PHYSIOLOGICAL ASPECTS OF SPORT TRAINING AND PERFORMANCE

Jay Hoffman, PhD

College of New Jersey

Human Kinetics

Library of Congress Cataloging-in-Publication Data

Hoffman, Jay, 1961-
    Physiological aspects of sport training and performance / by Jay
Hoffman.
        p. cm.
Includes bibliographical references and index.
    ISBN 0-7360-3424-2 (hbk.)
    1. Exercise--Physiological aspects. 2. Sprts--Physiological aspects.
    3. Physical education and training--Physiological aspects. I. Title.
    QP301 .H635 2002
    612'.44--dc21   2002004148

ISBN: 0-7360-3424-2

Acquisitions Editor: Michael S. Bahrke, PhD
Developmental Editor: Patricia A. Norris, PhD
Assistant Editor: Lee Alexander
Copyeditor: Patricia L. MacDonald
Proofreader: Sue Fetters
Indexer: Vita Richman
Permission Manager: Dalene Reeder
Graphic Designer: Robert Reuther
Graphic Artist: Dawn Sills
Photo Manager: Leslie A. Woodrum
Cover Designer: Keith Blomberg
Photographer (cover): David Madison/Newsport
Art Manager: Carl D. Johnson and Kelly Hendren
Illustrator: Tom Roberts (line drawings and mac art) and ICC (medical art)
Printer: Sheridan Books

Printed in the United States of America   10 9 8 7 6 5 4 3 2 1

Human Kinetics
Web site: www.humankinetics.com

United States: Human Kinetics
P.O. Box 5076
Champaign, IL 61825-5076
800-747-4457
e-mail: humank@hkusa.com

Canada: Human Kinetics
475 Devonshire Road Unit 100
Windsor, ON N8Y 2L5
800-465-7301 (in Canada only)
e-mail: orders@hkcanada.com

Europe: Human Kinetics
107 Bradford Road
Stanningley
Leeds LS28 6AT, United Kingdom
+44 (0) 113 255 5665
e-mail: hk@hkeurope.com

Australia: Human Kinetics
57A Price Avenue
Lower Mitcham, South Australia 5062
08 8277 1555
e-mail: liahka@senet.com.au

New Zealand: Human Kinetics
P.O. Box 105-231, Auckland Central
09-523-3462
e-mail: hkp@ihug.co.nz

A strong foundation must be the keystone for any great endeavor. My wife Yaffa, and children Raquel, Mattan, and Ariel have been the root of my strength. Their love, devotion, and unyielding support are cherished, and as always help turn my dreams to reality.

To my parents, who taught me that through hard work, intense desire, and persistence great goals can be accomplished.

To Dr. John Magel, Dr. Bill McArdle, Dr. Mike Toner, and Dr. Lawrence Armstrong who helped plant the seeds that laid the groundwork for what I have and will become.

To Dr. Carl Maresh, your friendship and guidance has far-outreached the confines of a laboratory. To Dr. Bill Kraemer, for giving me the opportunity and more importantly a path for which I can follow. It is the ability to be associated with such men of character that I am better for just having been in their presence.

To Coaches Jim Calhoun, Meir Kaminsky, and Eric Hamilton for helping me realize my dreams of bridging the gap between sport and science, and making their field or court my laboratory.

To my brother Richard and best friend Al Steinfeld, you know where I've been, and how far I've come – I love you guys.

To the men and women that over the years that I have worked and served with: Benjamin Davidson, Miki Bar-Eli, Leah Chapnik, Shmuel Epstein, Baraket Falk, Mark Goldberger, Jie Kang, Craig Landau, Ari Shamiss, Gershon Tenebaum, Itzik Weinstein, and Yoni Yarom – I thank you all, it has been an honor.

Finally, the people at Human Kinetics who have been tremendous in helping my dream of this textbook become a reality. Special thanks to Mike Bahrke, Pat Norris, and Lee Alexander for all your efforts.

CONTENTS

INTRODUCTION

The purpose of *Physiological Aspects of Sport Training and Performance* is to provide a focused and applied approach for those interested in exercise physiology and sports medicine. Sports medicine specialists and other health care providers may use this book as a primary reference that encompasses specific and practical advice related to the conditioning and performance of athletes. Chapters are fortified with figures and tables making it more attractive to the reader. Key terms are in bold print. In addition, some tables consisting of standardized performance data or specific athletic profiles (e.g., strength measures for collegiate football players) are included.

In comparison to other exercise physiology books on the market, *Physiological Aspects of Sport Training and Performance* provides an in-depth review of all components of an athlete's training program. A broad range of topic areas is covered, including environmental influences on performance, hydration status, nutritional concerns, and ergogenic aids. In contrast to other books, this one is primarily focused on training factors and how various conditions and situations affect exercise performance. Furthermore, practical applications are provided for so that exercise prescriptions can be developed for a number of different athletic populations.

Physiological Aspects of Sport Training and Performance is organized into five sections. The initial section examines physiological adaptation and the effects of various modes of training (aerobic, anaerobic and resistance) on biochemical, hormonal, muscular, cardiovascular, neural, and immunological adaptations. These adaptations are discussed as they relate to the training level of the athlete and their impact on sport performance.

The second section covers exercise training principles and prescription. Each chapter describes in detail the specific development of training programs for each mode of training (resistance, anaerobic, aerobic), including examples of exercises and training programs. The chapter on concurrent training explains how to incorporate several different modes of training into a periodized training program. In addition, some chapters include performance guidelines for various types of athletes to be used for comparison purposes. Expectations of performance improvements as athletes progress through their respective training programs are also included.

The third section discusses the relevance of nutrition, hydration status, and ergogenic aids to sports performance. The fourth section focuses on environmental factors and their influence on sports performance. Specific areas of focus include exercise in the heat, the cold, and at altitude, including the medical concerns related to exercise in these environments. The potential performance benefits of exercise at altitude are also mentioned.

Section five is focused on how certain medical and health conditions influence sport performance. Included in this section is a discussion of overtraining. Other chapters discuss conditions that are commonly encountered when working with athletes.

PART I

PHYSIOLOGICAL ADAPTATIONS TO EXERCISE

CHAPTER 1

NEUROMUSCULAR SYSTEM AND EXERCISE

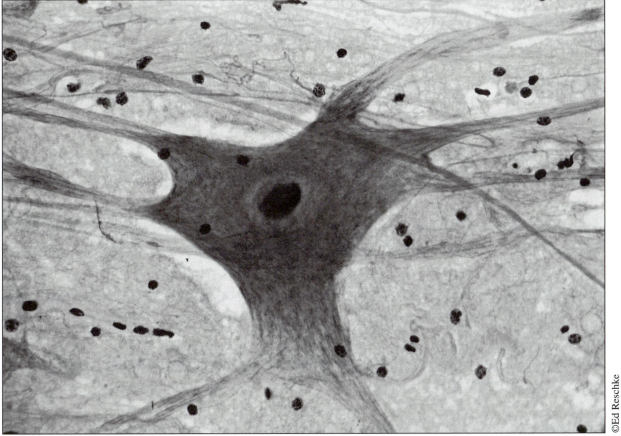

©Ed Reschke

The ability of the muscle to provide force and maintain physical activity is the basis for athletic performance. The common goal for all athletic training programs is to improve the functional capability of the exercising muscles. The specific goals of the athlete, however, may be different depending on the needs of the particular sport. This chapter provides a brief review of the gross structure of skeletal muscle and the mechanism of muscle contraction. Further discussion focuses on both neurological and skeletal muscle adaptations seen subsequent to athletic conditioning programs.

MUSCLE STRUCTURE

A skeletal muscle consists of thousands of cylindrical muscle cells called **fibers.** Muscle fibers are long, thin, and multinucleated. They lie parallel to one another, and the force of a muscle contraction is along the long axis of the fibers. A layer of **connective tissue** called the **epimysium** surrounds the entire muscle. The muscle tapers at both its proximal and distal ends to form a **tendon.** A tendon is a strong, dense connective tissue that serves to connect muscle to bone. Tendons, although much smaller in magnitude than muscles, have a greater tensile strength, which allows them to withstand the forces generated by the relatively large muscles. Below the epimysium are bundles of up to 150 fibers bound together. Each bundle is termed a **fasciculus** and is surrounded by another layer of connective tissue called the **perimysium.** The **endomysium** is an additional layer of connective

tissue that separates neighboring muscle fibers. A membrane called the **sarcolemma** surrounds the cellular contents that make up each muscle fiber. The number of fibers within each muscle varies considerably and depends on the size and function of the muscle. The fiber may extend the length of the muscle, or it may merge with another fiber. The structure of the muscle can be seen in figure 1.1.

A single muscle fiber is composed of many smaller units that lie parallel to the fiber. These units are called **myofibrils.** The myofibrils contain even smaller units, termed **myofilaments,** that consist mainly of two proteins, **actin** and **myosin.** The arrangement of these proteins gives the muscle fiber its striated appearance of light and dark bands (see figure 1.2). The light band is referred to as the **I band,** and the dark band is called the **A band.** The **H zone** is in the middle of the A band and at rest contains only the myosin filaments. During contraction the actin filaments are pulled into this zone causing this zone to be similar in appearance as the A band. The **Z line** bisects the I band and is connected to the sarcolemma to give stability to the entire structure. The repeating unit between two Z lines is called the **sarcomere,** which is the functional unit of the muscle cell.

MUSCLE CONTRACTION

The myosin filaments have projections, or **cross-bridges,** at the regions where the actin and myosin overlap. These projections, known as **myosin heads,** extend perpendicu-

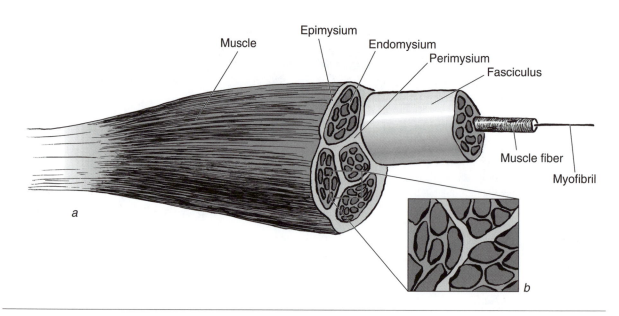

Figure 1.1   *(a)* Structure of muscle and *(b)* cross section.
Adapted, by permission, from J. Wilmore and D. Costill, 1999, *Physiology of sport and exercise* (Champaign, IL: Human Kinetics), 29.

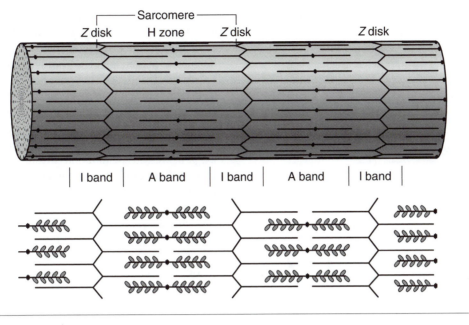

Figure 1.2 Sarcomere unit.
Adapted, by permission, from J. Wilmore and D. Costill, 1994, *Physiology of sport and exercise* (Champaign, IL: Human Kinetics), 32.

lar from the thick myosin filaments to the thinner actin filaments. The interaction between actin and myosin can be seen in figure 1.3.

Actin filaments contain two additional proteins, **troponin** and **tropomyosin,** that regulate the contact between the myofilaments during contraction. Tropomyosin appears to inhibit contact between actin and myosin by covering the **active sites** located on the actin molecule. Troponin triggers the contraction of these filaments by uncovering the active sites when stimulated by calcium.

Calcium ions are stored in interconnecting tubules that lie within the muscle fibers. These tubules are arranged parallel to the myofibrils and are called the **sarcoplasmic reticulum.** The sarcoplasmic reticulum terminate in large vesicles known as the **terminal cisternae** (see figure 1.4). The terminal cisternae abut an additional tubule system known as the **transverse tubules.** These tubules lie perpendicular to the myofibrils and are located in the lateral-most portion of the sarcoplasmic channels in the region of the Z line. The combination of the cisternae and the transverse tubule between them is referred to as the **triad.** The transverse tubules pass through the muscle fiber and form an opening to the inside of the muscle cell.

The transverse tubule system and triad function as the transportation network for the spread of the **action potential (depolarization)** from the outer membrane of the fiber inward. During depolarization, calcium ions are released from the sarcoplasmic reticulum and diffuse to the myofilaments. As mentioned previously, calcium ions bind to the protein troponin, causing a conformational

change in the tropomyosin molecule. Its inhibiting action is impeded, allowing the active sites on the actin filament to be uncovered. This permits the myosin head to connect to the active site, allowing muscle contraction to proceed (see figure 1.5 for depiction of cross-bridge activity with binding of Ca^{2+} to **troponin-tropomyosin complex**).

Muscle contraction is an active process that requires a constant influx of energy. For muscle contraction to continue, **ATP (adenosine triphosphate)** binds to its receptor on the myosin head. An enzyme (**myosin ATPase**) found on the myosin head splits ATP to yield **ADP (adenosine diphosphate)** and **Pi (inorganic phosphate).** This process causes the myosin head to disassociate from its active site on the actin molecule and ready itself for another cycle.

The **sliding filament theory** proposes that a muscle shortens or lengthens because the actin and myosin filaments slide past each other without the filaments themselves changing length (Huxley 1969). The major structural change occurring within the sarcomere is the pulling together of the Z bands, which decreases the region of the I band. During a muscle contraction, the cross-bridges do not move in a synchronous manner but may undergo repeated independent cycles of movement.

NEUROMUSCULAR SYSTEM

Muscle contraction is the result of stimuli processed by the central nervous system. Stimulation of the central nervous system initiates an electrical impulse that is

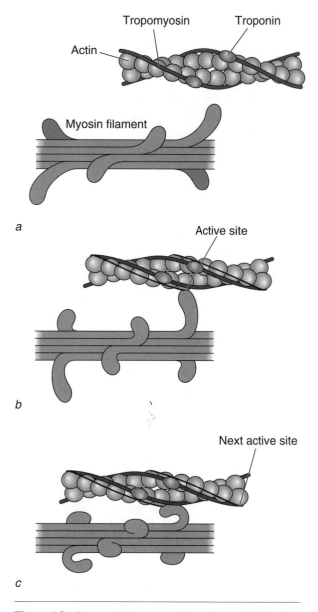

Figure 1.3 Interaction between actin and myosin. Muscle fiber *(a)* relaxed, *(b)* contracting, *(c)* fully contracted.
Adapted, by permission, from J. Wilmore and D. Costill, 1999, *Physiology of sport and exercise* (Champaign, IL: Human Kinetics), 36.

propagated along the length of a nerve cell, or **neuron.** The nerve cell consists of a cell body **(soma),** numerous short projections **(dendrites),** and a long projection **(axon)** that carry the electrical impulse from the soma toward the muscle. The connection between the nerve fiber and muscle is at the **neuromuscular junction.** The nerve fiber and the muscle that it innervates are known as the **motor unit.** Each motor unit is innervated by its own neuron, with its own specific contractile and metabolic characteristics. The individual fibers of a motor unit may be spread over a large portion of a muscle. Figure 1.6 is a schematic representation of the motor unit.

A nerve is composed of many nerve fibers. Nerve fibers covered by a **myelin sheath** are called **medullary fibers;** they ensure that neural impulses meant for a specific muscle action do not activate other muscle groups. This myelin sheath does not run continuously along the length of the nerve fiber but is interrupted by small spaces known as the **nodes of Ranvier.** Nerve impulses jump from one node to another, allowing a fast action or impulse conduction called **saltatory conduction.** In contrast, a nerve fiber without the myelin sheath is called **nonmedullary,** and the nerve impulse must travel the entire length of the nerve fiber.

Nerve impulses exist in the form of electrical energy. When there is no impulse, the inside of the nerve cell has a negative charge compared with the outside, basically because of a greater number of positive ions located outside of the cell (sodium) than inside the cell (potassium). This is termed the **resting membrane potential.** Excitation occurs when a nerve impulse causes the nerve cell membrane to become more permeable for both sodium and potassium. Sodium will move from an area of higher concentration to an area of lower concentration (from outside of the cell to inside), causing a positive charge on the inside of the cell in comparison with the outside. This is termed an **action potential** and lasts for a brief time. The action potential propagates along the length of the nerve, or motor neuron, until it reaches the muscle. The muscle fibers that the nerve innervates also have the capability of propagating an action potential and transmit the electrical impulse along its entire length.

The gap between the nerve fiber and muscle at the neuromuscular junction is called the **synapse.** The electrical impulse resulting from the action potential reaches the presynaptic side of the neuromuscular junction and causes the release of the neurotransmitter **acetylcholine.** Acetylcholine is stored in vesicles located at the presynaptic membrane and, on stimulation, diffuses across to the postsynaptic side of the neuromuscular junction. The interaction between acetylcholine with its receptor on the postsynaptic membrane causes an increase in the permeability of sodium and potassium ions. This propagation of an action potential will cause a release of calcium from the sarcoplasmic reticulum, resulting in muscle contraction. Figure 1.7 depicts a flow chart from the moment of the action potential to muscle contraction.

The muscle contraction is limited to the muscle fibers of the motor unit. When the nerve of a particular motor unit is activated, contraction of all the muscle fibers that it innervates occurs. This is known as the **all-or-none law.** However, it is important to understand that not all the fibers within a muscle are contracting. If this were true, there would be no ability to control the force output of the muscle. Gradations in force are accomplished by

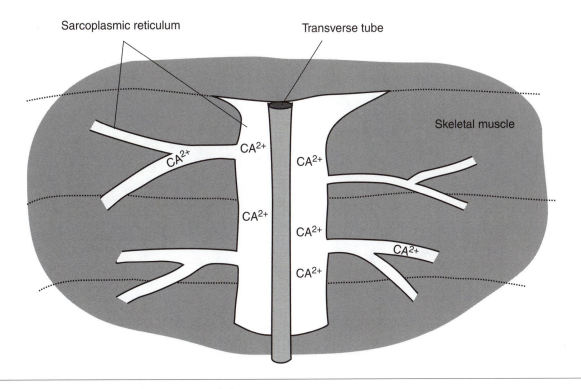

Figure 1.4 Sarcoplastic reticulum transverse tubules.

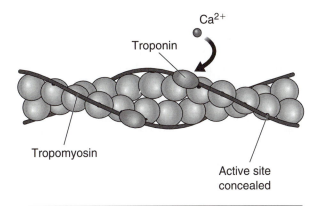

Figure 1.5 Binding of Ca^{2+} to troponin-tropomyosin complex.
Adapted, by permission, from J. Wilmore and D. Costill, 1999, *Physiology of sport and exercise* (Champaign, IL: Human Kinetics), 29.

the number of motor units recruited or stimulated. Maximal strength is accomplished when all the motor units of a muscle are activated.

MUSCLE FIBER TYPES

All motor units function in a similar manner although they may have different contractile and metabolic characteristics. Some motor units are more suited for

aerobic metabolism, whereas others are more appropriate for **anaerobic** activity. Two distinct fiber types have been identified and classified by their contractile and metabolic characteristics. These have been termed **slow-twitch** and **fast-twitch** fibers, also referred to as type I and type II, respectively. The characteristics of these fiber types are shown in table 1.1. As can be seen in the table, these fibers possess certain distinguishable characteristics that make them suited either for prolonged, low- to moderate-intensity activity (slow-twitch fibers) or short-duration, high-intensity activity (fast-twitch fibers).

For many years, muscle fiber classification was limited to these two classifications (type I and type II), with type II fibers being further subdivided into two distinct divisions: IIa and IIb. Type IIa fibers have a well-developed capacity for both aerobic and anaerobic metabolism and are commonly termed **fast oxidative glycolytic.** Type IIb fibers possess the greatest anaerobic capability and are termed **fast glycolytic.** In the last decade, with the improvement of muscle-staining techniques, additional subtypes within each fiber type have been reported (Staron et al. 1991; Fry, Allemeier, and Staron 1994). Subtypes of type I oxidative fibers have been labeled type I and type Ic. Type Ic fibers are thought to have less oxidative capacity than type I fibers. There have been five different fiber subtypes identified for type II fibers. These fibers, IIc, IIac, IIa, IIab, and IIb, represent a continuum

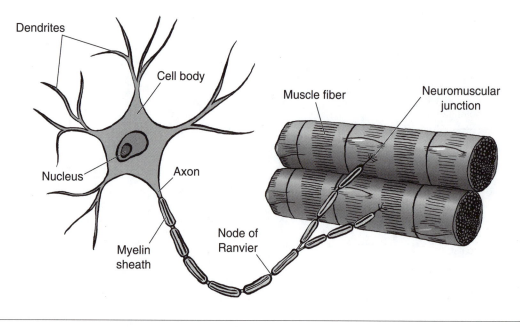

Figure 1.6   The motor unit.
Adapted, by permission, from S.J. Fleck and W.J. Kraemer, 1997, *Designing resistance training programs* (Champaign, IL: Human Kinetics), 46.

of aerobic and anaerobic characteristics (Pette and Staron 1990).

It is believed that genetics largely determine muscle fiber-type distribution and that it is set at birth or early in life. The average individual (man or woman) has an equal proportion of both slow- and fast-twitch fibers. The composition of muscle fiber types (percentage of type I to type II) is consistent among the major muscle groups in the body.

In elite athletes, the predominance of a particular fiber type appears to correspond to the metabolic requirements of their respective sport. Endurance athletes have a large percentage of type I fibers, whereas highly anaerobic athletes (e.g., sprinters) have a predominance of type II fibers. Elite endurance athletes may have 90% of their skeletal muscle made up of type I fibers, providing a large advantage for success in aerobic performance. Similarly, athletes with exceptional explosive power and speed would have a predominance of type II fibers.

MUSCLE RECRUITMENT

During muscle action, the muscle fibers are generally recruited according to the **size principle,** which suggests an orderly recruitment of the motor units. Motor unit recruitment patterns are related to the size of the neuron. Smaller motor units require a lower stimulus for activation. Thus, type I muscle fibers, which require a low stimulation, are initially recruited during muscle activation. As greater force in the muscle action is required, higher threshold units (type II fibers) are subsequently

recruited. However, there may be an exception to the size principle. Type II fibers appear to be recruited first during powerful, high-velocity activities.

The gradations seen in recruitment patterns suggest that fatigue-resistant, slow-twitch fibers are recruited primarily during low-intensity, long-duration activities. As greater intensity is needed, the higher threshold units (fast-twitch fibers) are stimulated. However, when muscle action requires immediate high-velocity, high-power movement (e.g., sprinting), the fast-twitch fibers may be recruited first. The benefit of the size principle is that it keeps the highly fatigable fast-twitch fibers in reserve until needed for a particular muscle action.

MUSCLE PROPRIOCEPTORS

Muscle action is finely controlled by the interaction between both motor and sensory activities. Sensory nerves, both **muscle spindles** and **golgi tendon organs,** which are located in the muscles, tendons, and fibrous capsules of joints, provide sensory feedback to the brain at a conscious and subconscious level on movement and position of the muscles and joints.

Muscle spindles are arranged parallel to the skeletal muscle fibers and are composed of several small **intrafusal fibers,** whose nerve endings are attached to the sheaths of the surrounding skeletal muscle fibers. Muscle spindles are responsible for monitoring the stretch and length of the muscle and also initiate contraction to reduce the stretch in the muscle. During

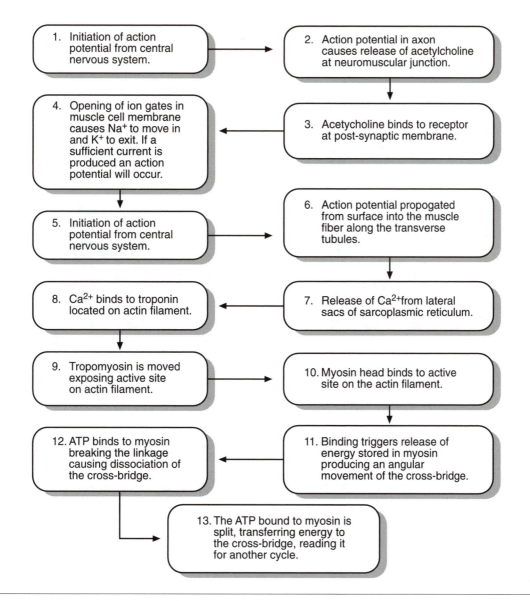

Figure 1.7 Sequence of events from action potential to skeletal muscle contraction.

Adapted, by permission, from S.J. Fleck and W.J. Kraemer, 1997, *Designing resistance training programs* (Champaign, IL: Human Kinetics), 34.

Table 1.1 Characteristics of Type I and Type II Muscle Fibers

	Type I Slow twitch	Type IIa Fast twitch	Type IIb Fast twitch
Force production	Low	Intermediate	High
Contraction speed	Slow	Fast	Fast
Myofibrillar ATPase activity	Low	High	High
Fatigue resistance	High	Moderate	Low
Glycolytic capacity	Low	High	High
Oxidative capacity	High	Medium	Low
Capillary density	High	Intermediate	Low
Mitochondrial density	High	Intermediate	Low
Myoglobin content	High	Intermediate	Low
Endurance capability	High	Moderate	Low
Glycogen storage capability	No difference	No difference	No difference

muscle contraction, force or tension develops as a result of direct stimulation of the motor unit and indirect stimulation through the muscle spindle.

Golgi tendon organs are sensory receptors located within muscle tendons. They are sensitive to tension in the **muscle-tendon complex** and form a protective mechanism to reduce the potential for injury. When tension in the activated muscle reaches levels that pose a potential risk of injury, the golgi tendon organ is stimulated. It inhibits contraction of the contracting, or **agonist,** muscles and activates the **antagonist** muscle groups to reduce the force of the muscle contraction. Disinhibition of the golgi tendon organ during training is thought to result in a greater strength expression by the exercising muscle.

NEUROMUSCULAR ADAPTATIONS TO EXERCISE TRAINING

There is a great deal of plasticity within the neuromuscular system. Participation in physical conditioning programs leads to improvements in muscle size or performance. These adaptations may occur fairly rapidly, depending on the training status of the indi-

vidual. In addition, the type of training program also affects physiological adaptation. Strength training may have a particular effect on muscle adaptation, whereas endurance training may cause a completely different type of adaptation. Likewise, if the training stimulus is removed, muscle tends to revert back to its pretraining state. An understanding of the type of alterations seen with a given training program will help the coach or athlete develop the most appropriate training program and set the most realistic training goals.

Neural Adaptation

Maximal strength expression is not determined solely by the quantity and quality of the muscle mass but also by the extent that the muscle mass is activated (Sale 1988). Maximal strength is achieved when the primary muscle group is fully activated and the **synergists** and antagonists are appropriately activated (Sale 1988). A better coordination in the activation of these muscle groups allows for a better expression of strength. The neural adaptation to resistance training is thought to be the initial adaptation leading to increased strength (Moritani and deVries 1979; Komi 1986).

Electromyographic Changes

The most common method for assessing neural activity of the muscle is through electromyograph (EMG) recordings. The EMG measures the electrical activity within the muscle and nerves and indicates the neural drive to a muscle. An EMG recording is generally performed with surface electrodes on a prime mover. Recording the EMG activity of a muscle before and after a resistance training program can measure the neural adaptation resulting from a resistance training program. Once the motor unit activity is recorded, the signal can be then integrated in a number of different fashions and is referred to as the integrated electromyogram (IEMG). Resistance training has been demonstrated to increase IEMG activity in both trained and untrained populations (Hakkinen and Komi 1983; Moritani and deVries 1979). Significant correlations between IEMG and increases in strength have been reported after strength training (Hakkinen and Komi 1986; Hakkinen, Komi, and Alen 1985). These findings support the idea that strength-trained subjects may be more capable of fully activating their primary muscles during maximal performance than untrained subjects. Further findings have shown that EMG activity will be significantly increased during the initial stages of a resistance training program but will decrease, or increase at a diminished rate, during the later stages of training (see figure 1.8). It is at this point in training that most strength increases are attributed to muscle **hypertrophy,** which is discussed later in the chapter.

Recruitment Patterns

Neural adaptations may also be recognized by a decrease in the electrical activity in the muscle with a corresponding increase in force output (Sale 1988). The greater strength expression seen with lower EMG responses may reflect a more efficient recruitment pattern of the muscles responsible for the force production. In addition, trained individuals may be able to recruit additional motor units as a consequence of training programs (Komi 1986).

Synchronization

An additional neural adaptation, which may also result in greater force production, is an increase in the **synchronization** of motor unit firing. Improved motor unit synchronization has been shown in a group of weightlifters after a short (6 weeks) training program (Milner-Brown, Stein, and Yemm 1975). A more synchronous pattern of motor unit firing simply means that the number of motor units active at any one time is increased, causing a greater number of fibers to be contracted at a given time, resulting in greater strength expression.

Inhibitory Mechanisms

As mentioned earlier in the chapter, an inhibition of muscle contraction by the golgi tendon organs is thought to limit force production. This protective mechanism appears to be especially active when maximal contractions are performed at slow speeds of contraction (Caizzo, Perrine, and Edgerton 1981; Wickiewicz et al. 1984). Resistance training is thought to cause an inhibition of these protective mechanisms. When the antagonist muscle group is contracted immediately before a lift, it is thought that this will partially inhibit the neural self-protective mechanism and allow for a more forceful contraction (Fleck and Kraemer 1997).

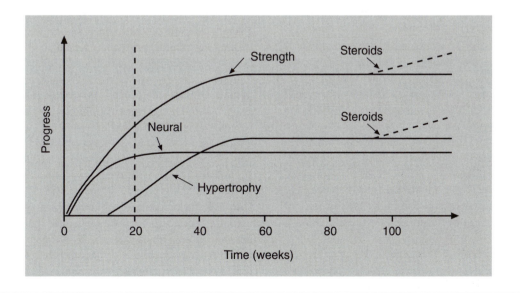

Figure 1.8 Relationship of neural and muscular adaptations to strength training.

Skeletal Muscle Adaptation

Skeletal muscle is dynamic in its response to training and can adapt to a wide range of functional demands. When skeletal muscle is forced to work at intensities exceeding 60-70% of its maximal force-generating capacity, adaptations occur that may result in an increase in muscle size and strength (MacDougall 1992).

The initial increases seen in muscle strength after resistance training have been attributed primarily to the neurological adaptations discussed earlier in this chapter. Further increases in skeletal muscle strength appear to be the result of a growth in muscle size. An increase in the size of preexisting muscle fibers (hypertrophy) or an increase in the number of fibers within the muscle (**hyperplasia**) may result in skeletal muscle growth.

Muscle Hypertrophy

Increases in muscle size are generally seen after 6 to 8 weeks of heavy resistance training. However, some evidence suggests that muscle growth may occur even earlier (Staron et al. 1991, 1994). Increases in muscle size have been attributed to increases in the cross-sectional area of existing muscle fibers (Alway, Grumbt, et al. 1989; MacDougall et al. 1984). This process of fiber growth appears to be related to the increased synthesis of contractile proteins (actin and myosin filaments) and to the increased formation of sarcomeres within the fiber (Goldspink et al. 1992; Alway, Grumbt, et al. 1989; MacDougall et al. 1979). The synthesis of these protein filaments, which constitute the contractile element of muscle fibers, may be related to the repeated trauma to the fibers from high-intensity resistance training. During recovery from the cellular damage caused by such training, an overcompensation of protein synthesis may occur, resulting in the noted anabolic effects (Antonio and Gonyea 1993).

Muscle hypertrophy occurs in both type I and type II fibers after resistance training programs. However, the type II fiber appears to undergo a greater relative hypertrophy (Staron et al. 1989). Since both type I and type II fibers are recruited during maximal contractions, the greater hypertrophy seen in the type II fiber may be related to the greater activation of high threshold units than normally activated during daily activity (MacDougall 1992). The magnitude of these increases varies considerably and depends on several factors, including the individual's responsiveness to training, the intensity of training, the duration of training, and the individual's training status.

Increases of 15.6%, 17.3%, and 28.1% in cross-sectional area have been reported in type I, type IIa, and type IIb muscle fibers, respectively, in novice female subjects after 6 weeks of high-intensity resistance training (Staron et al. 1991). As the length of the resistance training program increases (20 weeks), further increases in muscle size may be seen (15%, 45%, and 57% in type I, type IIa, and type IIb muscle fibers, respectively) (Staron and Johnson 1993). Similar increases in muscle hypertrophy have also been seen in untrained male subjects (Adams et al. 1993; Hather et al. 1991). Although gender differences in muscle growth do exist, these differences become apparent after longer periods of training.

As previously mentioned, the training status of the individual does have an important effect on the morphological changes seen in the muscle after resistance training. Experienced bodybuilders, both male and female, were examined during 24 weeks of training for competition. No significant improvements in muscle cross-sectional area were noted over the training period (Alway et al. 1992). This is consistent with what has been repeatedly reported in the literature concerning muscle growth in highly trained, experienced bodybuilders (Hakkinen, Komi, and Alen 1987; Hakkinen et al. 1988). It should be understood, however, that there might be a large difference between statistical significance versus practical significance in understanding muscle growth. Generally, studies examining muscle morphological changes do not have a large sample number. Thus, a great difference is needed to achieve statistical significance. Alway et al. (1992) reported a 3.6% increase in the cross-sectional area of the biceps in five experienced male bodybuilders after 24 weeks of training. Although, this was not statistically significant, it may represent an important component for success during competition. Muscle hypertrophy in experienced lifters may be attainable, but it requires a high-intensity training stimulus for a much longer training duration.

Fewer studies have compared muscle fiber size changes with strength, endurance, and combined strength/endurance training programs. Kraemer, Patton, et al. (1995) examined muscle fiber morphological changes in untrained subjects who exercised 4 days a week for 3 months in either a high-intensity resistance training program, an endurance training program, or a combined strength/endurance training program. All training programs were periodized (see chapter 11) to enhance recovery from exercise and prevent overtraining. Both strength and combined strength/endurance training programs resulted in significant muscle fiber hypertrophy. However, the inclusion of endurance training stunted the growth of both type I and IIc fibers (see figure 1.9). Endurance training alone caused a decrease in fiber size in the more oxidative fibers (type I and IIc). The attenuation of muscle hypertrophy with concurrent strength and endurance training has also been previously noted (Dudley and Djamil 1985). Interestingly, Kraemer,

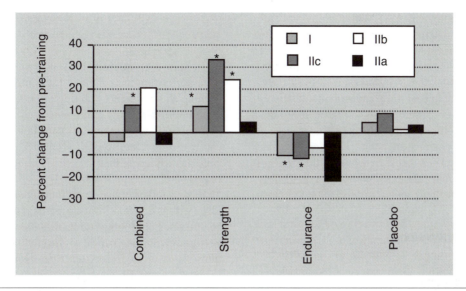

Figure 1.9 Muscle fiber changes as a consequence of strength, endurance, and combined strength/endurance training.
* = significant change.
Modified from Kraemer et al., 1995

Patton, and colleagues (1995) also included a study group that performed an upper-body resistance training program in addition to the endurance training program. Those subjects were able to mitigate the decreases in fiber size of the legs through upper-body training. This may have been caused by isometric contractions of the leg musculature during upper-body exercise.

Increases in fiber size do not appear to be accompanied by increases in mitochondrial number or in the capillary-to-fiber ratio. This lowering of the mitochondrial and capillary volume density in the fibers may not hinder strength or power performance, but it may have important implications for endurance capability in those muscles. This change might alter the oxygen kinetics within the muscle by delaying transport of oxygen from the vasculature to the exercising muscle. It is noteworthy to mention that endurance training decreases fiber size (see previous paragraph) while causing increases to both mitochondrial and capillary density, thus potentially improving the aerobic capability of the muscle. In contrast, sarcoplasmic reticulum and transverse tubule volume density increase in proportion to the change in myofibrillar volume (Alway, MacDougall, and Sale 1989), thus maintaining or improving contraction capabilities of the muscle.

Muscle Hyperplasia

It has been generally understood that muscle fiber number is fixed from birth and that skeletal muscle growth is a result of hypertrophy of existing muscle fibers. However, a number of studies have suggested that high-intensity resistance training may cause muscle hyperplasia (Gonyea 1980a, 1980b; Gonyea et al. 1986; Ho et al.

1980). Most of these earlier studies, which used animal models, met with criticism concerning the methodological means used in data analysis (Gollnick et al. 1981, 1983). However, later studies that accounted for those concerns were still able to demonstrate hyperplasia of skeletal muscle subsequent to muscle overload (Alway, Winchester, et al. 1989; Gonyea et al. 1986).

The use of an animal model invoked further criticism directed at proponents of skeletal muscle hyperplasia. The magnitude of hypertrophy seen in humans does not occur in many of the animal species (McArdle, Katch, and Katch 1996). Thus, for animals, muscle hyperplasia may be an important compensatory mechanism for combating muscle overload. Interestingly, MacDougall et al. (1982) and Tesch and Larson (1982) reported that elite bodybuilders had a greater number of muscle fibers than trained control subjects. These investigators suggested that the greater fiber number seen in the bodybuilders was attributable to years of high-intensity resistance training. However, these results were never duplicated in ensuing studies (MacDougall et al. 1984).

If muscle hyperplasia does occur, it is thought to be either through the development of new fibers from **satellite cells** (Antonio and Gonyea 1993; Appell, Forsberg, and Hollmann 1988) or through longitudinal splitting of existing muscle fibers (Antonio and Gonyea 1993; Ho et al. 1980; Gonyea et al. 1986). Satellite cells (located between the basement membrane and the plasma membrane) are thought to proliferate and grow to a **myoblast** and eventually **myotubes** that may develop into new muscle fibers. The myotube may also fuse with existing muscle fibers and remain incomplete along its length, leading to the wrong impression of a split fiber (Appell,

Forsberg, and Hollmann 1988). With longitudinal splitting, a hypertrophied muscle fiber that has reached some predetermined maximal ceiling of growth is thought to split into two or more smaller daughter cells through a process of lateral budding (Antonio and Gonyea 1994).

There does not appear to be any convincing support for the occurrence of muscle hyperplasia in humans. However, conflicting results still make this issue controversial and its potential appealing. Fleck and Kraemer (1997) have suggested that if hyperplasia does exist, it most probably occurs in a small portion of type II fibers when they reach their predetermined genetic growth limit.

Fiber-Type Conversions

As mentioned earlier in the chapter, the proportion of type I to type II muscle fibers appears to be genetically determined and their expression set early in life. A number of studies have examined whether conditioning programs can alter the proportion of type I to type II muscle fibers. Some studies have suggested that aerobic training may be able to increase the percentage of type I fibers (Howald et al. 1985; Simoneau et al. 1985), while others have reported increases in type II fiber proportion after sprint training (Jansson, Sjodin, and Tesch 1978; Jansson et al. 1990). However, the overwhelming majority of investigations have been unable to see any alterations in fiber-type composition as a consequence of conditioning programs. It is generally believed that only fiber-type transformations within a fiber type can be accomplished through training.

High-intensity resistance training appears to be a potent stimulus in causing a transformation of the type IIb to type IIa fiber subtype (Staron et al. 1989, 1991, 1994; Kraemer, Patton, et al. 1995). Most of the type IIb fibers have been reported to be converted to type IIa fibers after 20 weeks of resistance training (Staron et al. 1991). This is similar to the type II fiber conversions previously thought to be associated with aerobic exercise training

(Staron and Hikida 1992). Kraemer, Patton, and colleagues (1995) have also demonstrated skeletal muscle fiber subtype transformations from IIb to IIa in subjects performing high-intensity resistance training and in subjects performing a combined high-intensity resistance training and endurance training program. Subjects who were performing only endurance exercises also tended to increase the proportion of type IIa fibers but significantly elevated their type IIc fibers. This would be expected considering that the type IIc fibers are the most oxidative of the type II subtypes.

Fiber subtype transformations appear to occur rapidly (within 2 weeks) during participation in physical conditioning programs. These adaptations, however, may be transient. During periods of inactivity or detraining, a transformation of fast-twitch fiber subtypes from type IIa back to type IIb is observed (Staron et al. 1991). A return to training will result in a fiber-type transformation back to its trained state in a relatively shorter period of time. These studies highlight the dynamic nature of skeletal fiber transformations.

SUMMARY

This chapter presented a basic overview of muscle structure, motor units, muscle contraction, and muscle fiber types. In addition, neuromuscular adaptations to training are specific to the type of training program (strength or endurance training). Initial improvements in strength are primarily associated with neurological adaptations, whereas further increases in strength are more dependent on increases in the cross-sectional area of the muscle. These increases are thought to occur from either hypertrophy of existing muscle fibers or perhaps (although controversial) through the splitting of muscle fibers (muscle hyperplasia). Finally, fiber-type transformations may be possible within a subtype, but it is not possible to convert between fiber types (e.g., type I to type II).

CHAPTER 2

ENDOCRINE SYSTEM AND EXERCISE

©Michael Phillip Manheim

Hormones are chemical substances that circulate in the blood and interact with organs in the body to help combat various stresses. The primary role of hormones is to maintain internal equilibrium **(homeostasis).** Most hormones are synthesized in endocrine glands located throughout the body. Upon stimulation, the glands secrete their hormones into the surrounding extracellular space. The hormones then diffuse into the circulatory system and are transported to their respective target areas to perform their designated function.

OVERVIEW OF ENDOCRINE SYSTEM

Hormones influence the rate of specific cellular reactions by changing the rate of protein synthesis or **enzyme** activity and by inducing secretion of other hormones. In addition, hormones can facilitate or inhibit uptake of substances by cells. For example, **insulin** facilitates the uptake of **glucose** into the cell, and **epinephrine** inhibits glucose uptake to increase its concentration in the circulation.

Hormones can stimulate the production of enzymes or activate inactive enzymes. They can also combine with an enzyme to alter its shape **(allosteric modulation),** which will cause either an increase or decrease in the effectiveness of the enzyme.

Regulation of Hormone Secretion

For hormones to function properly, their secretion rate must be precisely controlled. A signal needs to be received that triggers the necessary steps for hormone secretion. The initial step is the detection of an actual or threatened homeostatic imbalance. This imbalance must be able to activate a **secretory apparatus** (e.g., the **endocrine gland**), resulting in hormone secretion. The circulating hormone interacts with its target organ or tissue and exerts its effect. Once the hormonal effect has occurred, the hormonal signal has to be turned off and the hormone needs to be removed from the circulation. Finally, the secretory apparatus must replenish the hormone in its secretory cells. The regulation of hormone secretion is depicted in figure 2.1.

The secretion of most hormones is regulated by **negative feedback,** meaning that some consequence of the hormone secretion acts directly or indirectly on the secretory apparatus to inhibit further secretion. This type of secretory mechanism is self-limiting. **Positive feedback** mechanisms are rare in endocrine regulation. During this type of regulation, some consequence of the hormonal secretion causes an augmented secretory drive. Rather than being self-limiting, the stimulus for triggering hormonal secretion becomes stronger. An example of positive feedback is the release of **oxytocin** from the posterior pituitary gland caused by dilation of the uterine cervix during childbirth. The oxytocin causes a greater dilation that in turn creates a greater stimulus for further oxytocin release.

Changes in Circulating Hormonal Concentrations

Increases in the concentration of hormones can be attributed to a number of different physiological mechanisms. Exercise or other physical or psychological stresses appear to be potent stimulators in elevating the secretory patterns of hormones. Fluid volume shifts, changes in clearance rates, and venous pooling of blood are additional mechanisms that may increase circulating concentration of hormones. Regardless of the mechanism, there

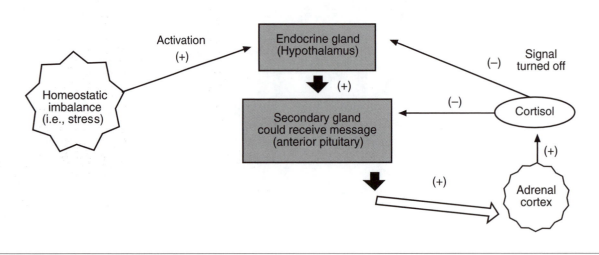

Figure 2.1 Regulation of hormone secretion.

is an increased potential for interaction with the receptor of the target tissue, leading to the desired cellular response.

Receptors are found in all types of cells within the body, and each hormone reacts with its specific receptor. The interaction of the hormone to its receptor has been called the **lock-and-key theory** (Kraemer 1994). The receptor is the lock and the hormone is the key. There is some **cross-reactivity,** meaning that there may be more than one hormone that can bind with the receptor (this is depicted in figure 2.2, where more than one hormone has an ability to cross-react with receptor C). When this occurs, the resulting biological actions are different from those induced by the primary hormone.

It is the hormone-receptor complex that results in a message being delivered to the cell nucleus for either inhibition or facilitation of **protein synthesis.** The number of receptors available for interactions with circulating hormones is considered another mechanism in initiating cellular action. Hormonal receptors are dynamic in that they also respond to physiological demand. They may increase in number to meet the demand of a rise in the circulating concentration of hormones. Such an increase in receptor number is termed **up-regulation.** Similarly, the number of receptors can be decreased if adaptation is no longer possible or to prevent an over-response by persistently increasing hormone levels. This adaptation is called **down-regulation.** This type of control on the part of the receptor is as dramatic as the changes in hormonal secretory patterns. The endocrine glands and the hormones they secrete are shown in table 2.1.

Types of Hormones

There are two main categories of hormones, **steroids** and **peptides.** Steroid hormones (e.g., **testosterone, cortisol**) are

Table 2.1 Endocrine Glands and Hormones

Gland	Hormone	Primary physiological action
Anterior pituitary	Growth hormone (GH)	Stimulates tissue growth, mobilizes fatty acids for energy, inhibits carbohydrate metabolism
	Proopiomelanocortin family includes: Adrenecorticotropic hormone (ACTH)	Stimulates secretion of glucocorticoids and other Adrenal hormones
	Endorphins, enkephalins, and dynorphins	Causes analgesia, and euphoria after exercise
	Luteinizing hormone (LH)	Act together to stimulate production of gonadal hormones (testosterone, estrogen and progesterone)
	Follicle-stimulating hormone (FSH)	
	Thyroid-stimulating hormone (TSH)	Stimulates production and secretion of the thyroid hormones from thyroid gland
	Prolactin	Stimulates milk production in mammary glands
Posterior pituitary	Antidiuretic hormone (ADH) also called vasopressin	Stimulates reabsorption of water by kidneys
	Oxytocin	Stimulates uterine contraction and milk secretion in lactating breasts
Thyroid gland	Thyroxine and triiodothyronine	Stimulates metabolic rate and regulates cell growth and activity
Parathyroid gland	Parathyroid hormone	Increases blood calcium and lowers blood potassium
Pancreas	Insulin	Promotes glucose transport into cell, protein synthesis
	Glucagon	Promotes hepatic glucose release, increase lipid metabolism
Adrenal gland (adrenal cortex)	Glucocorticoids (cortisol)	Promotes protein catabolism, stimulates conversion of protein into carbohydrates, promotes lipid metabolism
	Mineralcorticoids (aldosterone)	Promotes reabsorption of sodium and water by kidneys
Adrenal gland (adrenal medulla)	Catecholamines (epinephrine and norepinephrine)	Increases cardiac output, regulates blood vessels, increases glycogen catabolism and fatty acid metabolism
Liver	Insulin-like-growth factors	Increases protein synthesis in cells
Ovaries	Estrogen	Stimulates development of female sex characteristics
	Progesterone	
Testes	Testosterone	Stimulates growth and protein anabolism; development and maintenance of male sex characteristics

synthesized from circulating cholesterol. They are also fat soluble and can diffuse across the cell membrane. Once across the membrane, the steroid hormone binds with its receptor, most probably within the cytoplasm of the cell, to form the **hormone-receptor complex.** The hormone-receptor complex then transports itself to the nucleus of the cell, where its message is delivered, transcribed, and translated into action.

Peptide hormones (e.g., **growth hormone,** insulin) are made up of **amino acids** and bind to their receptors located on the cell membrane. Because peptide hormones are unable to cross the cell membrane, they must rely on secondary messengers to send their message to the cell nucleus. When the hormone binds with its receptor within the membrane, it triggers the production of **cyclic AMP** from ATP. This reaction is catalyzed by the enzyme **adenylate cyclase.** Cyclic AMP serves as the secondary messenger, activating a cascade of intracellular events and resulting in the cellular response.

HORMONES AND EXERCISE

Exercise has been shown to be a potent stimulus to the endocrine system. The hormonal response to an acute exercise session suggests that hormones may be involved in the recovery and remodeling processes that occur after exercise. The exercise stimulus has an important role in the hormonal secretion pattern. Variables such as intensity of exercise, volume of exercise, rest intervals, choice of exercise, and recovery status of the muscle appear to influence the hormonal response.

The mechanisms of hormonal interaction with the remodeling of muscle tissue are based on several factors. The acute increase in hormonal concentration caused by the exercise stimulus allows for a greater interaction between the hormone and its receptors. Since the adaptations to exercise (particularly resistance exercise) are **anabolic** in nature, the recovery mechanisms involve tissue repair and remodeling (Kraemer 1992b). In instances when training **intensity** or **volume** exceeds an individual's ability to recover, a possible situation of overtraining or overwork can occur, resulting in a greater **catabolic** effect. The hormonal response will either repair or remodel muscle tissue or perhaps impede this process.

The hormonal mechanisms may respond differently between trained and untrained individuals (Hakkinen et al. 1989). Furthermore, some hormonal mechanisms may not be operational in both males and females (e.g., testosterone). In addition, the effect of program design, genetic predisposition, fitness level, training experience, and adaptational potential all seem to affect the endocrine mechanism for maintaining or improving muscle size and strength (Kraemer 1992a; Hakkinen et al. 1989).

This section discusses both the acute and chronic training response of the hormones considered to be principally involved with the production of **muscle force** and

muscle hypertrophy. In addition, the role of the catabolic hormones in such mechanisms is discussed.

TESTOSTERONE

Testosterone is an **androgen,** a steroid hormone that has masculinizing effects. It is also anabolic because of its role in the maintenance and growth of muscle and bone tissue. Most of the circulating testosterone is produced in the testes, while small amounts are produced in the adrenal glands. Circulating testosterone binds to an androgen receptor located in the cytoplasm of skeletal muscle cells. This hormone-receptor interaction results in a migration of the hormone-receptor complex to the nucleus of the cell, resulting in an increase in protein synthesis (see figure 2.2).

Little testosterone is produced during childhood until ages 10 to 13, when a rapid increase occurs under the stimulus of **luteinizing hormone** (a gonadotrophic hormone secreted from the anterior **pituitary gland**). This testosterone surge is responsible for the distinguishable characteristics and maturity of the male sex organs and the development of secondary sexual characteristics. The physiological roles of testosterone are as follows:

- Increase in protein synthesis, resulting in muscle growth
- Development and maturation of male sex organs
- Development of secondary sexual characteristics
  - Increase in body hair
  - Development of masculine voice
  - Development of male pattern baldness
  - Development of libido
  - Control of spermatogenesis
  - Aggressive behavior
- Interaction on epiphyseal growth plates, contributing to the longitudinal growth of long bones and subsequent cessation of growth caused by fusion of the epiphyseal plates
- Increase in secretions of sebaceous glands, contributing to acne
- Possible role in glycogen synthesis

Acute Exercise Response

A single training session of resistance exercise has been demonstrated to significantly increase the peripheral concentration of testosterone above resting levels in males (Hakkinen et al. 1987, 1988; Kraemer, Marchitelli, et al. 1990). However, no significant difference in the response of testosterone to a single exercise session was seen in a cross-sectional comparison of both males and females with little resistance training experience compared with subjects (males and females) who had trained with weights for at least 2 years (Fahey et al. 1976). However, this may

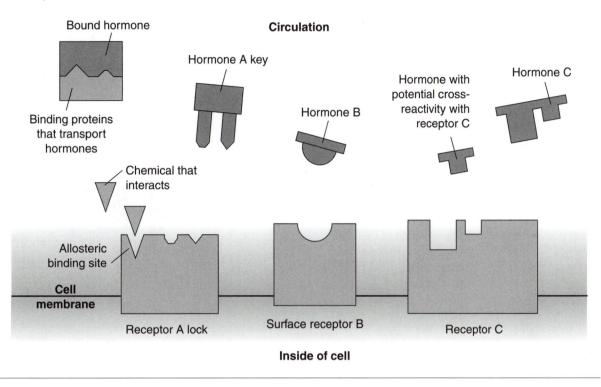

Figure 2.2   Hormone-receptor interaction.
Reprinted, by permission, T.R. Baechle and R.W. Earle, eds., 2000, *Essentials of Strength Training and Conditioning.* Champaign, IL: Human Kinetics), 96.

be dependent on length of training experience. Kraemer and colleagues (1992) reported that male weightlifters with more than 2 years of training experience had a significantly greater testosterone response to an exercise session than weightlifters with less than 2 years of lifting experience. Exercise response patterns of testosterone also appear to be related to the design of the exercise program. Single component variables (e.g., rest intervals, intensity) have been demonstrated to have a significant effect on the testosterone response to an acute bout of resistance exercise. Significantly higher testosterone concentrations have been observed when rest periods between sets are reduced (3 min versus 1 min between sets) or when the intensity of exercise is reduced (5 RM **[repetition maximum]** versus 10 RM) (Kraemer, Marchitelli, et al. 1990). This may help explain the variability in muscle hypertrophy seen with different resistance training programs and the combination of high volume and short rest periods in the training programs of bodybuilders. Further discussion of repetition maximum appears in Chapter 7.

There appears to be a biphasic response of testosterone to an acute bout of aerobic exercise, which appears dependent upon the duration of exercise. In studies performed primarily on male subjects, it has been shown that exercise of relatively short duration (e.g., 10-20 min) does not appear to increase plasma testosterone concentrations (Bottecchia, Borden, and Martino 1987; Galbo et al. 1977; Sutton et al. 1973). However, as exercise reaches 20-30

min in duration, significant elevations in testosterone are observed (Wilkerson, Horvath, and Gutin 1980; Hughes et al. 1996). As duration of exercise continues, a biphasic response is seen (see figure 2.3). Testosterone levels will continue to elevate as exercise is prolonged and then begin to decline toward baseline levels before exercise is completed. As exercise progresses past 3 h, significant declines (below resting levels) in testosterone have been reported (Dessypris, Kuoppasalmi, and Adlercreutz 1976; Guglielmini, Paolini, and Conconi 1984; Schurmeyer, Jung, and Nieschlag 1984; Urhaussen and Kindermann 1987). These reduced levels may continue for 48 h post exercise (Urhaussen and Kindermann 1987).

An increase in testosterone concentrations during an acute bout of anaerobic exercise appears to also depend on the duration of exercise. Increases in testosterone concentrations have been reported after both 90 s (Kindermann et al. 1982) and 2 min (Kuoppasalmi et al. 1980) of intermittent anaerobic exercise in male runners. However, exercise of shorter duration (a noncontinuous 15-s sprint) may not produce any change from resting levels (Kuoppasalmi et al. 1980). Interestingly, the immediate postexercise response of testosterone to anaerobic exercise may not be as important as its response during the recovery period. Decreases below resting levels have been seen after a 2-min bout of high-intensity anaerobic exercise (Kuoppasalmi et al. 1980). The physiological benefit or harm that may result from the depressed testosterone levels during recovery is not known.

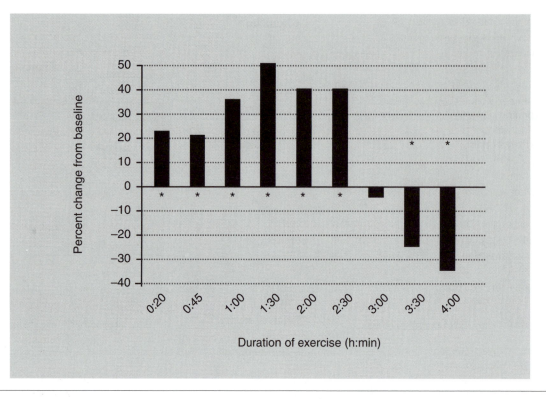

Figure 2.3   Response of testosterone to submaximal exercise of different durations. * = Significant differences from pre-exercise values, p < 0.05.
Data from Dessypris et al. 1976; Galbo et al. 1977; Guglielmini et al. 1984; Kuoppasalmi et al. 1980; Remes et al. 1980; Urhaussen et al. 1987; Wilkerson 1980.

Long-Term Response to Exercise

The relationships between strength, mass, and testosterone levels are not entirely understood. It is assumed that high resting concentrations of testosterone may enhance or facilitate the building of lean tissue (Hickson et al. 1994). This has primarily been the reason for the widespread use of anabolic steroids by power athletes and bodybuilders. The potential of weightlifters to alter androgen levels through prolonged resistance training is unclear. Six months of resistance training in noncompetitive lifters was unable to alter resting testosterone levels (Hakkinen et al. 1985). Even a full year of strength training was unable to alter resting testosterone concentrations in elite weightlifters, although strength increases were observed (Hakkinen et al. 1987). However, after 2 years of resistance training, elite weightlifters were able to significantly increase their resting testosterone concentrations while also improving their strength (Hakkinen et al. 1988). It is possible that changes in resting testosterone concentrations may be a reflection of an advanced adaptive strategy to increase force capability in subjects who have little potential for change in muscle hypertrophy (e.g., highly strength-trained athletes) (Kraemer 1992a).

    Low resting levels of testosterone are frequently observed in endurance-trained athletes (Ayers et al. 1985;

Hackney, Sinning, and Bruot 1988; Wheeler et al. 1984). Resting levels that are 31% (Hackney, Sinning, and Bruot 1988) and 40% (Ayers et al. 1985) below that of sedentary controls have been reported in endurance athletes. The depressed levels of testosterone seen in these athletes may be insufficient to stimulate skeletal muscle growth and may also make it difficult to counteract the catabolic effects of **glucocorticoids** on skeletal muscle.

GROWTH HORMONE

Growth hormone (GH) is a **polypeptide** hormone secreted from the anterior pituitary gland. Its secretion and release are controlled by neurotransmitters of the central nervous system. Physiological stimuli such as sleep, diet, and stress (including exercise) can all stimulate a GH response. The primary physiological role of GH is its involvement in the growth processes of skeletal muscle and other tissues in the body. The actions of GH are mediated to a certain extent by secondary hormones known as insulin-like growth factors (IGF). The basic physiological actions of GH are listed as follows:

- Increase in protein synthesis
- Increase in amino acid transport across cell membrane
- Growth and development of bones

- Reduction of glucose utilization
- Decrease in glycogen synthesis
- Increase in utilization of fatty acids
- Increase in lipolysis
- Metabolic sparing of glucose and amino acids
- Collagen synthesis
- Stimulation of cartilage growth

Growth hormone is secreted in burst-like, pulsatile fashion with its highest values occurring during sleep. These increases are thought to be involved with various tissue repair mechanisms (Kraemer 1992b). Thus, this pattern of GH secretion and release may influence the adaptations of the muscle and its subsequent expression of strength, especially during resistance training programs. It is important to note that GH treatment itself does not promote strength increases when not accompanied by the exercise stimulus (Rogol 1989).

Acute Response to Exercise

The acute GH response to a resistance exercise session is related to specific component variables of the training program. Both volume and intensity of training appear to be important factors in eliciting a GH response. At light exercise loads (28% of 7 RM), no changes in GH concentrations are observed (Van Helder, Radomski, and Goode 1984). When a more moderate exercise intensity is employed (10 RM), a significant increase in GH is observed (Kraemer, Marchitelli, et al. 1990). This increase is significantly greater than that seen after a resistance training program of higher intensity (5 RM). The most dramatic increases in GH were observed when exercise intensity was moderate (10 RM) and the rest intervals were short (1 min) (Kraemer, Marchitelli, et al. 1990). The GH response, similar to what was previously mentioned in the discussion on testosterone, appears to be heightened when training programs of moderate intensity and short rest periods are utilized.

The volume of training also appears to be a potent stimulus in the GH response to exercise. Hakkinen and Pakarinen (1993) demonstrated a greater GH response (4.5-fold difference) in a high-volume, moderate-intensity training protocol (10 sets at 10 RM) versus a low-volume, maximal-intensity training protocol (20 sets at 1 RM). The greater fatigue (greater blood **lactate** concentrations) observed in the high-volume training program most likely contributed to the elevated GH response. The importance of training volume in eliciting significantly greater GH responses has been confirmed in a number of other studies (Kraemer, Marchitelli, et al. 1990; Craig and Kang 1994). Mulligan et al. (1996), controlling for the intensity of exercise, showed that a multiple-set training program (3 sets at 10 RM) elicits a greater GH response during an acute exercise session than a single-set design (1 set at 10 RM).

The GH response to resistance exercise in females appears to be sensitive to changes in acute program variables (e.g., rest, intensity, volume of training) (Kraemer, Gordon, et al. 1991; Kraemer, Fleck, et al. 1993). Those changes do appear to be different from those observed in males and may be related to higher resting concentrations of GH in females during the early follicular stage of the menstrual cycle (Kraemer, Gordon, et al. 1991). Considering the lack of any significant testosterone response in women, GH may have a more primary role in the anabolic adaptations in female skeletal muscle (Hakkinen et al. 1992; Kraemer, Gordon, et al. 1991; Kraemer, Fleck, et al. 1993).

Elevations in GH concentrations are typically reported during aerobic exercise and these elevations are positively related to both the duration and intensity of exercise (Bunt et al. 1986; Chang et al. 1986; Hartley et al. 1972; Karagiorgos, Garcia, and Brooks 1979; Sutton and Lazarus 1976; Van Helder et al. 1986). However, research examining exercise intensity and increases in GH concentrations remains equivocal. Some researchers have suggested that a minimum exercise intensity may be needed to elicit increases in GH concentrations (Chang et al. 1986). Others have suggested that elevations in GH concentrations can occur without any change in blood lactate levels (Hansen 1973), and similar GH responses have been reported between continuous and intermittent exercise despite a greater lactate response during intermittent exercise (Lugar et al. 1992). It has been suggested that exercise above the lactate threshold needs to be for a minimum duration (10 min), but the blood lactate levels cannot predict the amplitude and duration of the GH response (Felsing, Brasel, and Cooper 1992; Weltman et al. 1997).

Long-Term Adaptations to Exercise

Significant increases in the GH response to exercise have been seen after resistance training programs (Hakkinen et al. 1985; Kraemer et al. 1990, 1993). However, resistance training does not appear to alter resting GH concentrations (Hakkinen et al. 1985; Kraemer et al. 1992). A year of exercise training at an exercise intensity above the lactate threshold has been shown to amplify the pulsatile release of GH at rest (Weltman et al. 1992). However, the GH response to an acute exercise stimulus in trained subjects is ambiguous. Training has been reported to decrease (Hartley et al. 1972; Koivisto et al. 1982; Weltman et al. 1997), increase (Bunt et al. 1986), or not affect (Kjaer et al. 1988) the GH response to an exercise bout. A reduction in the GH response to exercise may occur within 3 weeks of training (Weltman et al. 1997). However, these changes may be related to a lower relative intensity used for the postexercise period. When trained subjects exercise at the same relative intensity, accounting for improvements in performance, a greater GH response to the exercise stimulus is seen (Bunt et al. 1986).

INSULIN-LIKE GROWTH FACTORS

Many of the effects of GH appear to be mediated through small polypeptide hormones called **insulin-like growth factor (IGF)** (Kraemer 1992b). Insulin-like growth factors (also called **somatomedins**) are secreted in both hepatic and nonhepatic tissue. Originally, the liver was thought to be the only source of circulating IGF; however, it appears that IGF may also be secreted to a lesser extent in various tissues, including skeletal muscle tissue (Deschenes et al. 1991). Somatomedins are thought to be an important stimulator of the anabolic processes in skeletal muscle, and they may also be involved in the growth processes of bone and connective tissue (Kraemer 1992b).

Exercise Response

Little information is available concerning the IGF response to exercise. Acute increases in IGF have been shown after high-intensity aerobic exercise (Cappon et al. 1994), while both increases (Kraemer, Marchitelli, et al. 1990; Kraemer, Gordon, et al. 1991) and no change (Kraemer, Aguilera, et al. 1995) from preexercise concentrations have been reported after resistance training. Increases in IGF concentrations after resistance training do not appear to be significantly affected by changes in acute program variables (Kraemer, Marchitelli, et al. 1990; Kraemer, Gordon, et al. 1991). In addition, the acute temporal increases of IGF during both aerobic and resistance training do not relate to any increase in the GH response to exercise. Kraemer, Aguilera, et al. (1995) have suggested that this lack of relationship between IGF and GH may be related to the relatively long latency period (3-9 h) reported between GH-stimulated messenger RNA (mRNA) synthesis in hepatic tissue and the peak increase in IGF.

Effect of Training

The effect of training on the IGF response is largely unknown. Kraemer, Aguilera, et al. (1995) have suggested that the lack of change in the IGF response to an acute bout of resistance training in recreationally trained lifters may reflect an ability of such lifters to tolerate an advanced bodybuilding workout. In addition, these individuals may have already reached an upper limit of IGF concentrations in the blood (Kraemer, Aguilera, et al. 1995).

INSULIN

Insulin is a protein hormone secreted by the β-**cells** of the **islets of Langerhans** within the pancreas. The major function of insulin is to regulate glucose metabolism in all tissues except the brain. This is accomplished by facilitating an increase in the rate of glucose uptake into both muscle tissue and fat cells. Glucose that is not used is converted into **glycogen.** If glycogen stores are full, excess **carbohy-**drates are stored as **triglycerides** in adipose tissue. Insulin appears to also increase the rate of amino acid uptake by skeletal muscle and other tissue (Hedge, Colby, and Goodman 1987). However, its role in muscle remodeling may become more prevalent when functioning to decrease the rate of protein degradation within muscle tissue (Deschenes et al. 1991) or to provide the muscle with sufficient nutrients to stimulate muscle growth (Florini 1985).

Exercise Response

Exercise appears to decrease the circulating concentrations of insulin. This is likely the result of the inhibitory effect of **catecholamines** on the β-cells of the pancreas. The reduction of insulin appears to be a function of the duration of exercise. As exercise duration lengthens, a greater decrease in insulin concentrations is seen (Galbo 1981; Koivisto et al. 1980). Insulin levels also appear to decrease during both mild and moderate exercise intensities. However, as the exercise stimulus approaches maximum intensity (e.g., 90% of $\dot{V}O_2$ max), insulin concentrations may not decline (Galbo 1985). The decrease in insulin concentrations during exercise is likely controlled by other hormonal interactions (e.g., catecholamines). The action of insulin becomes more prevalent after exercise. A single bout of exercise enhances insulin sensitivity and skeletal muscle responsiveness to glucose uptake in exercised muscles (Richter et al. 1989). Thus, exercise-induced increased insulin sensitivity of glucose uptake serves to replenish depleted glycogen stores during the postexercise meal.

Effect of Training

Training appears to increase the sensitivity to insulin of both the skeletal muscle and the liver (Devlin et al. 1986; Rodnick et al. 1987). Thus, less insulin is required to regulate blood glucose in trained individuals. Trained individuals also appear to have a less pronounced insulin reduction during exercise than untrained individuals (Bloom et al. 1976).

CORTISOL

Cortisol, a steroid hormone synthesized and released from the **adrenal cortex** of the adrenal gland, is the primary glucocorticoid hormone found in humans. Its synthesis is stimulated by adrenocorticotropic hormone (ACTH), which is secreted by the anterior pituitary gland. The primary physiological function of cortisol is the stimulation of **gluconeogenesis** and the mobilization of fatty acids from body stores. Increases in cortisol appear to be stimulated by stress, diet, immobilization, inflammation, high-intensity exercise, and disease. The following is a list of physiological functions of cortisol:

- Conversion of amino acids to carbohydrates
- Increase in proteolytic enzymes

- Inhibition of protein synthesis
- Increase in protein degradation in muscle
- Stimulation of gluconeogenesis
- Increase in blood glucose concentrations
- Facilitation of lipolysis

Because cortisol is involved with protein degradation of skeletal muscle mass, it is considered to be a catabolic hormone. Much interest has been generated in comparing its response with that of testosterone. A ratio between these two hormones (testosterone/cortisol) has been used in an attempt to examine the anabolic/catabolic status of the body and to relate these measures to changes in performance (Kuoppasalmi and Adlercreutz 1985). Elevations of cortisol are considered to be a marker of catabolic activity in the muscle as well as a marker of physiological stress.

Exercise Response

The acute response of cortisol to a resistance training session appears to be related to the volume of training. In elite weightlifters performing 20 sets × 1 RM, no increase in cortisol concentration was seen from its resting level (Hakkinen and Pakarinen 1993). However, when volume of training was increased (10 sets × 10 RM), a significant increase in cortisol was observed. Other studies have shown that postexercise cortisol concentrations are elevated in the initial stage of a resistance training program in novice weightlifters but remain at resting levels after several weeks of training (Hickson et al. 1994; Potteiger et al. 1995). The elevated cortisol levels seen in elite lifters may reflect an ability of these athletes to push themselves maximally during each training session.

Response to Training

Prolonged aerobic exercise appears to be a potent stimulator of the adrenocortical system. Increases in cortisol appear to be proportional to the intensity of exercise (Farrell, Garthwaite, and Gustafson 1983). However, cortisol levels may not change from baseline levels during exercise at mild to moderate intensities; only when exercise is greater than 70% of $\dot{V}O_2$ max is a consistent increase in cortisol observed (Few 1974). Significant increases in cortisol concentrations also occur during a short bout (1 min) of exercise as long as the exercise is performed at maximal intensity (Buono, Yeager, and Hodgdon 1986). Training appears to lower the cortisol response during prolonged endurance exercise (Tabata et al. 1990). These changes appear to reflect a better maintenance of blood glucose levels in these individuals.

CATECHOLAMINES

The catecholamines (epinephrine, **norepinephrine,** and **dopamine**) are secreted by the **adrenal medulla** of the

adrenal gland and are controlled entirely by sympathetic nervous input. Epinephrine makes up 80% of the total catecholamine secretion from the adrenal medulla (Hedge, Colby, and Goodman 1987), while most of the circulating norepinephrine is derived from sympathetic neurons. Catecholamines are stimulated by hypoglycemia, physical or psychological trauma, circulatory failure, stress, exercise, illness, hypoxia, and cold exposure. The most potent of these stimuli is hypoglycemia. Decreases in blood glucose may elevate adrenal medulla secretion 10- to 50-fold (Hedge, Colby, and Goodman 1987). The direct and indirect actions of catecholamines (as suggested in Kraemer 1992b) on muscle function are as follows:

- Increase in force production
- Increase in contraction rate
- Increase in blood pressure
- Increase in energy availability
- Augmentation of secretion rates of other hormones

Catecholamines may be the most important hormones for the acute expression of strength (Kraemer 1992b). In addition, acute increases in catecholamines may be involved with the potentiation of other hormonal mechanisms (e.g., testosterone, IGF).

Exercise Response

Catecholamine concentrations appear to be elevated during both endurance and resistance exercises. These increases seem to reflect the acute demands and stresses of the training programs (Kjaer 1989; Kraemer et al. 1987). The intensity rather than the duration of exercise appears to be the primary determinant of the catecholamine response (Brooks et al. 1990; Jezova et al. 1985). Even short-duration sprints (several seconds) of maximal intensity appear sufficient to elevate both epinephrine and norepinephrine concentrations (Brooks et al. 1988; Kraemer, Dziados, et al. 1990). Before high-intensity exercise, an anticipatory rise in plasma epinephrine has been reported and is thought to be representative of some preparatory mechanism in advance of the exercise (Kraemer, Gordon, et al. 1991).

During exercise at submaximal intensities, there may be a differential catecholamine response that appears to depend on the duration of exercise. During such exercise, increases in norepinephrine concentrations are seen within 15 min without any increase in epinephrine. As exercise duration increases, epinephrine concentrations may then increase above resting levels (Pequignot et al. 1979). This most likely reflects a greater need for substrate mobilization (hepatic glucose production) during exercise of longer duration.

Response to Training

Training does not appear to alter resting catecholamine concentrations (Kjaer and Galbo 1988; Kraemer et al.

1985). However, trained individuals do appear to have a greater capacity to secrete epinephrine (Kjaer 1989; Kjaer and Galbo 1988) and perhaps norepinephrine as well (Kraemer et al. 1985).

THYROID HORMONES

The **thyroid gland** secretes three hormones. Calcitonin is secreted by parafollicular cells and is involved with the regulation of calcium balance. The thyroid gland also secretes **thyroxine (T_4)** and **triidothyronine (T_3).** These hormones are made up of both iodide and the amino acid tyrosine. Secretion of these hormones is stimulated by **thyroid-stimulating hormone** released by the anterior pituitary gland. T_4 is secreted in greater quantity than T_3, but T_3 is the active and more potent form of the hormone. T_4 can be metabolized to T_3 in numerous tissues in the body, including skeletal muscle and the liver. In fact, 80% of the circulating T_3 is formed by extrathyroidal metabolism of T_4.

The primary function of the thyroid hormones is to increase the basal metabolic rate. Thyroid hormones may also potentiate the glucose uptake caused by insulin. However, an oversecretion of thyroid hormone, which is often seen in hyperthyroid patients, results in an increase in liver glycogen depletion. All aspects of lipid metabolism are stimulated by the thyroid hormones. Although thyroid hormones are required for normal growth, high levels will have a catabolic effect on skeletal muscle.

Exercise Response

Relatively little research is available concerning the response of thyroid hormones to both an acute exercise stress and prolonged training. The response of both T_3 and T_4 to acute exercise is inconclusive. These hormones have been reported to increase (Balsam and Leppo 1975) or not change from resting levels (Galbo et al. 1977) during acute exercise at varying intensities. However, the influence of acute exercise may not be detectable until several days after the exercise session (Galbo 1981).

Response to Training

Several investigations have examined the response of resting thyroid hormones to prolonged exercise training and as a possible hormonal indicator of overtraining (Alen et al. 1993; Hoffman, Epstein, Yarom, et al. 1999; Pakarinen et al. 1988). During prolonged periods of high-intensity training (specifically resistance training), decreases in resting levels of T_4 and T_3 may be observed (Alen et al. 1993; Pakarinen et al. 1988). Further research on the response of the thyroid hormones to prolonged training appears to be needed.

FLUID REGULATORY HORMONES

Maintenance of body fluid and **electrolyte balance** is regulated through the endocrine system and is very important during prolonged exercise, or exercise in a hot environment, when individuals are at a great risk of **dehydration** (loss of body water through excessive perspiration). In an attempt to conserve water, as well as electrolytes, the body will stimulate hormonal action in both the renal and circulatory systems. The principle hormones used in regulating blood volume and electrolyte balance are **antidiuretic hormone (ADH), aldosterone, and angiotensin II.** Figure 2.4 depicts the fluid regulatory hormone response to dehydration.

Antidiuretic Hormone

Antidiuretic hormone (ADH), or arginine vasopressin (AVP), is the hormone principally responsible for regulating fluid balance. ADH acts on the collecting ducts within the nephrons of the kidney, making them more permeable to water, and on arterioles in skeletal muscle and skin to produce its vasoconstrictor effects. It is synthesized and secreted in the posterior pituitary gland. The most important stimulus for ADH secretion is change in plasma **osmolality.** As long as plasma osmolality remains at or below 280 mOsm $\cdot$ L^{-1}, little to no ADH will be secreted. However, ADH is sensitive to changes in the plasma osmolality and will respond proportionately.

ADH may be the most potent constrictor of vascular smooth muscle, perhaps 10 times more active than norepinephrine or angiotensin II for stimulating arteriole contractions (Goodman and Fray 1988). A change in blood volume will be sensed by receptors in both the arterial and venous circulations. However, because of an extensive buffering capacity of the baroreceptors, changes in arterial pressure are seen only with large decreases in blood volume. At normal osmolality, ADH secretion is minimal as long as blood volume remains close to its physiological set point. A decrease in blood volume of 10-15% appears to be the minimum threshold to stimulate ADH secretion. ADH is considered to be an emergency response rather than a fine tuner of blood volume (Goodman and Fray 1988). Because ADH responds to two different inputs (osmolality and blood volume), its sensors need to integrate the signals and respond appropriately. Volume depletion and increased osmolality, as might be seen during dehydration, result in a heightened sensitivity and the stimulation of ADH release.

Aldosterone

Aldosterone is a steroid hormone secreted by the adrenal cortex of the adrenal gland. It is primarily involved with salt and water balance and is stimulated mainly by

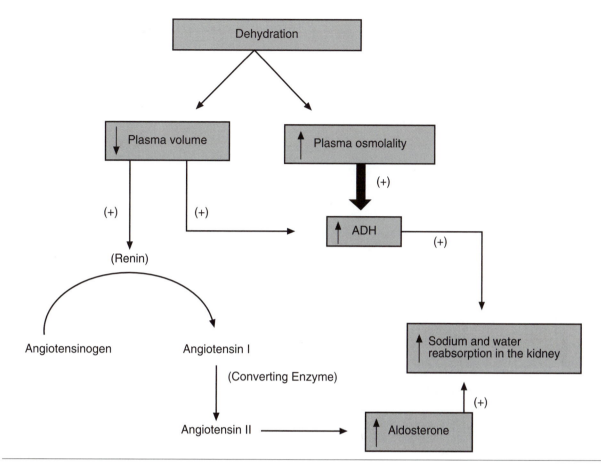

Figure 2.4 Fluid regulatory response to dehydration.

angiotensin II as well as ACTH and high concentrations of potassium. Aldosterone is also stimulated by decreases in blood volume as detected by lowered blood pressure. Its response time is relatively slow (approximately 30 min lag time) compared with ADH (instantaneous). Its principle site of action is in the nephrons of the kidney, specifically on the cortical portion of the renal collecting ducts, to promote absorption of sodium and excretion of potassium. Under the influence of aldosterone, sodium is reabsorbed in exchange for potassium. This exchange of ions is not a one-to-one trade; a much greater concentration of sodium is reabsorbed compared with the amount of potassium lost. Aldosterone also affects the ratio of sodium to potassium in both sweat and saliva.

Angiotensin II

Angiotensin II is a peptide hormone whose primary role is maintaining salt and water balance. It accomplishes this by being the primary stimulator of aldosterone secretion. In addition, it is also a powerful constrictor of vascular tissue and acts centrally to excite sympathetic vasomotor outflow, reinforcing its vasoconstrictor action. It causes elevations in blood pressure, and it is considered to be the most potent pressor agent known. The vasoconstricting action of angiotensin II is not uniform to all vascular beds, thereby causing a redistribution of blood to the brain, heart, and skeletal muscle at the expense of visceral organs and the skin.

Angiotensin II is formed in the blood through a two-step process. Angiotensinogen is secreted by the liver and converted to angiotensin I by the enzyme **renin.** Renin is produced in the kidneys and catalyzes the proteolytic cleavage of angiotensinogen to angiotensin I. Angiotensin I is inactive and is converted to angiotensin II (active form) through additional proteolytic cleavage catalyzed by a converting enzyme. The rate-limiting step of this conversion process is the conversion of angiotensinogen to angiotensin I. Thus, the secretion of aldosterone is regulated by the secretion of renin by the kidneys (Goodman 1988).

Fluid Regulatory Hormones' Response to Exercise

Because the fluid regulatory hormones respond to similar stimuli and all act to defend fluid balance within the body, their responses to exercise will be discussed together. Increases in ADH, aldosterone, and plasma renin activity (PRA, indicator of angiotensin II activity) as a result of exercise are well documented (Convertino et al. 1981; Convertino, Keil, and Greenleaf 1983; Melin et al.

1980; Wade and Claybaugh 1980). As previously mentioned, these hormones conserve body fluid during periods of exercise when fluid and electrolyte loss occurs as a result of sweating. The responses of these hormones appear to be similar between males and females (Maresh, Wang, and Goetz 1985; DeSouza et al. 1989), but aldosterone concentrations may be significantly elevated in the midluteal phase of the menstrual cycle (DeSouza et al. 1989).

The fluid regulatory hormones respond in a graded fashion to the level of hypohydration (loss of body fluid), and the response is further magnified when exercise is performed in a hot environment (Francesconi et al. 1985; Montain et al. 1997). Exercise intensity is also a potent stimulator of AVP, aldosterone, and PRA secretion patterns. A significantly greater response of these hormones is observed when comparing high-intensity exercise with low-intensity exercise (Freund et al. 1991; Montain et al. 1997). Rehydration with water during exercise reduces or abolishes the response of ADH and PRA, but aldosterone levels decline only when an isotonic solution is consumed (Brandenberger et al. 1986).

Fluid Regulatory Hormones' Response to Training

Training reduces the fluid regulatory hormone response when exercise is performed at the same pretraining exercise load (same absolute load). When comparing trained versus untrained individuals performing exercise at the same relative intensity, a greater response of these hormones is seen (Convertino, Keil, and Greenleaf 1983). It is well understood that prolonged training results in an increase in the fluid volume reserve, primarily because of plasma volume expansion. Thus, trained individuals are able to maintain a higher circulating volume even though they have a greater absolute loss in plasma volume because of an increased sweat rate.

OPIOIDS AND EXERCISE

Several endogenous peptides secreted by the anterior pituitary gland produce analgesia upon binding to their receptors in the brain. **Proopiomelanocortin (POMC)** is a large molecule from which other active molecules are split by enzymatic cleavage. POMC is the source of ACTH and endogenous **opioids.** These opioids can be subdivided into three groups (**endorphins, enkephalins,** and **dynorphins)** stemming from three major precursor molecules. Opioid receptors are seen in both the central and peripheral nervous systems (Akil et al. 1984). However, peripheral stimulation of these receptors is thought to be more prevalent when opioids are used in pharmacological dosages.

The primary function of the opioids is controlling pain. Stimulation of opioid receptors causes a reduction in the pain response to harmful stimuli, while pain relief is re-versed when naloxone (an opioid antagonist) interacts with the opioid receptor. Opioids can also act as neurotransmitters by inhibiting the gonadotrophic hormones LH and FSH (Kraemer et al. 1992) and stimulating GH and prolactin release (McArdle, Katch, and Katch 1996).

Increases in endogenous opioids, primarily β-**endorphins,** have been consistently reported during endurance exercise (Donevan and Andrew 1987; Farrell et al. 1987; Schwarz and Kindermann 1989). Elevations in β-endorphins have also been reported during both interval sprint (Fraioli et al. 1980) and resistance exercise (Kraemer, Dziados, et al. 1993). However, the β-endorphin response during these modes of exercise (sprint and resistance) depends on single component variables of the exercise program.

The magnitude of the β-endorphin response during and after exercise depends on the intensity of the exercise stimulus (Farrell et al. 1987; Goldfarb et al. 1990). During supramaximal exercise of very short duration, β-endorphins show no significant change from resting levels (Kraemer et al. 1989). A minimum duration of exercise appears to be necessary to stimulate β-endorphin release. An increase in β-endorphins is related to elevations in the stress hormones (e.g., ACTH and cortisol) and apparently requires a certain degree of fatigue to be reached. Similarly, during resistance exercise, elevations in endorphin levels are noted when the exercise paradigm calls for moderate intensity (10 RM) of training with short rest periods (1 min) (Kraemer, Dziados, et al. 1993). Resistance training programs of greater intensity (5 RM to 8 RM) or of longer rest periods (3 min)—less fatiguing programs—do not appear to be of sufficient stress to cause an endorphin response (Kraemer, Dziados, et al. 1993; Pierce et al. 1994).

The physiological significance of elevated endorphin levels after exercise is unclear. However, it is associated with the "high" or euphoria felt after exercise. In addition, endorphins are also thought to be involved in pain tolerance, improved appetite control, and reduction in anxiety, tension, anger, and confusion—all of which are considered to be the psychological benefits of exercise (Morgan 1985; McArdle, Katch, and Katch 1996; O'Connor and Cook 1999).

SUMMARY

The primary function of hormones is to influence the rate of cellular reactions. Both physical and psychological stresses are potent stimulators in elevating the secretory patterns of hormones. The hormonal response to exercise is influenced by single component variables of the exercise program. These components (intensity of exercise, volume of exercise, choice of exercise, rest intervals, and recovery status) need to be well designed to optimize the desired endocrine response.

CHAPTER 3

METABOLIC SYSTEM AND EXERCISE

©Human Kinetics

Many of the physiological adaptations during prolonged training relate to an improved ability to generate more **energy** and to use this energy more efficiently. The adaptations seen as a consequence of training are specific to the mode of training performed. **Endurance training** results in both **metabolic** and **morphological** changes that enhance the ability to bring nutrients to the muscle, allow the muscle to more efficiently use these nutrients, and increase the ability of the muscle to generate more energy. **Anaerobic training** causes metabolic adaptations that are much different from what is commonly seen during **aerobic training** and that are more specific to the needs of the particular athlete. These adaptations are primarily aimed at enhancing the ability of the muscle to generate energy specific for that energy system and at improving the athlete's tolerance to muscle acid/base imbalances through an improved buffering capacity.

The primary focus of this chapter is the metabolic adaptations subsequent to exercise training that enhance the athlete's ability to perform. The following brief review on **bioenergetics** will allow the reader to better appreciate the metabolic adaptations discussed during the remainder of the chapter. These adaptations will be discussed specific to the type of training program performed (i.e., aerobic or anaerobic).

Bioenergetics is primarily concerned with the source of energy for muscular contractions. There are three physiological systems in the body that yield energy. Two of these systems can function without the benefit of oxygen and are termed anaerobic. Specifically, these energy systems are called the **phosphagen energy system (ATP-PC)** and the **glycolytic energy system.** The third system requires oxygen for its energy production and is termed aerobic, or the **oxidative energy system.**

The energy for all cellular functions is derived from the **metabolism** of various substances stored within the muscle (e.g., **glycogen** or **triglycerides**) or within storage sites in the body (e.g., **adipose** tissue). A set of metabolic reactions occurs within each cell to produce a potential store of chemical energy. This energy drives all cellular processes that do not otherwise proceed spontaneously. The principal energy component that governs all cellular functions is adenosine triphosphate (ATP). The ATP molecule has three inorganic phosphate (Pi) groups attached to an adenosine molecule. The enzyme ATPase causes the removal of a phosphate group from ATP to form adenosine diphosphate (ADP) and Pi plus the release of a large amount of energy. The process by which ATP can be formed is called **phosphorylation** and can proceed as mentioned previously through three different sources.

ATP-PC ENERGY SOURCE

ATP-PC is stored within muscle and is available for immediate use. Like ATP, PC (**phosphocreatine**) has a phosphate group and a high-energy bond attached to a **creatine** molecule. Unlike ATP, in which the breakdown of ATP to ADP produces energy for direct use in cellular function, the Pi group removed from creatine, facilitated by the enzyme **creatine kinase,** can only be used to combine with an ADP molecule to reform ATP.

The ATP-PC energy system is the simplest of the three energy systems. Oxygen is not required to release the energy from ATP-PC, thus it is considered an anaerobic energy source. However, only a limited amount of ATP and PC is available within the muscle, and during maximal exercise, the supply will be exhausted within 30 s. Although the ATP-PC energy system is available for a relatively short period of time, there are several advantages in its use as an energy source. Basically, it's the energy source that is readily available for immediate use. It also has a large power capacity, providing the muscle with a large amount of energy within a short period of time. These characteristics make the ATP-PC energy source ideal for short-duration, high-intensity events (e.g., 100-m sprint, shot put, long jump).

An interesting and important question is whether the ATP content of the cell ever reaches a level at which the force-generating capacity of the muscle or the cycle rate of the actin/myosin cross-bridge is compromised. From a number of different studies, it appears that ATP-PC cellular concentrations do not reach such a critical level. Fatigue produced from other factors appears to reduce the ATP utilization rate before ATP concentrations become self-limiting (Bergstrom and Hultman 1988). In fact, ATP concentration within skeletal muscle may not fall below 70% of its resting level even during extreme cases of fatigue (Fitts 1992).

The decline in ATP utilization during maximal exercise is related to a large extent to the decrease in PC concentrations within the cell, as well as to an increase in cellular H^+ concentrations generated from the highly anaerobic event. Although ATP concentrations do not seem to be completely expended during maximal exercise, PC levels decline rapidly to the point of complete exhaustion as it is used to replenish diminished ATP levels. This relationship between ATP and PC concentrations in skeletal muscle during maximal exercise is depicted in figure 3.1.

The ability to perform repetitive, highly intensive exercise (e.g., football or basketball) requires a rapid regeneration of phosphagens. The resynthesis of PC is biphasic in nature. Initially, there is a rapid recovery with a $t_{1/2}$ (half-time) of 20-30 s, followed by a slower phase of recovery that may require up to 20 min (Harris et al.

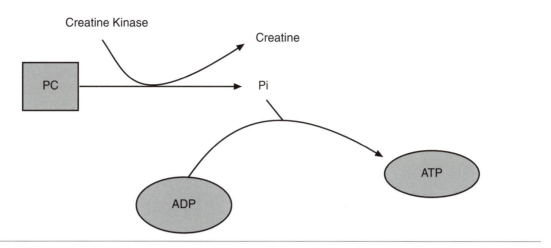

Figure 3.1 Relationship between ATP and PC.

1976). However, most of the PC is regenerated by 3 min postexercise.

GLYCOLYTIC ENERGY SOURCE

There is an additional energy source that produces ATP through the breakdown of a glucose molecule. This process of metabolizing glucose is called **glycolysis** and results in the net production of 2 or 3 ATP, depending on where the glucose molecule was derived from. Glycolysis results in the release of energy and the breakdown of a glucose molecule, through a chain of chemical reactions, into a compound called **pyruvic acid.** A description of glycolysis can be seen in figure 3.2. Because this energy system can produce ATP without the need of oxygen, it is also considered an anaerobic energy source.

The glucose metabolized during glycolysis comes from the blood either through the digestion of carbohydrates or from the breakdown of glycogen (storage form of glucose) from the liver. Glucose can also be metabolized from glycogen stored within the working muscle cells. The process of metabolizing glycogen into glucose is called **gluconeogenesis.** In all tissues, glycogen is metabolized into **glucose-1-phosphate** by the enzyme **phosphorylase** and is further broken down into **glucose-6-phosphate.** Once glucose-6-phosphate is formed, the process of glycolysis can begin. The importance of the phosphorylated glucose molecules should not be lost. The phosphate molecule attached to each glucose molecule prevents it from diffusing out of the cell. However, the liver possesses a specific phosphatase enzyme that degrades glucose-6-phosphate into glucose and Pi. This allows the glucose molecule to diffuse into the circulation and reach tissues that require additional glucose. No other tissue has the ability to dephosphorylate glucose, allowing it to be transported to tissues in need.

Metabolizing glucose into pyruvic acid is a 10-step process. If glycolysis begins with the breakdown of stored

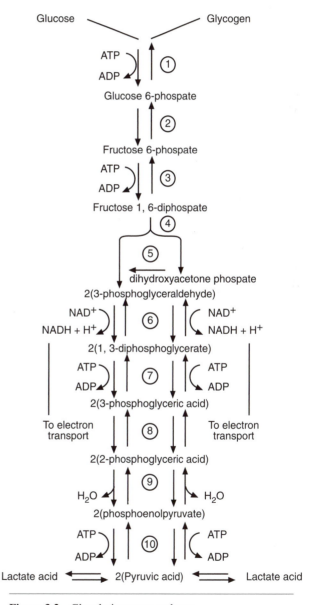

Figure 3.2 Glycolytic energy pathway.

Adapted, by permission, from J. Wilmore and D. Costill, 1999, *Physiology of sport and exercise* (Champaign, IL: Human Kinetics), 122.

glycogen, a net of 3 ATP is produced from its complete metabolism. However, if glycolysis begins from glucose, then only a net total of 2 ATP molecules is produced, because 1 ATP is used for the conversion of glucose to glucose-6-phosphate. Because no oxygen is present during anaerobic glycolysis, the pyruvic acid will be converted to **lactic acid.**

The accumulation of lactic acid within the muscle has several harmful effects that interfere with muscle function. An increase in lactic acid lowers **muscle pH** (becomes more acidic) and is often felt as a burning sensation in the muscle after high-intensity exercise. The burning sensation within the exercising muscle during such exercise reflects the activation of sensory receptors **(nociceptors)** that are sensitive to elevations in H^+. Increases in lactic acid also interfere with chemical processes affecting the production of ATP, as well as hinder muscle contraction capability by preventing the binding of calcium to troponin.

Despite these unwanted side effects, glycolysis can produce a larger amount of energy than ATP-PC. However, it cannot supply as much energy per unit time and, therefore, is not as powerful as the ATP-PC energy source. Glycolysis is the primary energy source for high-intensity exercise lasting 1-3 min.

OXIDATIVE ENERGY SOURCE

The oxidative energy system utilizes oxygen in the production of ATP and is therefore referred to as an aerobic energy source. The oxidative production of ATP occurs within the **mitochondria** of all cells. In skeletal muscle, the mitochondria are located adjacent to the myofibrils and throughout the sarcoplasm. The oxidative production of ATP cannot produce enough ATP per unit time to provide energy to sustain highly intense activity. However, because of the abundance of stored fat and carbohydrates in the body, the oxidative energy system can provide ample energy for prolonged periods of submaximal exercise. Thus, this is the primary energy system used for long duration aerobic events.

Aerobic metabolism begins in the same way as glycolysis, with the breakdown of glycogen into glucose and the subsequent conversion of glucose to pyruvic acid. However, in the presence of oxygen, the pyruvic acid is converted into **acetyl coenzyme A** and enters into a series of chemical reactions called the **Krebs cycle** and the **electron transport chain** (see figure 3.3a & b).

The Krebs cycle is a series of chemical reactions that produces **carbon dioxide** (expired through the lungs) and hydrogen. The hydrogen combines with the coenzymes nicotinamide adenine dinucleotide (NAD) and flavin adenine dinucleotide (FAD) and transports them from the cell cytoplasm to the mitochondria where they enter the electron transport chain. The hydrogen atoms involved in the electron transport chain are split into **protons** and **electrons.** The hydrogen protons combine with oxygen to form water, while the electrons pass through a series of reactions that phosphorylate ADP to form ATP. This process is also known as oxidative phosphorylation.

Oxidative metabolism uses primarily carbohydrates and fat. However, during periods of carbohydrate depletion, starvation, or prolonged exercise, significant amounts of protein can be metabolized for energy as well. While at rest, the body derives most of its energy from stored fat. However, the body begins to metabolize a greater percentage of stored carbohydrate during exercise. The oxidative metabolism of 1 molecule of glycogen produces a net gain of 39 ATP (see table 3.1).

The use of stored fat as an energy source relies exclusively on the breakdown of triglycerides stored within both adipose sites and muscle. The process of breaking down fat for energy is called **lipolysis** and results in triglycerides being metabolized into a **glycerol** molecule and three **free fatty acids.** It is the free fatty acids that are used as the primary energy source. As the free fatty acids enter the mitochondria, they undergo a further catabolic process called **β-oxidation.** This involves the enzymatic cleavage of a free fatty acid molecule into acetyl-CoA. The acetyl group then enters the Krebs cycle through the **citrate synthase reaction** and is oxidized in the same way that carbohydrates are oxidized through aerobic glycolysis. The energy production from the oxidation of 1 molecule of a fatty acid such as **palmitic acid** yields 129 ATP (see table 3.1).

INTERACTION OF THE ENERGY SOURCES

Although one energy source may be the predominant system working at any given time, all three sources of energy supply a portion of the needed energy (ATP) for exercise at all times. Thus, the ATP-PC source also provides energy at rest, and the oxidative energy source is also used during maximal exercise. The more intense the exercise, the greater the portion of ATP derived from anaerobic energy sources. As intensity of exercise decreases and exercise duration increases, energy production is primarily from aerobic metabolism. There is no exact point at which one energy source drops off and another energy source begins to provide more energy. Rather there is a gradual transition from one energy source to another.

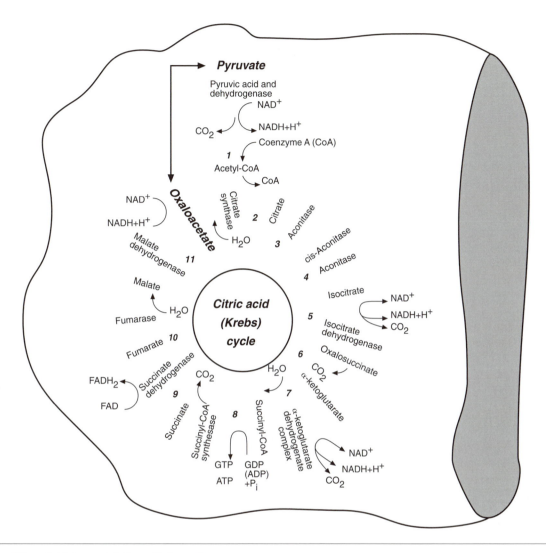

Figure 3.3a Oxidative metabolism: Krebs cycle.
Adapted, by permission, from J. Wilmore and D. Costill, 1999, *Physiology of sport and exercise* (Champaign, IL: Human Kinetics), 122.

METABOLIC ADAPTATION TO ENDURANCE TRAINING

Endurance training, such as prolonged running or cycling, results in profound physiological adaptations that cause significant improvements in exercise capacity. Improved endurance performance is the result of adaptations in a number of different physiological systems (e.g., cardiovascular, neuromuscular). This section examines the metabolic adaptations to endurance training.

Capillary Density

Endurance-trained men have been shown to have a 5-10% greater **capillary density** than sedentary control subjects (Hermansen and Wachtlova 1971; Ingjer 1979). Other studies examining highly trained endurance athletes have reported an even larger disparity in capillary density (37-50% differences) in comparison with untrained individuals (Jansson, Sylven, and Sjodin 1983; Saltin and Rowell 1980). Although the greater capillary density of endurance athletes may be a function of genetics, a 15% increase in the capillary content of skeletal muscle has been reported after an endurance training program (Ingjer 1979). The greater capillary content after endurance training is an adaptation that specifically enhances aerobic exercise performance, allowing for a greater exchange of gases, heat, waste, and nutrients between the blood and exercising muscle. These increases appear to occur within several weeks or months after the

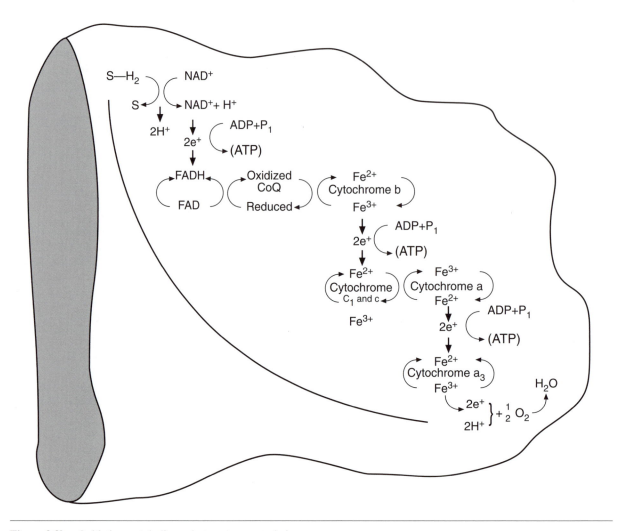

Figure 3.3b Oxidative metabolism: electron transport chain.
Adapted, by permission, from J. Wilmore and D. Costill, 1999, *Physiology of sport and exercise* (Champaign, IL: Human Kinetics), 33.

Table 3.1 Energy Production (ATP) from Glycolysis and Oxidative Metabolism

Energy source	Glycolysis	Oxidative phosphorylation (Krebs cycle and electron transport)
Glucose	2	38
Glycogen	3	39
Palmitic Acid	–	129

onset of an endurance training program (Wilmore and Costill 1999).

Myoglobin Content

Myoglobin is the oxygen-transporting and storage protein of muscle. It shuttles the oxygen molecules from the capillaries to the mitochondria. Several animal studies have shown that the myoglobin content of skeletal muscle can be increased during prolonged endurance training (Hickson 1981; Froberg 1971). However, similar results have not been achieved in humans. Several studies have failed to see any changes in myoglobin concentration from baseline levels after an endurance training program (Coyle et al. 1985; Jansson, Sylven, and Sjodin 1983). The specific role of myoglobin in improving aerobic capacity in humans remains unclear.

Mitochondrial Function and Content

The oxidative energy-producing capability of skeletal muscle is remarkably improved after endurance training. These improvements have been partially attributed to an enhancement in mitochondrial function. Increases in the size and number of mitochondria have been reported after endurance exercise programs (Holloszy 1988; Holloszy and Coyle 1984). In a study of rats, mitochondrial content was shown to increase by 15%, while the size of the mitochondria increased by 35% during 27 weeks of endurance training (Holloszy et al. 1970).

Oxidative Enzymes

An important metabolic adaptation seen subsequent to endurance training is an increase in concentration of the enzymes involved in the Krebs cycle and electron transport chain and the enzymes responsible for the activation, transport, and β-oxidation of free fatty acids.

The increase in these enzymes allows for a more efficient metabolic system for oxidizing nutrients to form energy (ATP). In addition, the greater concentration of the oxidative enzymes is also thought to spare muscle glycogen and reduce the production of lactate during exercise of a given intensity (Holloszy and Coyle 1984).

In untrained individuals, the concentration of mitochondrial enzymes appears to be twice as high in type I (slow-twitch) fibers than in type II (fast-twitch) fibers (Holloszy and Coyle 1984). Mitochondrial density may also be higher in type I versus type II fibers. During endurance training, the oxidative enzymes appear to increase at a greater rate in the type II oxidative fibers, making the difference between enzyme concentrations of type I and type II fibers negligible or even nonexistent in highly trained aerobic athletes (Holloszy 1988).

Succinate dehydrogenase (SDH) and citrate synthase are enzymes of the Krebs cycle that are often measured to provide a quantitative analysis of the improvement in oxidative potential of endurance-trained individuals. In a rodent model, a twofold increase in these enzymes has been reported after endurance training (Holloszy 1975; Holloszy et al. 1970). Similar findings have also been seen in studies on humans (Gorostiaga et al. 1991; Green et al. 1995). A moderate amount of daily exercise (20 min per day) appears to be an adequate stimulus to significantly increase oxidative enzymes (see figure 3.4). Training for a longer duration (60-90 min per day), as might be experienced by endurance athletes, may cause an even greater increase in the oxidative enzymes. However, these changes appear to occur primarily during the initial stages (first few months) of training and may plateau even when training volume is further elevated (see figure 3.5).

Increases in oxidative enzymes do not correlate well with changes in maximal aerobic capacity ($\dot{V}O_2$ max), suggesting that other factors (e.g., circulation) may have a greater influence on improving aerobic capacity (Gollnick et al. 1972). Increases in the concentration of oxidative enzymes in endurance athletes may be more important for allowing exercise at a higher intensity (e.g.,

©Human Kinetics

running at a faster pace) than for improving aerobic capacity (Wilmore and Costill 1999).

Glycolytic Enzymes

Unlike the oxidative enzymes, the enzymes of glycolysis do not appear to be significantly affected by prolonged endurance exercise (Holloszy and Booth 1976). Endurance training places a greater demand on the oxidative energy system and, therefore, is more effective in increasing the mitochondrial enzymes.

Training's Effect on Carbohydrate Utilization

As previously mentioned, the benefit of an increase in oxidative enzymes and mitochondrial density appears to

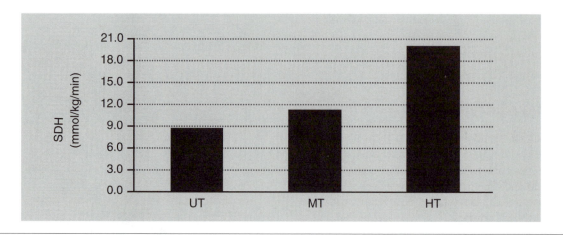

Figure 3.4   Succinate dehydrogenase activity of the gastrocnemius muscle in untrained (UT) and moderately trained (MT) joggers and highly trained (HT) marathon runners.
Adapted, by permission, from J. Wilmore and D. Costill, 1999, *Physiology of sport and exercise* (Champaign, IL: Human Kinetics), 190.

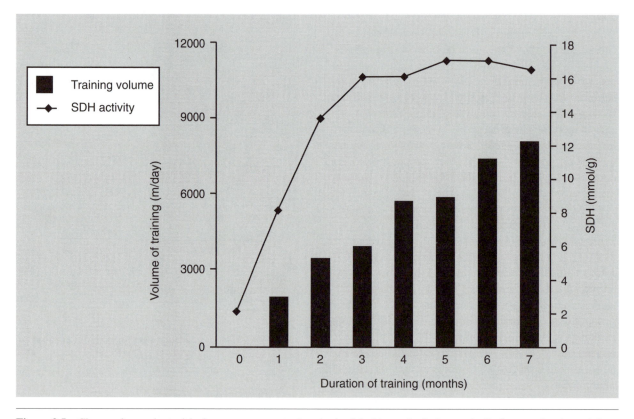

Figure 3.5   Changes in succinate dehydrogenase concentrations in the deltoid muscle during gradually increased swim training.
Adapted, by permission, from J. Wilmore and D. Costill, 1999, *Physiology of sport and exercise* (Champaign, IL: Human Kinetics), 189.

be in the sparing of muscle glycogen. This enhances the ability to transport **pyruvate** produced during glycolysis into the larger mitochondrial volume and through the process of oxidative phosphorylation. As a result, there is a reduced buildup of lactic acid and a lowering of the ATP to ADP ratio during submaximal exercise. Thus, there will be a reduction in carbohydrate utilization (gly-

cogen degradation) and an improvement in exercise tolerance at submaximal intensities.

Training's Effect on Fat Utilization

After endurance training programs, there is an increased reliance on stored fat as a source of energy during submaximal exercise (Gollnick and Saltin 1988), largely dependent on the mechanisms previously discussed concerning the glycogen-sparing effect. In addition, the increase in capillary density after endurance training provides an enhanced opportunity for the exchange of free fatty acids from adipose tissue to the exercising muscle.

It has traditionally been accepted that fat diffuses across the plasma membrane of the cell. However, recent research has suggested that fat may also be transported across the plasma membrane through a carrier-mediated transport system (Turcotte 2000). This transport system, made up of proteins embedded in the plasma membrane, appears to respond to chronic endurance training (Turcotte et al. 1999). Thus, increases in fat utilization may also be related to changes in its uptake mechanism into the muscle cell.

Endurance training may also cause a greater increase in free fatty acid concentrations in plasma. This is thought to reflect a greater release from adipose storage sites, contributing to the sparing effect on muscle glycogen (Costill et al. 1977). However, elevated free fatty acid concentrations have not been consistently demonstrated in all studies (Gollnick and Saltin 1988), which may be related to the greater uptake and oxidation of fats seen in endurance-trained individuals compared to sedentary controls.

METABOLIC ADAPTATIONS TO ANAEROBIC EXERCISE

High-intensity training such as sprinting (running, swimming, or cycling) or competing in high-intensity sports such as basketball or hockey causes adaptations that are specific to the anaerobic energy system. As described earlier, the anaerobic energy system is made up of both the ATP-PC and glycolytic systems. This section discusses the metabolic adaptations seen in exercise programs that emphasize these energy systems.

ATP-PC System's Adaptations to High-Intensity Training

High-intensity training appears to cause little or no change in resting ATP or PC concentrations (Karlson, Dumont, and Saltin 1971; MacDougall et al. 1977; Troup, Metzger, and Fitts 1986). Whether the resting concentrations of the enzymes (creatine kinase and **myokinase**) that catalyze the phosphagen energy system can be positively altered by high-intensity training is unclear. Costill, Coyle, and colleagues (1979) were unable to

demonstrate any change in these enzymes after 6 s of maximal exercise (knee extensions). In contrast, another study using a similar exercise protocol (5 s of knee extensions) showed significant elevations in these enzymes (Thorstensson 1975). Other research using a different mode of exercise (high-intensity cycle ergometer training) also met with conflicting results. Parra et al. (2000) reported significant elevations in creatine kinase after 2 weeks of daily sprint training. However, when training was prolonged to 6 weeks with longer rest intervals between workouts, similar increases in resting enzyme concentrations were not seen. This was similar to other studies that examined high-intensity cycle ergometer exercise for 15 weeks and failed to see any changes in creatine kinase levels after training (Simoneau et al. 1987).

Even when exercise duration is increased, changes in the ATP-PC enzymes are still not consistent. When exercise consisted of 30 s of continuous knee extensions, significant elevations in both creatine kinase and myokinase were seen (Costill, Coyle, et al. 1979). In contrast, Jacobs et al. (1987) were unable to find any significant change in muscle creatine kinase levels after 6 weeks of high-intensity training (15- and 30-s maximal sprints on a cycle ergometer). The lack of any consistency in the ATP-PC enzyme response to exercise is difficult to explain. There is minimal data available concerning high-intensity, short-duration exercise programs and changes in these specific enzymes, and the available studies have used a variety of exercise protocols. It is possible that a longer duration of training may be needed to stimulate changes in resting creatine kinase levels or that there is an upper limit to creatine kinase concentrations within the muscle that cannot be altered with training. This latter hypothesis is explored further when creatine kinase supplementation is discussed in chapter 16.

Glycolytic System's Adaptations to High-Intensity Training

As high-intensity exercise is prolonged, energy is derived primarily from the glycolytic energy system. Studies examining exercise training using bouts of exercise of 30 s or more have shown significant elevations in the glycolytic enzymes (**phosphofructokinase,** phosphorylase, **lactate dehydrogenase)** (Costill, Coyle, et al. 1979; Houston, Wilson, et al. 1981; Jacobs et al. 1987). The increase in the concentrations of these enzymes might enhance the glycolytic capacity, allowing the muscle to maintain a high intensity of exercise for a longer period of time. Increases between 10 to 25% in these glycolytic enzymes have been seen with dynamic high-intensity training programs.

The increase in glycolytic enzymes appears to depend on the mode of exercise. In the previously mentioned studies, the training involved either high-intensity running, cycling, or swimming. In contrast, studies examining the effect of resistance training on changes in glycolytic enzyme concentrations have been unable to show any significant alterations in these enzymes (Sale et al. 1990a, 1990b; Tesch, Komi, and Hakkinen 1987). Apparently, resistance training alone is unable to stimulate any metabolic adaptation in the glycolytic enzymes. These studies suggest that athletes training for anaerobic sports with a large strength component (e.g., football) need to include both resistance training and sprint or interval exercises in their conditioning programs in order to maximize their physiological adaptation for the sport.

Oxidative Enzymes' Adaptations to High-Intensity Exercise

High-intensity exercise that stimulates increases to glycolytic enzymes also appears to significantly increase mitochondrial enzyme activity (oxidative enzymes) (Dudley, Abraham, and Terjung 1982; Troup, Metzger, and Fitts 1986). However, it appears that these increases are more prevalent when the duration of high-intensity exercise exceeds 3 min (Fitts 1992). In addition, the in-crease does not reach the magnitude typically seen after prolonged endurance training. The implications of an increase in oxidative enzymes from anaerobic training programs suggest that individuals who train anaerobically may still be able to generate some improvements in aerobic capacity. Table 3.2 highlights the changes in selected muscle enzymes from aerobic or anaerobic training programs.

High-Intensity Exercise and Buffering Capacity

High-intensity exercise (e.g., sprinting, cycling, swimming) results in an accumulation of lactic acid within the exercising muscle. Elevation in lactic acid causes a lowering of muscle pH and the onset of muscle fatigue. Training programs that stress the anaerobic energy system improve the buffering capacity within the muscle, changing the ability of the muscle to tolerate high concentrations of lactic acid. Buffers such as **bicarbonate** and muscle phosphates combine with the H^+ released from lactic acid to maintain **acid/base** balance within the exercising muscle. This prevents the onset of fatigue and also allows the individual to exercise with a higher concentration of lactic acid within the muscle. Change in buffering capacity has been demonstrated to increase 12-50% within 8 weeks of high-intensity exercise on a cycle ergometer

Table 3.2   Changes in Muscle Enzyme Concentrations ($mmol \cdot g^{-1} \cdot min^{-1}$) from Aerobic and Anaerobic Training Programs

	Untrained	Anaerobically trained	Aerobically trained
AEROBIC ENZYMES			
Succinate dehydrogenase	8.1	8.0	20.8*
Malate dehydrogenase	45.5	46.0	65.5*
ANAEROBIC ENZYMES			
ATP-PCr system			
Creatine Kinase	609.0	702.0*	589.0
Myokinase	309.0	350.0*	297.0
GLYCOLYTIC SYSTEM			
Phosphorylase	5.3	5.8	3.6*
Phosphofructokinase	19.9	29.2*	18.9
Lactate dehydrogenase	766.0	811.0	621.0

* = Significantly different from untrained value

Data from Wilmore and Costill 1999.

(Sharp et al. 1986), whereas blood lactate concentrations have been shown to increase 9.6% after 6 weeks of high-intensity cycling (Jacobs et al. 1987).

Lactate Production

High-intensity training programs increase the tolerance of trained individuals to large concentrations of lactic acid because of an improved buffering capacity within the skeletal muscle. However, endurance training does not appear to stimulate an improvement in the buffering capacity of muscle. As might be expected, the effects of aerobic training on lactate production are different from the effects of anaerobic training.

During exercise of submaximal intensity, both blood and skeletal muscle lactate concentrations are lower in the trained state than in the untrained state (Holloszy and Booth 1976; Hurley et al. 1984). Comparisons of blood lactate concentrations in sedentary men before and after a 12-week endurance exercise training program can be seen in figure 3.6. Common belief is that lower lactate production in the trained individual is related to a greater oxygenation of the exercising muscle, possibly from the increase in blood volume and capillary density seen after endurance training programs. However, some evidence suggests that during submaximal exercise the blood flow per gram of muscle may actually be lower in the trained state compared with the untrained state (Holloszy 1988). This is thought to evolve from a reduced diversion of blood flow from the periphery (skin) or internal organs (liver) to the muscles during submaximal exercise. The exercising muscles in the trained state compensate for the lower blood volume by extracting more oxygen, resulting in a larger arteriovenous oxygen difference.

Lower lactate production in trained individuals during submaximal exercise may also be related to greater reliance on fat as the primary energy source for ATP generation. In addition, the greater mitochondrial content in the muscle after endurance training programs reduces the available pyruvate for conversion to lactate. A greater proportion of the pyruvate generated through glycolysis is channeled into the mitochondria for oxidative metabolism. Another mechanism that may contribute to reduced lactate concentrations in the blood (which reflects muscle lactate production) is an increase in the rate of lactate removal (Holloszy 1988).

SUMMARY

Metabolic adaptations are specific to the type of exercise training program employed. Endurance training results in improved capacity to generate energy through oxidative metabolism. This is reflected by increases in capillary density, mitochondrial size and content, and oxidative enzymes. Anaerobic training results in elevations in the glycolytic enzymes as well as enhanced buffering capacity within skeletal muscle. In addition, although high-intensity training does not rely on oxidative metabolism as its primary energy source, the increase in oxidative enzymes during high-intensity exercise programs suggests that slight improvements may also be seen in aerobic capacity during such training programs.

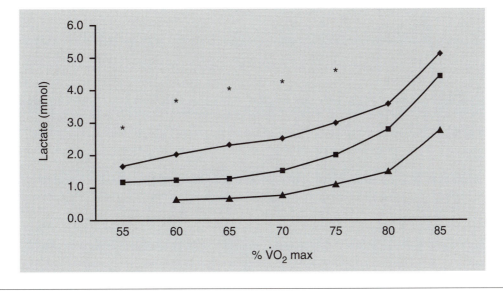

Figure 3.6 Blood lactate concentrations at the same relative exercise intensities. Sedentary men studied before (♦) and after (■) a 12-week endurance training program. Competitive long-distance runners (▲) regularly training.

Data from Hurley et al. 1984.

CHAPTER 4

CARDIOVASCULAR SYSTEM AND EXERCISE

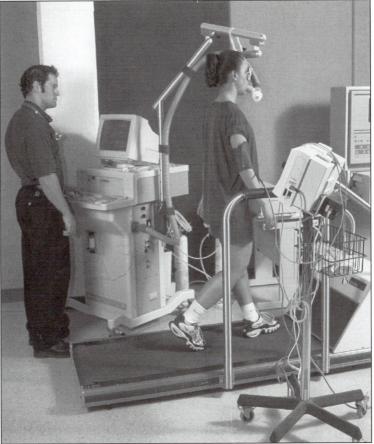

©Human Kinetics

At rest, the heart provides approximately 5 L of blood per minute to meet the energy demands of the average person. As the metabolic demands increase, as might be expected during exercise, the heart is able to compensate by increasing the volume of blood that it pumps into circulation. Cardiac output during exercise can increase more than fourfold in an average person, and in an elite endurance athlete, cardiac output may reach 40 L · min⁻¹. Like any other muscle, the heart will adapt to the increased demands placed on it during prolonged exercise training. These adaptations are specific to the type of exercise stimulus that is presented. This chapter reviews the cardiovascular adaptations that are seen during acute exercise and prolonged endurance and resistance training. The close relationship between the cardiovascular and respiratory systems and the effect that exercise has on improved cardiorespiratory function are also discussed.

OVERVIEW OF CARDIOVASCULAR SYSTEM

The cardiovascular system consists of an elaborate network of **vessels** (the circulatory system) and a powerful pump (**the heart**). It is responsible for delivering oxygen and nutrients to active organs and muscles and removing the waste products of metabolism. The heart is a four-chambered muscular organ located in the midcenter of the chest cavity. Its anterior border is the sternum, and its posterior border is the vertebral column. The diaphragm is inferior to the heart, and the lungs are situated on the lateral borders. Approximately two-thirds of its mass lies to the left of the body's midline. The longitudinal axis of the heart from its base to its apex is directed anterior-inferior and 45° to the left of the midline.

Morphology of the Heart

The heart muscle, referred to as the **myocardium,** is similar in appearance to striated skeletal muscle. However, the fibers of the myocardium are multinucleated and interconnected end to end by **intercalated disks.** These disks contain **desmosomes,** which maintain the integrity of the cardiac fibers during contraction, and **gap junctions** that allow for a rapid transmission of the electrical impulse that signals for contraction. The structure of the myocardium can be thought of as three separate areas: **atrial, ventricular,** and **conductive.** The atrial and ventricular myocardium function similarly to skeletal muscle in that they will contract in response to electrical stimuli. However, an electrical stimulus of only a single cell in either chamber will result in an **action potential** being rapidly spread to the other cells of the atrial and ventricular myocardium, resulting in a coordinated contractile mechanism. In addition, the cardiac fibers in each of these areas can function separately. The conductive tissue that is found

between these chambers provides a network for the rapid transmission of conductive impulses, allowing for coordinated action of both the atrial and ventricular chambers.

The structural detail of the heart can be seen in figure 4.1. There is a striking difference in the anatomy and physiology of the right and left sides of the heart that relates to their specific functions. The right side of the heart (right **atrium**) receives blood from all parts of the body, and the right **ventricle** pumps deoxygenated blood to the lungs through the **pulmonary circulation.** The left atrium receives oxygenated blood from the lungs and pumps this blood from the left ventricle into the **aorta** and through the entire **systemic circulation.** The left ventricle is an ellipsoidal chamber surrounded by thick musculature that provides the power to eject the blood through the entire body. The right ventricle, however, is crescent shaped with thin musculature, reflecting the reduced ejection pressures seen in this ventricle (25 mmHg) compared with approximately 125 mmHg in the left ventricle at rest. A thick solid muscular wall, or **interventricular septum,** separates the left and right ventricles.

Blood flow from the right atrium to the right ventricle goes through the **tricuspid valve** (consisting of three cusps or leaflets that allow only a unidirectional flow of blood). The **bicuspid** or **mitral valve** allows blood flow between the left atrium and left ventricle. The **semilunar valves,** located on the arterial walls of the outside of the ventricles, prevent blood from flowing back into the heart between contractions. During **systole,** the cusps lie against their arterial wall attachments; during **diastole** or during retrograde flow, the cusps fall passively inward, sealing the **lumen.**

Cardiac Cycle

The **contraction phase** in which the atria or ventricles expel the blood in their chambers is called systole. The **relaxation phase** in which these chambers refill with blood is referred to as diastole. The **cardiac cycle** is the total time spent in one complete revolution of systole and diastole. At rest, the heart spends most of its time (approximately 60%) filling with blood (diastole) and less time (approximately 40%) expelling the blood (systole). However, during exercise this situation is reversed, with most of the cardiac cycle spent in systole. During systole, the tricuspid and mitral valves are closed. However, blood flow from pulmonic and systemic circulation continues into the atria. As systole ends, the atrioventricular valves rapidly open and the blood that has accumulated in the atria flows quickly into the ventricles, accounting for 70-80% of the ventricular filling. This period of rapid filling accounts for one-third of diastole. The middle one-third of diastole is characterized by very little blood flow into the ventricle and is referred to as **diastasis.** During the last one-third of diastole, ventricle filling is completed with an additional 20-30% of

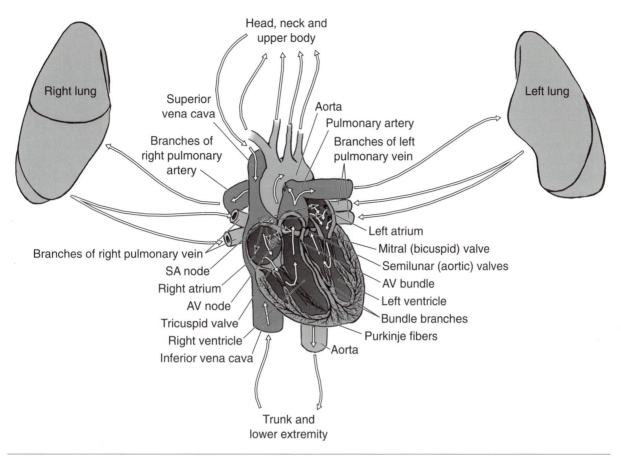

Figure 4.1 Structural detail of the heart.

blood pumped into the ventricle as the result of atrial systole.

The volume of blood in the ventricle at the end of diastole is called the **end-diastolic volume** (EDV). Two main phases occur during systole: **pre-ejection and ejection.** The pre-ejection phase includes an electromechanical lag, which is the time delay between the beginning of ventricular excitation (**depolarization**) and the onset of ventricular contraction and **isovolumic contraction.** Isovolumic contraction is the phase in which intraventricular pressure is raised before the onset of ejection. This part of the pre-ejection phase occurs between the closure of the mitral valve and the opening of the aortic semilunar valve During the ejection phase, the blood within the ventricle is pumped into the systemic circulation through the opening of the semilunar valve. This phase ends with the closing of the semilunar valve. The blood remaining in the ventricle at the end of ejection is referred to as **end-systolic volume (ESV).** The difference between EDV and ESV is called the **stroke volume (SV).** The proportion of the blood pumped out of the left ventricle with each beat is called the **ejection fraction (EF)** and is determined by SV ÷ EDV. The ejection fraction averages about 60% at rest. This sim-

ply means that 60% of the blood in the left ventricle at the end of diastole will be ejected with the next contraction. Figure 4.2 depicts a complete cardiac cycle.

Heart Rate and Conduction

A unique feature of the heart is its ability to contract rhythmically without either neural or hormonal stimulation. This autorhythmicity is due to a specialized intrinsic conduction system that consists of the **sinoatrial node (SA node), internodal pathways, atrioventricular node (AV node),** and **Purkinje fibers.**

The SA node is located in the right atrium and is a collection of specialized cells that are capable of generating an electrical impulse. Because of this distinctive ability, it is appropriately nicknamed the pacemaker of the heart. Once an impulse leaves the SA node, it propagates leftward and downward, spreading through the **atria synctium** of first the right and then left atria along internodal pathways to the AV node, which is located toward the center of the heart on the lower right atrial wall.

The **AV node,** or the AV junction (made up of the AV node and the **bundle of His**), delays transmission of the

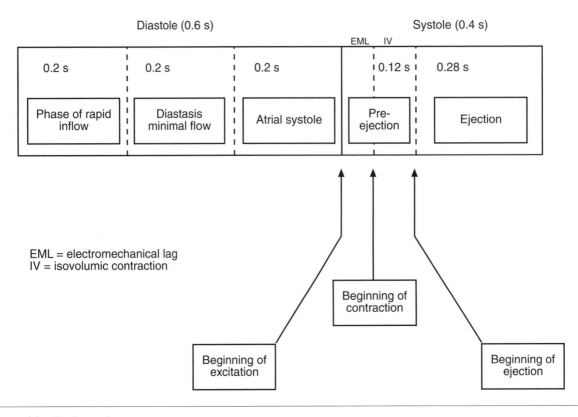

Figure 4.2   Cardiac cycle.

impulse for 0.1 s. This slight delay of ventricular excitation and contraction allows the atria to contract and also permits a limitation in the number of signals that are transmitted by the AV node. This appears to serve as a protective mechanism for the ventricles from atrial tachyarrhythmias. The bundle of His is found distally in the AV junction and divides into a right and left segment (**bundle branches**) that transmit the electrical impulses to the right and left ventricles, respectively. The Purkinje fibers are found on the distal tips of the right and left bundle branches and extend into the walls of the ventricles, accelerating the conduction velocity of the impulse to the rest of the ventricle. The conduction velocity of the Purkinje fibers may increase fourfold compared with the bundle of His.

As mentioned earlier, the SA node, AV node, and Purkinje fibers have the inherent ability for spontaneous initiation of the electrical impulse. However, the autonomic nervous system can also influence the rate of impulse formation (**chronotropy**), contractile state of the myocardium (**inotropy**), and the rate of spread of the excitation impulse. The **sympathetic** and **parasympathetic** nervous systems, as well as certain hormones, can influence cardiac contractility. The atria are well supplied with both sympathetic and parasympathetic neurons, whereas the ventricles are primarily innervated by

sympathetic neurons. Sympathetic stimulation releases the catecholamines epinephrine and norepinephrine from sympathetic neural fibers. These neural hormones accelerate heart rate by increasing SA node activity, and they increase both atrial and ventricular contractile force. Increases in heart rate are termed **tachycardia.**

Parasympathetic stimulation through the **vagus nerves** releases the neurohormone **acetylcholine,** which depresses SA node activity and decreases atrial contractile force. Decreases in heart rate are termed **bradycardia.** Sympathetic stimulation may increase heart rate by over 120 beats · min⁻¹ and strength of contraction by 100%, whereas maximal vagal stimulation may decrease heart rate by 20-30 beats · min⁻¹ and lower strength of contraction by approximately 30% (Adamovich 1984).

Cardiac Output

Cardiac output is the product of heart rate and stroke volume. It generally refers to the amount of blood that is pumped by the heart in 1 minute. Cardiac output responds to the energy demands of the body and varies considerably between people. On average, the total blood volume pumped out of the left ventricle is approximately 5 L · min⁻¹ for an adult male. This volume is similar for trained and sedentary males. In the untrained male, these 5 L of blood are sustained with a heart rate of 70 beats · min^{-1}. Thus, stroke volume would need to be approximately 71 ml · beat^{-1}. In the endurance athlete, heart rate is generally much lower at rest because of greater vagal tone and reduced sympathetic drive. If the heart rate of the endurance athlete was 50 beats · min^{-1}, the stroke volume would be 100 ml · beat^{-1}. A comparison of the cardiac output in trained and sedentary males can be seen in table 4.1. The mechanism that drives this particular adaptation is not entirely clear but is likely related to the increased vagal tone seen after endurance training and to some of the morphological adaptations of the heart, which will be discussed later.

Vasculature

The **vascular system** is composed of a series of vessels that carry oxygenated blood away from the heart to the tissues **(arterial system)** and return deoxygenated blood from the tissues back to the heart **(venous system).** The heart has its own coronary vascular system responsible for supplying the myocardium with oxygen and nutrients. The arterial system receives the blood pumped from the left ventricle of the heart and distributes it throughout the body via a network of **arteries, arterioles** (small arterial branches), **metarterioles** (smaller branches), and **capillaries.** The left ventricle pumps the blood from the heart into a thick elastic vessel called the aorta. The blood is then circulated throughout the body through the previously mentioned arterial network. The walls of the arteries are strong and thick to withstand the rapid transport of blood under high pressure to the tissues. Their thickness prevents any exchange of gases between the arterial vessels and the surrounding tissues. In addition, these vessels are richly innervated by the sympathetic nervous system, which allows them to be effectively stimulated for regulating blood flow. As the blood reaches the tissues, it becomes diverted to smaller branches of the arterial system. At the end of the metarterioles (the smallest arterial vessels) are the capillaries, which consist of a single layer of **endothelial cells.** The capillaries are microscopic in size (approximately 0.01 mm in diameter) but may contain approximately 5% of the total blood volume at rest. Be-

cause of this small diameter, the rate of blood flow decreases as the blood circulates toward and into the capillaries. In addition, extensive branching of the capillary microcirculation creates a large surface area between the capillary vasculature surrounding the tissue. This large surface area, the slow rate of blood flow, and the thin layer of endothelial cells make the capillaries an ideal place for gas exchange between the blood and the tissues.

As the blood leaves the capillaries it enters the venous circulation. The venous system comprises vessels increasing in size as they get closer to the heart. Deoxygenated blood leaving the capillaries enters **venules** (small veins), and the rate of blood flow is increased (because of the smaller cross-sectional area of the venous system in comparison with the capillary system). The blood is transported back to the heart via the **superior vena cava** (venous blood returning from areas above the heart) and the **inferior vena cava** (venous blood returning from areas below the heart). The deoxygenated blood then enters the right atrium, goes through to the right ventricle, and is pumped to the lungs to be reoxygenated and subsequently transported back into the left side of the heart to be circulated through the arterial circulation.

During rest, blood flow is controlled by the **autonomic nervous system** and is primarily distributed to the liver, kidneys, and brain. However, during exercise there is a redistribution of the blood flow to the exercising muscles. The muscles may receive 75% or more of the available blood at the expense of the other organs. In combination with a greater cardiac output, the exercising muscles may receive up to a 25-fold increase in blood flow. The flow of blood to the muscles and organs at rest and during exercise can be seen in figure 4.3.

Blood Pressure

In a resting state, as the blood is pumped into the aorta from the left ventricle during contraction (systole), the pressure within the aorta increases to approximately 120 mmHg. This measurement is referred to as the **systolic blood pressure** and represents the strain against the arterial walls during ventricular contraction. Since the pumping action or contraction of the left ventricle of the heart is pulsatile in nature, the arterial pressure fluctuates between its high level during systole to a lower level seen during the relaxation phase of the heart (diastole). **Diastolic blood pressure** is approximately 80 mmHg at rest and provides an indication of the **peripheral resistance,** or ease at which blood flows into the capillaries. As blood flows through the systemic circulation, the pressure continues to progressively fall to approximately 0 mmHg as it reaches the right atrium. The decrease in arterial pressure during each segment of the systemic circulation is directly proportional to the vascular resistance

Table 4.1 Cardiac Output at Rest in Sedentary and Endurance Trained Males

	Cardiac output = heart rate · stroke volume
Sedentary	4,970 mL = 70 beats · minute^{-1} · 71 mL
Trained	5000 mL = 50 beats · minute^{-1} · 100 mL

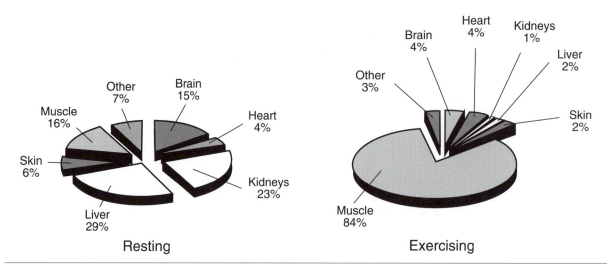

Figure 4.3   Distribution of cardiac output while resting and exercising.

in that segment. Changes in the resistance of the systemic circulation are important for the regulation of blood flow.

OVERVIEW OF RESPIRATORY SYSTEM

The coordination between the cardiovascular and respiratory systems provides the body with an efficient means to transport oxygen to the tissues and remove carbon dioxide. During respiration, air is breathed in (**inspiration**) though the nasal cavity or mouth. From there the air travels through the **pharynx, larynx, trachea,** and into the lungs. Once in the lungs, the air flows through an elaborate system of branches termed **bronchi** and **bronchioles** that expand the surface area for gas exchange (see figure 4.4). From the bronchioles the air reaches the **alveoli,** the smallest respiratory unit, where gas exchange with the pulmonary circulation occurs. The lungs are located in the chest cavity (**thorax**) but do not have any direct attachment to the ribs or any other bony structure. Instead, they are suspended by **pleural sacs** that connect to both the lungs and thoracic cavity. A fluid is present between the pleural sacs and lungs to prevent friction during respiration.

Inspiration and Expiration

During inspiration, the muscles of the thoracic cavity (**diaphragm** and **external intercostal**) contract, causing the thorax to expand and the lungs to stretch and fill with additional air. The lung expansion causes a reduced pressure gradient and the pressure within the lungs is reduced to levels below that on the outside, causing air to rush in. During exercise, additional muscles (e.g., pectorals, sternocleidomastoid) may be recruited, causing a

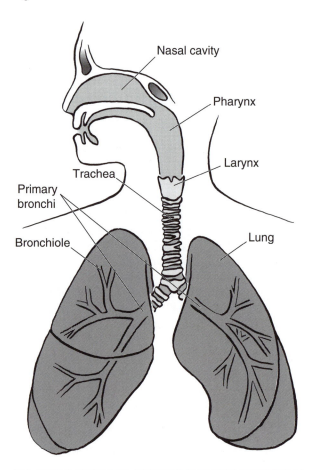

Figure 4.4   Anatomy of the respiratory system.

greater movement of the thorax and creating an even larger lung expansion.

When air is breathed out (**expiration**), the inspiratory muscles relax. In the case of forced expiration, contraction of the **internal intercostal** and abdominal muscles causes the thorax to return to its normal position. As a result, the pressure within the lungs expands to levels above that outside and expiration occurs.

Change in pressure is the primary reason for air and gases to flow into and out of the lungs and through the entire respiratory and circulatory systems. For **ventilation** (the process of inspiration and expiration) to occur, only small changes in pressure between the lungs and the outside environment are required. For instance, standard atmospheric pressure is 760 mmHg and only a slight change in **intrapulmonary pressure** (pressure within the lungs) causes air to be inhaled. This process is not as simple at altitude and will be explained in much greater detail in chapter 20.

Pressure Differentials in Gases

In addition to changes in pressure that cause inspiration and expiration, pressure differentials in the air also result in both oxygen and carbon dioxide exchange. The air we breathe is a mixture of gases. Each gas exerts a pressure in proportion to its concentration in the gas mixture, known as its **partial pressure.** Air is made up of 79.04% nitrogen, 20.93% oxygen, and 0.03% carbon dioxide. Thus, at sea level in which atmospheric pressure is 760 mmHg, the partial pressure of oxygen is 159.1 mmHg (20.93% of 760 mmHg) and carbon dioxide is 0.2 mmHg (0.03% of 760 mmHg).

As the air reaches the alveoli, the partial pressures of the gases in the alveoli and the partial pressures of the gases in the blood create a pressure gradient. This is the basis of gas exchange. If the partial pressures of the gases on either side of the membrane are equal, no gas exchange occurs. The greater the pressure gradient, the faster the gases will diffuse across the membrane. As the inspired air moves into the alveoli, the partial pressure of oxygen (PO_2) is between 100 and 105 mmHg (due to mixing of air within the alveoli). However, alveolar gas concentrations remain fairly stable (Wilmore and Costill 1999). The pressure gradient between the capillaries and the alveoli is depicted in figure 4.5. At the pulmonary capillary, blood has been stripped of most of its oxygen by the tissues. Typically, the PO_2 at the pulmonary capillary is between 40 and 45 mmHg. As you can see, the pressure gradient favors oxygen going from the alveoli to the capillary. In addition, the pressure gradient of carbon dioxide favors exchange from the capillary to the alveoli, where it can be exhaled from the body during expiration. The pressure gradient for carbon dioxide is not as great at the capillary-alveoli membrane as it is for oxygen. Nevertheless, carbon dioxide diffuses easily across the membrane, despite the low pressure gradient, because of a greater membrane solubility than oxygen.

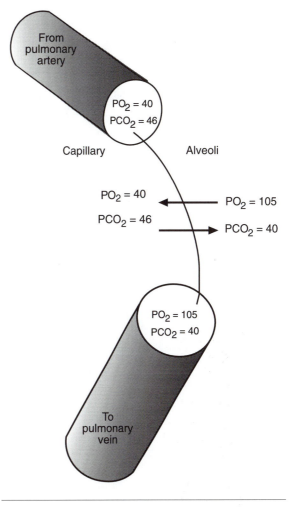

Figure 4.5 Pressure gradient between the capillary and alveoli within the lungs.

Oxygen and Carbon Dioxide Transport

Oxygen is transported in the blood either combined with **hemoglobin** (98%) or dissolved in blood plasma (2%). Each molecule of hemoglobin can carry four molecules of oxygen. The binding of oxygen to hemoglobin depends on the PO_2 in the blood and the affinity between oxygen and hemoglobin. The greater the PO_2, the more saturated the hemoglobin molecules are with oxygen. In addition, the temperature and pH of the blood (see figure 4.6) also affect the affinity between oxygen and hemoglobin. As the pH of the blood decreases, the affinity that hemoglobin has to oxygen is decreased and oxygen is released. The rightward shift of the curve is known as the **Bohr effect** and is important during exercise when a greater amount of oxygen is needed in the exercising tissues. On the other hand, when the pH is high, as it would be in the lungs, there is a greater affinity between oxygen and hemoglobin. This is important in order to saturate the hemoglobin molecules with oxygen.

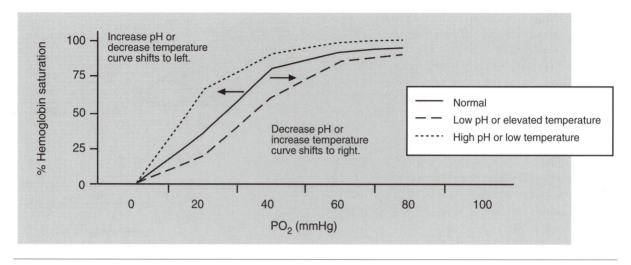

Figure 4.6   Oxygen saturation curve.

In the average male, there is approximately 14-18 g of hemoglobin in each 100 ml of blood. In the female, the concentration of hemoglobin ranges from 12 to 16 g per 100 ml of blood. Each gram of hemoglobin can bind 1.34 ml of oxygen. Thus, for males, the oxygen-carrying capacity of hemoglobin fully saturated with oxygen is approximately 18-24 ml per 100 ml of blood, while in females the range is approximately 16-22 ml per 100 ml of blood. At rest, normal oxygen saturation is approximately 95-98% (Pruden, Siggard-Anderson, and Tietz 1987).

Carbon dioxide transport in the blood occurs primarily in the form of **bicarbonate ions** (approximately 60-70%). Carbon dioxide is also transported dissolved in the **plasma** (7%-10%) or bound to hemoglobin. When bound to hemoglobin it forms the molecule **carbaminohemoglobin.** However, it does not compete with oxygen since it has its own binding site on the **globin molecule.** In contrast, the binding site for oxygen is on the **heme molecule.** As carbon dioxide diffuses from the muscle to the blood, it combines with water to form **carbonic acid.** This very unstable acid quickly dissociates, releasing a hydrogen ion (H^+) and forming a bicarbonate ion (HCO_3^-). The H^+ binds to hemoglobin and causes the Bohr effect to occur, whereby hemoglobin loses its affinity for oxygen and increases the rate of diffusion of oxygen into the tissues. An example of this action is depicted in figure 4.7.

CARDIOVASCULAR RESPONSE TO ACUTE EXERCISE

Oxygen consumption ($\dot{V}O_2$) is elevated during acute exercise to meet the higher energy needs of the exercising muscle. As exercise intensity increases, a greater demand for energy is met by an increase in the cardiac output or by a greater oxygen extraction from the vasculature [a greater $(a\text{-}\bar{v})O_2$ difference]. During the early stages of exercise, rapid increases in both heart rate and stroke volume bring about elevations in cardiac output. Figure 4.8 demonstrates the effects of varying intensities of exercise on heart rate, stroke volume, and cardiac output.

Cardiac Output During Acute Exercise

Cardiac output at rest is approximately 5 L. However, during maximal exercise, cardiac output may increase up to 20 L in young, sedentary males, and in young, endurance-trained male athletes, cardiac output may reach up to 40 L. In examining this considerable difference in cardiac output, we can see that the maximal heart rate for individuals from both these groups (assuming that both men are 20 years old) is approximately 200 beats · min^{-1} (maximal heart rate = 220 – age). Thus, a difference in stroke volume must account for the large differences seen in cardiac output. In our example, the stroke volume of the sedentary male is approximately 100 ml · $beat^{-1}$, whereas the stroke volume in the endurance-trained athlete may reach 200 ml · $beat^{-1}$.

The importance of a large cardiac output for the endurance athlete is reflected by the linear relationship seen between cardiac output and oxygen consumption (Lewis et al. 1983). This relationship is seen not only in adults but also in children and adolescents (Cunningham et al. 1984) and between trained and untrained individuals (Hermansen and Saltin 1969).There appears to be a 6:1 ratio between maximal cardiac output and $\dot{V}O_2$max (McArdle, Katch, and Katch 1996).

Heart Rate During Acute Exercise

Heart rate elevation during exercise is primarily controlled by sympathetic stimulation from the higher somatomotor centers of the brain. The heart rate response is directly

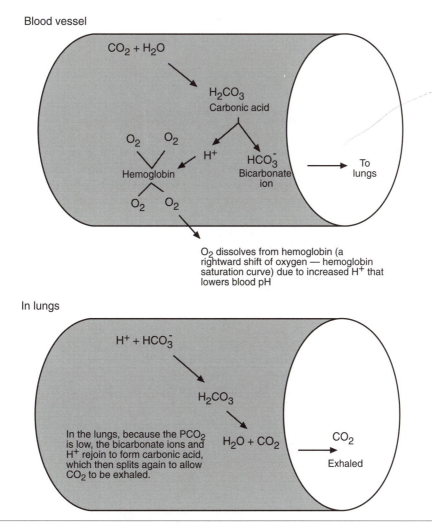

Blood vessel

$CO_2 + H_2O$

H_2CO_3
Carbonic acid

O_2 O_2
Hemoglobin

H^+ HCO_3^-
Bicarbonate
ion

O_2 O_2

To
lungs

O_2 dissolves from hemoglobin (a
rightward shift of oxygen — hemoglobin
saturation curve) due to increased H^+ that
lowers blood pH

In lungs

$H^+ + HCO_3^-$

H_2CO_3

In the lungs, because the PCO_2
is low, the bicarbonate ions and
H^+ rejoin to form carbonic acid,
which then splits again to allow
CO_2 to be exhaled.

$H_2O + CO_2$

CO_2

Exhaled

Figure 4.7   Effect of carbon dioxide release into the blood.

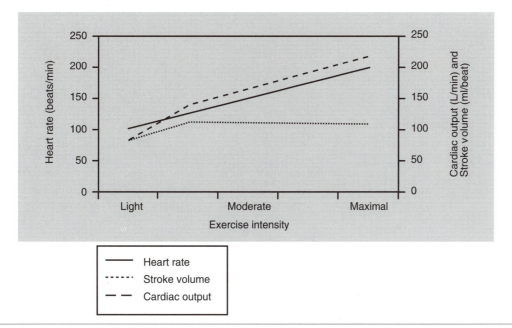

Figure 4.8   Effect of exercise intensity on heart rate, stroke volume, and cardiac output.

proportional and linear to the intensity of exercise. As intensity of exercise increases, the heart rate will continue to increase until exercise reaches maximal intensity (see figure 4.9). At maximal intensity, the heart rate will plateau, indicating that the individual is reaching his or her maximal level.

Initial increases in heart rate are also related to a withdrawal of parasympathetic input. This occurs during low-intensity exercise. As exercise continues in duration, or increases in intensity, a greater sympathetic stimulation becomes the driving force in elevating heart rate. Sympathetic activation occurs from feedback mechanisms in both peripheral mechanical and chemical receptors that monitor changes in pH, hypoxia, temperature, or other metabolic variables that can alter sympathetic drive.

During certain activities, an increase in heart rate can be seen before the onset of exercise. This anticipatory rise in heart rate appears to be primarily related to sprint or anaerobic-type events (McArdle, Katch, and Katch 1996). As the length of the exercise event increases [from a 60-yard (55-m) sprint to a 2.0-mile (3.2-km) run], the preexercise heart rate becomes lower. This pattern of an anticipatory heart rate response to high-intensity exercise may be a "feed-forward" mechanism to provide for a rapid mobilization of bodily reserves, controlled by the central command center in the medulla of the brain (McArdle, Katch, and Katch 1996). Such a mechanism does not appear warranted for longer duration events.

The more times that the heart beats per minute, the greater the volume of blood pumped into the circulation. However, there is a limit to this effect. As the heart rate rises above a certain level, the strength of each contrac-tion may decrease because of metabolic overload. More important, the greater rate of contraction results in less time spent in diastole. The time between contractions becomes so reduced that there is not sufficient time for the blood to flow from the atria to the ventricles. Thus, the total volume of blood made available to the circulation is reduced. This is why during artificial electrical stimulation the heart rate will only be elevated to between 100 and 150 beats · min^{-1}. However, sympathetic stimulation results in a stronger systolic contraction, decreasing the time during systole and thereby allowing a greater time for filling during diastole. Elevations in heart rate from sympathetic stimulation result in a heart rate between 170 and 250 beats · min^{-1}.

Stroke Volume During Acute Exercise

Increases in stroke volume are accomplished early during exercise primarily through an increase in left ventricular end-diastolic volume (EDV). This rapid augmentation of stroke volume is due to the **Frank-Starling mechanism,** which is related to the increased volume of blood that returns to the heart during exercise. With a greater volume of blood returning to the heart, the ventricles become stretched to a greater extent than normal and respond with a more forceful contraction. This stronger contraction results in a greater volume of blood entering the systemic circulation with each heart beat. This mechanism appears to occur early during exercise and at a relatively low level of exercise intensity. The Frank-Starling mechanism may cause an approximate 30-50% increase in stroke volume (Bonow 1994). As exercise continues, increases in EDV reach a plateau

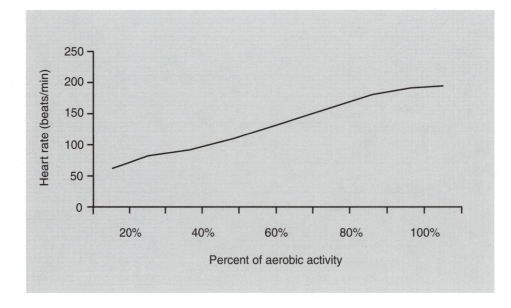

Figure 4.9   Relationship between heart rate and exercise intensity.

while exercise intensity is still submaximal. Further increases in stroke volume are attributed to the enhanced left ventricular contractile function (controlled by enhanced sympathetic stimulation), resulting in a greater decrease in ventricular end-systolic volume.

There are two mechanisms that appear to be responsible for the increase in EDV during exercise. The initial mechanism uses the exercising muscles as a pump to increase the rate of return of blood to the heart. This would be expected to increase the pressures within the ventricular cavity during filling, thereby raising diastolic pressure. However, this does not occur in the healthy heart and, in contrast, the relaxation seen in the left ventricle reduces the ventricular pressure below that of the left atrium. This causes the mitral valve to open and the onset of ventricular filling. As mentioned earlier, the enhanced sympathetic response during exercise increases the relaxation time during diastole. During this time, the increase in size of the left ventricle causes a further reduction in pressure, creating a suctioning effect that draws additional blood into the chamber. This facilitation of the suctioning mechanism by sympathetic drive is the secondary mechanism that contributes to the increased stroke volume and is crucial in the recruitment of the Frank-Starling mechanism (Bonow 1994).

Cardiac Drift

As exercise duration is prolonged, or when exercise is performed in a hot environment, a gradual increase in heart rate and a decrease in stroke volume may occur even when exercise intensity is maintained. This has been referred to as **cardiac drift** and may also be accompanied by decreases in arterial and pulmonary pressures (Wilmore and Costill 1999). The drift is thought to occur from a greater percentage of circulating blood being diverted to the skin in an attempt to dissipate body heat caused by an increased core temperature. The greater concentration of blood in the periphery and a loss of some plasma volume to sweat results in a reduced blood return to the heart. This decrease in EDV results in a reduced stroke volume. Heart rate is elevated to compensate for the change in stroke volume and to maintain cardiac output.

Arteriovenous Oxygen Difference During Exercise

At rest, a person with normal hemoglobin concentration has approximately 200 ml of oxygen in every 1 L of blood. With a normal cardiac output of $5 L \cdot min^{-1}$ at rest, a potential of 1 L of oxygen is available to the body. However, only 250 ml, or 25%, of the available oxygen is extracted from arterial blood during rest, leaving the remaining 750 ml of oxygen available for reserve. This is called the $(a-\bar{v})O_2$ difference.

During exercise, the amount of oxygen extracted from the arterial blood is increased. Up to 75% of the available oxygen may be used by the exercising muscles. The increase in oxygen extraction appears to be related to the intensity of exercise and may be further enhanced after endurance training programs. The ability to extract oxygen from the blood and the total blood volume available to the muscles is critical for determining the aerobic capacity of the individual. This is reflected by the **Fick equation:**

$$\dot{V}O_2max = \text{maximal cardiac output} \times \text{maximal } (a-\bar{v})O_2 \text{ difference}$$

There may be very little difference between moderately trained individuals and endurance athletes in the ability to extract oxygen, despite large differences in $\dot{V}O_2max$. Therefore, the primary factor determining aerobic capacity appears to be cardiac output (McArdle, Katch, and Katch 1996).

Distribution of Cardiac Output During Exercise

As discussed earlier and depicted in figure 4.3, there is up to a 25-fold increase in blood flow during exercise. However, most of the blood is diverted to the exercising muscles. The extent of this shunting depends on the environmental condition and possibly other factors including type of exercise and fatigue. The shunting of blood is generally accomplished by diverting blood to the exercising muscles from organs or areas of the body that can tolerate a reduction in blood flow. However, certain organs such as the heart cannot function without a normal blood flow and do not compromise their blood supply during exercise.

Blood Pressure Response to Acute Exercise

Blood pressure typically increases during dynamic exercise such as walking, jogging, or running. In the healthy individual, this increase is seen only in the systolic response. Increases in systolic blood pressure are linear in nature and may exceed 200 mmHg during maximal exercise (either endurance or resistance). However, these increases do not parallel the four- to eightfold changes that may be seen in cardiac output. The systolic blood pressure appears to be buffered to a large extent by the decrease in peripheral resistance caused by vasodilation in the vasculature of the exercising muscles (MacDougall 1994). The decrease seen in peripheral resistance also appears to account for the minimal to no change observed in diastolic pressures. Diastolic pressure may also decrease during higher intensity bouts of exercise. The blood pressure response to dynamic endurance exercise can be seen in figure 4.10.

During exercise that involves the upper body only, both systolic and diastolic blood pressures are higher than when exercise is performed with only the legs (Toner, Glickman, and McArdle 1990). This is thought to occur because of the relatively smaller muscle mass and vasculature of the arms. Even when these vessels are maximally dilated, they do not have the same effect on peripheral resistance as lower-body exercise. Other possible explanations include a greater involvement of a **Valsalva** (forced expiration against a closed or partially closed glottis) or partial Valsalva maneuver or that a given absolute power output with arm exercise represents a greater relative exercise intensity than performing similar exercise with the lower body (MacDougall 1994). Regardless of the mechanism, the higher pressor response seen with upper-body exercise has important implications for determining the exercise prescription for individuals with coronary heart disease.

During resistance exercise, large increases in both systolic and diastolic blood pressures are evident (MacDougall et al. 1985, 1992; Sale et al. 1993). During maximal efforts that involve a large muscle mass, intra-arterial blood pressures exceeding 350/250 mmHg in healthy young men have been reported (MacDougall et al. 1985). The large pressor response during resistance training is the result of a compression of the vasculature within the contracting muscles and a Valsalva maneuver. The magnitude of the pressor response is also related to the relative size of the muscle mass involved and the intensity of the effort. Blood pressure increases with each repetition in a set and then drops rapidly to below resting levels after the last repetition (MacDougall et al. 1985; Sale et al. 1993). This transient decrease is likely related to the large vasodilation of the vasculature that was occluded during muscle contraction and may contribute to the dizziness sometimes experienced after an intense exercise session.

A major portion of the large pressor response to resistance training is attributed to the Valsalva maneuver (MacDougall 1994). The Valsalva maneuver does not appear to take effect until exercise intensity reaches 80-85% of 1 RM or when fatigue sets in (MacDougall et al. 1985, 1992). During a Valsalva maneuver, a rapid increase in intrathoracic pressure causes an immediate and direct effect on the arterial tree, resulting in an increase in both systolic and diastolic blood pressures (MacDougall et al. 1992). However, if the Valsalva maneuver is maintained, the systolic and diastolic pressures will begin to drop within several seconds because of the reduced diastolic filling caused by impaired venous return. Although often contraindicated during resistance exercise, the Valsalva maneuver may in fact be beneficial and may have a protective effect in healthy resistance-trained individuals (MacDougall 1994; McCartney 1999). The increase in intrathoracic pressure during the Valsalva maneuver stabilizes the spinal column and has also been shown to decrease left ventricular transmural pressure **(afterload)** (Lentini et al. 1993), in contrast to the high afterload normally expected when systolic pressures are elevated. In addition, the increase in intrathoracic pressure is also transmitted to the cerebral spinal fluid, which reduces the transmural pressures of the cerebral vessels and prevents vascular damage at the time of peak peripheral resistance (McCartney 1999).

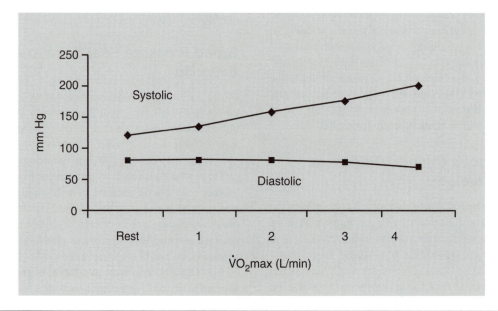

Figure 4.10 Blood pressure response to progressive endurance exercise.

PULMONARY VENTILATION DURING EXERCISE

During exercise, an increase in ventilation results from several chemical and neural stimuli that appear to occur simultaneously (Eldridge 1994; Whipp 1994). Changes in ventilatory patterns appear to occur in three phases (see figure 4.11). Initially, there is a rapid increase in ventilation rate followed by a brief plateau. This initial increase in ventilation is thought to be the result of a central command (cerebral cortex in the brain) as well as feedback from the active muscles and the effects of increased K^+ concentrations (Eldridge 1994). After approximately 20 s, ventilation continues to increase as a result of these same stimuli; however, the neural component appears to increase as a result of an increasing drive from medullary short-term potentiation. As steady-state exercise is reached, all the mechanisms that controlled the increase in ventilatory pattern stabilize. During this phase there can be added input from peripheral sources (e.g., chemoreceptors and core temperature) that can fine tune the ventilatory response (McArdle, Katch, and Katch 1996). During recovery, the abrupt decrease in ventilation is the result of the removal of the central command drive and the afferent input from the active muscles. During the latter stages of recovery, the slower return to resting ventilatory levels represents the gradual return to normal metabolic, thermal, and chemical levels.

During submaximal exercise, ventilation increases linearly with oxygen uptake. The increase in oxygen consumption is primarily the result of an increase in **tidal volume** (amount of air inspired or expired during a normal breathing cycle). As exercise intensity is elevated, the increase in oxygen consumption may rely more on increasing the breathing rate. During steady-state exercise, **minute ventilation** (liters of air breathed per minute) plateaus when the demand for oxygen is met by supply (see figure 4.11). The ratio of minute ventilation to oxygen consumption is termed the **ventilatory equivalent** and is symbolized by $\dot{V}_E/\dot{V}O_2$. During submaximal exercise, the ventilatory equivalent in healthy individuals is approximately 25:1 (Wasserman, Whipp, and Davis 1981). That is, 25 L of air is breathed in for every 1 L of oxygen. This ratio may be slightly higher in children (Rowland and Green 1988) and also may be affected by the mode of exercise (swimming versus running) (McArdle, Glaser, and Magel 1971). However, during maximal exercise, minute ventilation increases disproportionately in relation to oxygen uptake, and the ventilatory equivalent may reach as high as 35-40 L of air per liter of oxygen consumed in the healthy adult.

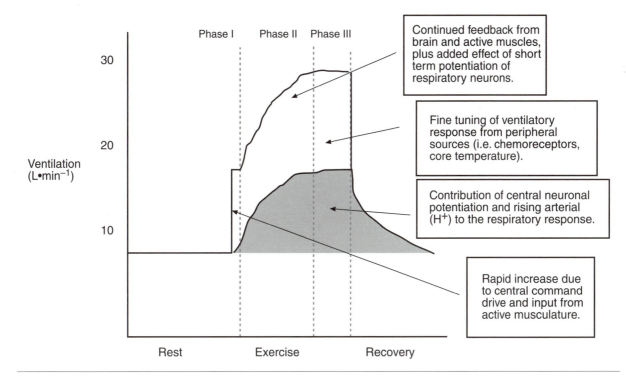

Figure 4.11 Dynamic phases of exercise hyperpnea.

Adapted, by permission, from Elderidge, 1994, "Central integration of mechanisms in exercise hypernea," *Medicine and Science in Sports and Exercise* 26:319-327.

CARDIOVASCULAR RESPONSE TO TRAINING

Long-term physical conditioning results in a number of cardiovascular adaptations specific to the type of exercise program. In general, endurance training and resistance training are the exercise programs that are often compared. These modes of training represent distinctly different physiological demands placed on the cardiovascular system. Although many of the cardiovascular adaptations observed in these training programs are similar, others are very different. A summary of these adaptations is discussed in this section and can be seen in table 4.2.

Training's Effect on Cardiac Output

Increases in $\dot{V}O_2$max are characteristic of endurance training programs. These increases are generally accompa-

nied by increases in cardiac output and improved extraction capability of skeletal muscle [increase in $(a-\bar{v})O_2$]. The improvement in oxygen extraction is related to the greater perfusion capabilities of the exercising muscle. The increase in cardiac output is primarily the result of improved stroke volume. Maximal heart rates are unaffected by training and will not differ between elite endurance athletes and age-matched sedentary individuals. Thus, improvements in cardiac output are directly related to the increase in stroke volume.

Training's Effect on Stroke Volume

Endurance training has been consistently demonstrated as a potent stimulus for increasing stroke volume at rest and during maximal exercise. Endurance-trained athletes have been shown to have a 60% greater stroke volume than sedentary control subjects, which is consistent with the relative difference in $\dot{V}O_2$max seen between these individuals (McArdle, Katch, and Katch 1996). The im

Table 4.2 Cardiovascular Adaptations to Prolonged Endurance and Resistance Training

	Endurance training	Resistance training
RESTING ADAPTATIONS		
Heart rate	↓	↓ or –
Stroke volume	↑↑	↑ or –
Cardiac output	–	–
Blood pressure		
Systolic	↓ or –	↓ or –
Diastolic	↓ or –	–
EXERCISE ADAPTATIONS		
Heart rate	–	–
Stroke volume	↑↑	↑ or –
Cardiac output	↑↑	↑ or –
Blood pressure		
Systolic	↓ or –	↓ or –
Diastolic	↓ or –	↓ or –
MORPHOLOGICAL ADAPTATIONS		
Left ventricular mass	↑	↑
Left ventricular diameter	↑↑	↑ or –
Wall thickness		
Left ventricle	↑	↑↑
Septum	↑	↑↑

↑ = increase; ↓ = decrease; – = no change

proved stroke volume is related to an enlarged ventricular chamber (referred to as **eccentric hypertrophy)** caused by chronic increased ventricular filling common to endurance exercise. This increased preload is thought to relate to the expanded plasma volume associated with such training (Carroll et al. 1995; Convertino 1991).

Resistance training results in little to no change in maximal aerobic capacity. As such, minimal changes would be expected in cardiac output. Significantly greater stroke volumes have been reported in elite-level weightlifters compared with recreational lifters (Pearson et al. 1986). However, when this study and others were examined by a meta-analysis (Fleck 1988), the increase in stroke volume seen in these athletes appeared to be more of a factor of a larger body size than a training adaptation.

Training's Effect on Heart Rate

A decrease in resting heart rate and a relative decrease in heart rate at any given submaximal $\dot{V}O_2$ is a commonly found adaptation in endurance training programs (Blomqvist and Saltin 1983; Charlton and Crawford 1997). However, the magnitude in heart rate reduction during endurance training may be much smaller than that reflected in some of the cross-sectional studies comparing elite endurance athletes with sedentary controls (Wilmore et al. 1996). Resistance training may or may not result in any significant change in resting heart rate. Several studies have reported significant decreases in resting heart rate after resistance training programs (Goldberg, Elliot, and Kuehl 1994; Kanakis and Hickson 1980; Stone, Nelson, et al. 1983), whereas others studies have failed to see any significant changes (Lusiani et al. 1986; Ricci et al. 1982; Stone, Wilson, et al. 1983). The mechanism regulating training-induced bradycardia is not thoroughly understood but is likely related to a change in the balance between sympathetic and parasympathetic activity. In addition, a decrease in the intrinsic rate of firing of the SA node after long-term training has also been suggested to be a factor in the bradycardic response to long-term training (Schaefer et al. 1992).

Training's Effect on Blood Pressure

Endurance training has been shown to attenuate the blood pressure response to exercise so that after training, the same absolute level of intensity will result in a lower systolic pressure response than before the exercise program (MacDougall 1994). However, this is likely related to the initial conditioning level of the individual, and higher intensities and longer duration of training may be needed to see such adaptations in well-conditioned endurance athletes.

In normotensive individuals, resting systolic or diastolic pressures are generally unresponsive to endurance training programs (MacDougall 1994). However, based on a number of epidemiological studies and other investigations examining exercise and hypertension, it appears that exercise is a potential stimulus for reducing both systolic and diastolic blood pressure in hypertensive individuals (Seals and Hagberg 1984). The reduction in resting blood pressure appears to occur during endurance exercise programs of three to five sessions per week, at least 30 min in duration, and between 50-70% of $\dot{V}O_2$max (Fagard 2001; Seals and Hagberg 1984).

The majority of the investigations examining resistance-trained athletes have reported average to slightly below average systolic and diastolic blood pressures (Fleck 1992). Several resistance training studies have also reported no change or slight decreases in resting blood pressures (Goldberg, Elliot, and Kuehl 1994) and significant decreases in the blood pressure response during resistance exercise at the same absolute load (McCartney et al. 1993; Sale et al. 1994). The decreases in resting blood pressure after resistance training are likely the result of a decrease in body fat and a possible reduction in the sympathetic drive to the heart (similar to what may drive the reduction in blood pressure during endurance exercise) (Fleck and Kraemer 1997).

Training's Effect on Cardiac Morphology

The size of the heart in athletes is large in comparison with healthy individuals who participate only in recreational activities or are sedentary in nature. The enlarged heart in athletes resembles the size seen in many pathological conditions of the heart, and for a while some controversy existed whether it was indeed a consequence of pathological or physiological processes (Shapiro 1997).With the advent of M-mode echocardiographic studies in the 1970s, a better understanding of the physiological adaptations from prolonged training could be studied.

Ventricular Wall Thickness and Internal Diameter
During exercise, the hemodynamic demands on the heart are predominantly pressure and volume. During prolonged training, the heart will adapt to match the workload that is placed on the left ventricle in order to maintain a constant relationship between systolic cavity pressure and the ratio of wall thickness to ventricular radius (Shapiro 1997). Adaptations to the morphology of the heart are governed by the law of LaPlace, which states that wall tension is proportional to pressure and the radius of curvature (Ford 1976). During a pressure overload, common to resistance exercise programs, the septum and posterior wall of the left ventricle increase in size to normalize myocardial wall stress. During a

volume overload, common to endurance training programs, the increase is predominantly in the internal diameter of the left ventricle (increasing the size of the cavity), with a proportional increase in both the septum and posterior wall of the ventricle. Both endurance training and resistance training are at either ends of the spectrum concerning the volume and pressure stresses placed on the heart. However, most sports have a parallel impact on both cavity dimension and wall thickness (Spirito et al. 1994). In these sports, athletes perform some combination of aerobic and anaerobic training, resulting in cardiovascular adaptations associated with both an enlarged diastolic cavity dimension and a larger wall thickness. In the sports that primarily emphasize a single form of training, the morphological changes of the heart may be more extreme.

Endurance-trained athletes have been shown to have a greater than normal left ventricular internal diameter, with normal to slightly thicker walls (Maron 1986; Morganroth et al. 1975; Pelliccia et al. 1991; Spirito et al. 1994). These changes are in accordance with the law of LaPlace in that there is a compensatory thickening of the walls of the ventricle in response to the greater internal diameter. This type of left ventricular hypertrophy is termed eccentric hypertrophy and is considered to be a normal physiological response to a volume overload (greater end-diastolic volumes) consistent with prolonged endurance training.

Resistance-trained athletes, on the other hand, have normal internal diameters but significantly thicker ventricular walls (Fleck, Henke, and Wilson 1989; Menapace et al. 1982; Morganroth et al. 1975; Pearson et al. 1986). This is referred to as **concentric hypertrophy** and at times may approach levels that are seen in hypertrophic cardiomyopathy (a disease of the myocardium associated with great thickening of the septum and posterior wall at the expense of cavity size, significantly impairing left ventricular function). The concentric hypertrophy seen in the resistance-trained athlete does not impede the internal diameter of the ventricle. In addition, the type of hypertrophy seen in cardiomyopathy is usually asymmetric, whereas in resistance-trained or power athletes, the change in wall size is generally symmetrical.

It appears that the type of resistance training program employed may determine the extent of cardiac morphological changes. Bodybuilders have been reported to have greater than normal cavity dimensions (Deligiannis, Zahopoulou, and Mandroukas 1988) but, when examined relative to body surface area or lean body mass, no significant differences between bodybuilders or weightlifters were observed. However, right ventricular and left atrial volumes have also been shown to be greater in bodybuilders than in weightlifters. These changes were still evident even when examining these differences relative to body surface area or lean body mass (Deligiannis, Zahopoulou, and Mandroukas 1988; Fleck, Henke, and Wilson 1989). The high-volume bodybuilding programs appear to have the greatest potential to affect cardiac chamber size (Fleck and Kraemer 1997).

Left Ventricular Mass

Left ventricular mass is, on average, 45% greater in highly trained athletes than in age-matched control subjects (Maron 1986). This increase in mass is related to the increases in left ventricular internal diameter and ventricular wall thickness. When examined relative to changes in body mass or body surface area, the significantly greater ventricular mass is still present. Some studies have suggested that differences are more prevalent in elite athletes than in athletes of lesser caliber (Fleck 1988).

RESPIRATORY ADAPTATIONS TO TRAINING

For the most part, the respiratory system is not a limiting factor in providing sufficient oxygen to the exercising muscles. However, similar to most other physiological systems in the body, the respiratory system can also adapt to physical exercise in order to maximize its efficiency. In general, lung volume and capacity change very little as the result of physical exercise. It does appear that **vital capacity** may increase slightly during maximal exercise, but this may be related to the slight decrease seen in **residual volume** (amount of air remaining in the lungs after a maximal expiration) (Wilmore and Costill 1999).

TRAINING EFFECTS ON RESPIRATORY RATE, MINUTE VENTILATION, AND VENTILATORY EQUIVALENT

Training may cause slight decreases in both respiratory rate and minute ventilation during submaximal exercise. This likely reflects the improved exercise efficiency resulting from prolonged endurance training. However, endurance training does appear to cause both respiratory rate and minute ventilation to increase during maximal exercise. During maximal exercise, minute ventilation is thought to increase in relation to increases in $\dot{V}O_2$max (McArdle, Katch, and Katch 1996). In untrained subjects, Wilmore and Costill (1999) reported that minute ventilation can increase from 120 L · min⁻¹ to about 150 L · min⁻¹ after training. In addition, minute ventilation in highly trained endurance athletes may increase to 180 L · min⁻¹ and has been reported to be as high as 240 L · min⁻¹ in elite rowers (Wilmore and Costill 1999).

Endurance training does appear to reduce the ventilatory equivalent during submaximal exercise (Andrew, Guzman, and Becklake 1966; Girandola and Katch 1976; Yerg et al. 1985). In other words, a reduced amount of air is inspired at a particular rate of oxygen consumption. Thus, the oxygen cost of exercise attributable to ventilation is reduced. The benefit during exercise may be realized by a reduction of fatigue of the ventilatory musculature and a greater oxygen availability to the exercising muscles (Martin, Heintzelman, and Chen 1982).

TRAINING'S EFFECT ON BLOOD VOLUME AND RED BLOOD CELLS

Endurance training appears to be a potent stimulus for causing **hypervolemia** (increases in blood volume). This has been demonstrated in both young and old populations (Carroll et al. 1995; Convertino 1991). During the initial 2-4 weeks of training, plasma volume expansion is thought to account for the hypervolemia (Convertino 1991). As training progresses, blood volume expansion appears to be the result of both continued plasma volume expansion and an increase in the number of red blood cells.

The increase in plasma volume is believed to be the result of increases in antidiuretic hormone and aldosterone (see chapter 2), which increase fluid retention by the kidney. In addition, exercise causes an increase in plasma proteins, primarily albumin (Yang et al. 1998). This increase in plasma proteins within the blood causes a greater osmotic pull, causing fluid to be retained in the blood.

Increases in blood volume do appear to be the result of both plasma volume expansion and an increase in red blood cell number. However, plasma volume expansion seems to be a greater contributor to hypervolemia (Green et al. 1991). Figure 4.12 shows the effect of prolonged endurance training on blood volume expansion and the contributions of plasma volume expansion and increases in red blood cell number. Although, both plasma volume and red blood cell volume increase, they do not increase proportionally. Thus, **hematocrit** (% of red blood cells to total blood volume) decreases as a response to training. A high hematocrit could be dangerous because of an increased blood viscosity. However, if hematocrit is reduced, the viscosity of the blood will decrease, which may facilitate the blood flow through the circulation. Reductions in hematocrit do not appear to cause a concern for low hemoglobin concentrations. In fact, hemoglobin concentrations in endurance-trained athletes are typically above normal and provide an ample capacity of oxygen to meet the needs of the body during exercise.

SUMMARY

This chapter demonstrated the effects of acute exercise on cardiac function and how the heart compensates for the increased energy demands of exercising muscles. This compensation is manifested by changes in cardiac output, which is regulated in part by enhanced sympathetic

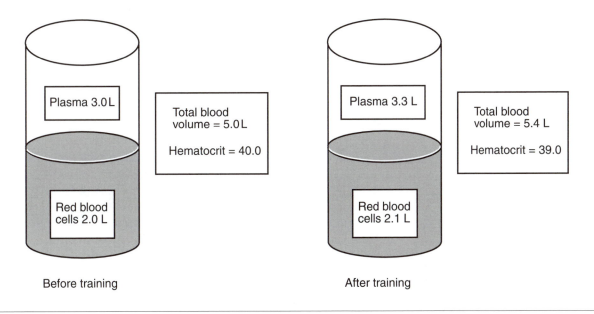

Plasma 3.0 L

Total blood volume = 5.0 L

Hematocrit = 40.0

Red blood cells 2.0 L

Before training

Plasma 3.3 L

Total blood volume = 5.4 L

Hematocrit = 39.0

Red blood cells 2.1 L

After training

Figure 4.12 Effect of endurance training on blood volume, plasma volume, and hematocrit.

drive and increased venous return. In addition, blood flow is diverted to exercising muscles from non-exercising muscles and non-essential organs to provide for greater oxygen delivery. Differences in the acute cardiac response between endurance and resistance training programs were reviewed. The chapter also discussed the effects of pro-longed training on cardiovascular adaptations and how these adaptations are dependent on the type of training program employed. Finally, this chapter reviewed the coordinated relationship between the cardiovascular and respiratory systems and the effects of both acute and pro-longed training on the respiratory system.

CHAPTER 5

IMMUNOLOGICAL SYSTEM AND EXERCISE

©Mary Langenfeld

Interest in the immune response to exercise has gained much popularity and importance because of the implication that intense training increases the athlete's susceptibility to infection, especially upper respiratory tract infections. However, regular moderate exercise is also known to be beneficial to people's health and well-being and is thought to reduce the risk of many diseases and illnesses. Exercise appears to be beneficial to a certain point, at which time either an increase in intensity or duration may impair the immune response. The focus of much of the research on exercise and the immune response has been directed at better understanding the intensity and volume of training that best optimizes immune function.

This chapter briefly reviews the immune system, including the various types of cells and their functions. Further discussion focuses on the immune response to exercise and its implications for performance.

The immune system has two functional divisions: the **innate** immune system and the **adaptive** immune system. Innate immunity is the body's natural response and its first line of defense against infectious agents. This system combats all invading microorganisms seen for the first time. Natural immunity does not improve with additional exposures to these same microorganisms. The innate immune system includes the skin, which prevents the infectious agents from penetrating, mucus membranes, pH of body fluids (e.g., stomach acid), and bodily secretions. The **complement** system, **lysozymes, phagocytes,** and **natural killer (NK) cells** are also part of the innate immune system response.

If the initial line of defense is not sufficient to kill invading pathogens, the adaptive immune system is activated. This immune system produces a specific reaction for each infectious agent. Normally, the adaptive system is successful in destroying the invading pathogen. The adaptive system is also capable of generating a memory of this exposure and producing **antibodies** or other immune cells to respond more effectively and quickly on subsequent exposures to that specific infectious agent.

CELLS OF THE IMMUNE SYSTEM

Immune cells arise from **stem cells** through two lines of differentiation: the **lymphoid lineage,** which produces **lymphocytes,** and the **myeloid lineage,** which produces phagocytes and other cells. Two types of lymphocytes, **T cells** and **B cells,** have receptors for **antigens.** The T cells develop in the thymus, whereas the B cells develop in the bone marrow. These lymphocytes are produced at a very high daily rate (10^9). They migrate into the circulation and into secondary **lymphatic tissue,** such as the spleen, lymph nodes, tonsils, and unencapsulated lymphoid tissue. The lymphocytes represent approximately 20% of the total **leukocyte** (white blood cell) population in an adult.

Types of Leukocytes

There are three major types of leukocytes: **granulocytes, monocytes,** and lymphocytes (see figure 5.1). Granulocytes (also referred to as polymorphonuclear leukocytes) make up 60-70% of the total leukocytes and include **neutrophils, eosinophils,** and **basophils.** The granulocytes, considered part of the initial response to foreign **pathogens,** are primarily involved in **phagocytosis** and destruction of infectious organisms. Neutrophils are the most common granulocyte found in the circulation. They are rapidly attracted by **chemotactic** factors to sites of infection or injury. Neutrophils are short-lived and act primarily by releasing **proteases** and **phospholipases** to eliminate the infectious organism. They are also involved

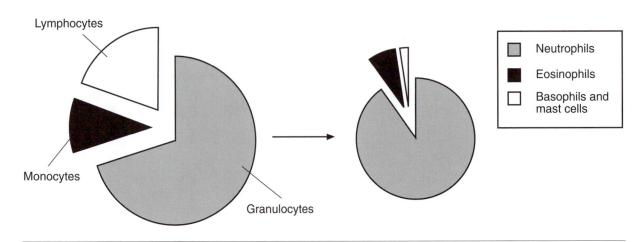

Figure 5.1 Leukocyte distribution.

in generating toxic molecules, such as **oxygen radicals.** Neutrophils are very active during tissue injury and inflammation and are believed to be involved in the degradation of damaged tissue. Eosinophils, basophils, and **mast cells** make up a very small percentage of the leukocytes. Eosinophils appear to be the most active in resistance to parasitic infection, whereas basophils and mast cells are primarily involved in allergic and inflammatory reactions.

Monocytes are also involved in phagocytosis during the early stages of the immune response. Once the monocytes enter the affected area, they differentiate further into **macrophages.** The monocytes/macrophages release proteases from **lysosomes** and generate oxygen radicals and **nitric oxide,** which are lethal to infectious agents. Monocytes also produce **cytokines,** which activate lymphocytes and stimulate the inflammatory process.

Phagocytosis

The phagocytic cells (e.g., neutrophils, monocytes, and macrophages) are brought to sites of infection or inflammation by chemotactic agents. The phagocytes have receptors on the surface that allow them to have a nonspecific affinity to a variety of microorganisms. At-

tachment to these microorganisms may be enhanced if the microorganism has been opsonized by the **C3b** component of the complement system (Roitt, Brostoff, and Male 1993). Opsonization alters bacteria in a manner that allows them to be more readily and efficiently engulfed by phagocytes. The phagocytes have receptors that bind specifically to the C3b component, enhancing the phagocytic recognition of the infectious agent. After attachment, the phagocytes engulf and destroy the micro-organism (see figure 5.2).

LYMPHOCYTES

The lymphocytes account for 20-25% of the total leukocyte population. They are made up of several subpopulations, each possessing a specific function. Lymphocytes are part of the initial immune response and are responsible for producing cytokines, antibodies, **cytotoxicity,** and memory of previous infection. There are three primary subpopulations of lymphocytes (T cells, B cells, and NK cells). They differ in regard to their size and morphological structure. The T cells and B cells are the major effectors of adaptive immunity, and natural killer (NK) cells have innate immunity capability.

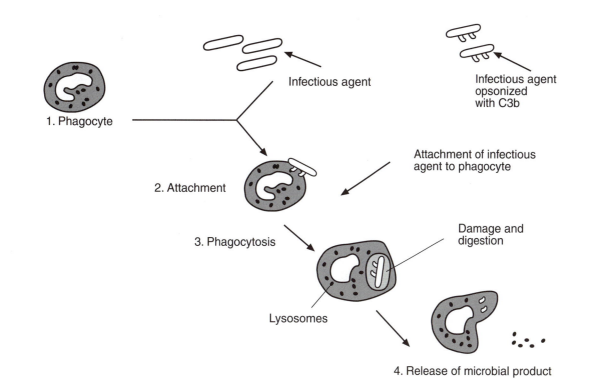

Figure 5.2 Phagocytic action.

T Cells

T lymphocytes are relatively small and agranular. The primary function of these cells is to initiate and regulate most of the immune response to infection and injury. This includes activation of B cells as well as use of their cytotoxic capability. All T cells are able to recognize their target by foreign antigen fragments attached to cell surface receptors on infected cells. These cells are called **major histocompatibility complex (MHC)** molecules (Janeway and Travers 1996). T cells have two distinct morphological patterns that differ in function and cell surface proteins.

T cells that have cytotoxic properties are referred to as cytotoxic T cells. These cells have a surface protein known as **CD8.** T cells with a **CD4** surface protein are further subdivided into T helper and T inflammatory cells. T helper cells activate B cells, and T inflammatory cells stimulate monocyte/macrophage phagocytic and antibacterial activities (Mackinnon 1999). CD4 cells also cause the production of several cytokines. The **clusters of differentiation (CD)** are briefly discussed in the following paragraph.

B Cells

B cells are stimulated to action by T helper cells. On stimulation, they differentiate into plasma cells that produce and secrete large amounts of antibody. Each antibody recognizes a single antigen. The B cells are capable of memory, meaning that they will have a faster and more effective response to future exposures to the same antigen.

Natural Killer (NK) Cells

NK cells are large granular lymphocytes with many cytoplasmic granules. They function as part of the initial immune reaction to defend against infectious agents. NK cells possess an ability to kill a wide variety of targets by releasing toxic substances similar to those released by cytotoxic T cells. They also have the ability to release some cytokines in response to stimulus by T cells.

Clusters of Differentiation

Leukocytes possess a number of different molecules on their surfaces. Some appear at particular stages of cell differentiation or are characteristic of different cell lineages (Roitt et al. 1993). These molecules can be used as markers to identify particular cells through monoclonal antibodies binding to surface antigens of the cells. The numerical listing of the CD system has no specific meaning and is primarily related to the order in which they were discovered and described.

IMMUNOGLOBULINS

Immunoglobulins, or antibodies, are a group of **glycoproteins** found in all bodily fluids. They are produced in large amounts by plasma cells that have developed from precursor B cells. All antibodies are immunoglobulins but not all immunoglobulins demonstrate antibody activities. Antibodies are critical for the recognition of previous exposures to an antigen. They combat infectious agents through both direct and indirect means. They act directly by binding to antigens on the microorganism to prevent it from entering host cells. Immunoglobulins may bind to both the antigen and to the host tissue, including cells of the immune system. They act indirectly by stimulating recognition by other phagocytic cells that kill the invading organism. The latter method of action appears to be more prevalent.

There are five distinct classes of immunoglobulins (**IgG, IgA, IgM, IgD,** and **IgE**). Each differs from the others in size, amino acid components, and carbohydrate content. Their functions and properties can be seen in table 5.1

CYTOKINES

The cytokines regulate growth factors and mediate the inflammatory response. They are involved in all aspects of the immune response and may have a role on nonimmune cells as well. There are several different types of cytokines. Although each cytokine acts on a number of target cells, the target cell will have its own specific response. The cytokines are part of a group of soluble factors found in bodily fluids that mediate immune function. Other soluble factors include complement and acute phase proteins, as well as the previously discussed immunoglobulins.

The cytokines consist of a group of different cells with various functions. Table 5.2 lists the primary producers and major functions of the cytokines frequently reported in the exercise science literature. The **interleukins (IL)** are a group of unrelated cytokines that are produced primarily by T cells but also by monocytes/macrophages and NK cells. The interleukins appear to have a wide array of roles, including inflammatory mediation, stimulation of further cytokine production, and enhancement of phagocytic function. The **interferons (IFN)** are released by both leukocytes and **fibroblasts** and are involved in preventing or limiting the spread of a virus between infected and noninfected cells. **Tumor necrosis factor (TNF)** is produced by both macrophages and T cells. TNF defends against both viral and bacterial infection by stimulating cytotoxic activity of leukocytes or acting synergistically with them to kill these microorganisms.

Table 5.1 Immunoglobulins and Their Functions

Immunoglobulin	% of total immunoglobulin pool	Function
IgG	70%–75%	Distributed evenly between intravascular and extravascular spaces and is the major antibody of secondary immune response.
IgA	15%–20%	Is seen predominantly in saliva, breast milk, respiratory, genitourinary, and gastrointestinal secretions. It acts as a defense against infectious agents entering through mucosal secretions.
IgM	10%	Primarily found in the mucosal secretions, and is seen early in the immune response.
IgD	<1%	Found on the membrane of circulating B cells and may be involved in antigen-triggered lymphocyte differentiation.
IgE	Trace	Found on surface membrane of mast cells and basophils. Associated with immediate sensitivity to asthma and hay fever.

Table 5.2 Cytokines Frequently Reported in the Exercise Science Literature

Cytokine	Primary producer	Major function
IL-1α, IL-1β	Macrophages	Both have same exact function. Primarily involved as an inflammatory mediator (activates NK cells, TNF, and induces fever). They also potentiate the response of lymphocytes.
IL-2	T cells	Promotes T cell division and B cell growth. It is also involved with activating monocytes and NK cells.
IL-3	Activated T cells	Stimulate production of neutrophils and monocytes in bone marrow.
IL-4	Activated T cells	Activates T helper cells and B cells. Is also involved in IgE expression.
IL-5	Activated T cells	Eosinophil differentiation and maturation.
IL-6	Activated T cells, macrophages	T and B cell growth. Acts synergistically with IL-1 and TNF to stimulate acute phase protein response. It is pyrogenic.
IFN-α IFN-β	Leukocytes Fibroblasts	Both have similar anti-viral activity, activate NK cells, and enhance antigen recognition by increasing expression of MHC class I molecules.
IFN-γ	T, NK Cells	Anti-viral and anti-bacterial activity. Inhibits viral replication. Enhances antigen recognition by increasing expression of MHC I and II molecules. Stimulates phagocytic and cytotoxic activities.
TNF-α	Macrophages and NK cells	Stimulates cytotoxic activity of CD8 T cells, NK cells, and macrophages. Increases vascular permeability and migration of leukocytes to areas of inflammation and infection. Acts synergistically with IL-1 and IL-6 to stimulate acute phase proteins.
TNF-β	CD4 T cells	Acts synergistically with IFN-α to increase direct killing of infectious agents. It also acts as a chemotactic agent drawing macrophages to sites of infection and inflammation.

Data from Mackinnon 1999.

COMPLEMENT SYSTEM

The complement system is a group of proteins found in the blood whose primary function is to initiate and amplify the inflammatory response. The biological activity of the complement system is recruitment of macrophages and neutrophils to sites of injury or infection, **lysis** of target cells (most likely bacteria), and **opsonization** of pathogens. Lysis refers to the rupture of the cell membrane and resultant loss of cytoplasm. Opsonization is a process that deposits an **opsonin** (e.g., antibody or C3b-complement component) on bacteria, which enhances recognition by phagocytes. The complement system is made up of a number of components whose nomenclature is related to the manner in which they were discovered.

The complement system is activated via one of two pathways, classical or alternative. Activation of these pathways occurs through the interaction of certain classes and subclasses of antibodies with antigens. The alternative pathway is thought to provide a nonspecific natural immunity, whereas the classical pathway is thought to represent an adaptive mechanism (Roitt et al. 1993). Both of these pathways result in the production of C3b, the central component of the complement system. C3b in turn activates the membrane attack complex (C5-C9), which provides a direct killing action on bacteria. C3a and C5a are **anaphylatoxins** that stimulate **chemotaxis** of leukocytes and **degranulation** of basophils and mast cells.

ACUTE PHASE PROTEINS

Acute phase proteins, part of the innate immune response, are synthesized in the liver and circulate in the vascular fluid. These proteins increase rapidly (> 100-fold) during both infection and inflammation. The **C-reactive protein** is the primary acute phase protein. Its role during inflammation is thought to be opsonization of pathogens or damaged cells. **Ceruloplasmin** and **A_1-antitrypsin** are other acute phase proteins thought to act away from the site of injury as a protective mechanism, neutralizing oxygen radicals and proteases (Evans and Cannon 1991).

EXERCISE AND IMMUNE RESPONSE

Exercise has a significant effect on immune function. The changes seen in the circulating concentration of the immune cells and their release pattern and distribution are governed to a large extent by changes in hormonal concentrations (e.g., catecholamines and cortisol) and cytokines (IL-1). This section examines the changes in immune cell function as they relate to acute exercise and to long-term training programs. Where possible, the response patterns are differentiated between different types of exercise programs (e.g., aerobic, anaerobic, or resistance exercise).

Acute Exercise and Leukocyte Response

Circulating leukocyte concentrations are consistently reported to increase after a wide range of exercise stresses,

ranging in duration from several seconds to several hours (Gabriel, Urhausen, and Kindermann 1992; Gray et al. 1993; Nieman, Berk, et al. 1989; Nieman, Hensen, et al. 1995; Ndon et al. 1992; Shek et al. 1995). The magnitudes of these increases appear dependent on the intensity and duration of exercise. As the intensity or duration of exercise increases, a greater leukocyte concentration can be seen (Mackinnon 1999; McCarthy and Dale 1988).

During short-duration, high-intensity exercise (e.g., sprints or resistance exercise), significant elevations in leukocyte concentrations (150-180% above baseline levels) are seen (Gabriel, Urhausen, and Kindermann 1992; Gray et al. 1993; Nieman, Hensen, et al. 1995). These increases remain elevated during recovery, and the rate of decline appears to be related to the volume of the exercise protocol. Following brief maximal exercise, leukocyte concentrations begin to decline toward baseline levels after 30-60 min. As maximal exercise is maintained for a longer period of time (as might be seen with repeated interval training or multiple sets during resistance exercise), leukocyte concentrations may remain elevated for up to 2 h postexercise (Gray et al. 1993; Nieman, Hensen, et al. 1995). As duration of exercise continues (up to 30 min at either moderate or high intensity), elevations in leukocyte concentrations may also remain for up to 2 h postexercise (Mackinnon 1999; Ndon et al. 1992).

During endurance exercise or prolonged exercise (up to 3 h), leukocytes are reported to increase rapidly and continue to rise for the duration of exercise (Mackinnon 1999). Total leukocyte concentrations may increase 2.5- to 3-fold higher than resting levels (Nieman, Berk, et al. 1989; Shek et al. 1995) and remain elevated for up to 6 h post-exercise (Nieman, Berk, et al. 1989). The magnitude of leukocyte elevation during endurance exercise may be a function of the intensity of exercise. Exercise performed at the anaerobic threshold produces a greater leukocyte response than exercise performed below the anaerobic threshold (Mackinnon 1999).

Considering that neutrophils predominate the leukocyte family, it is not surprising to see their response to exercise parallel the changes generally seen in total leukocyte concentrations. During recovery from exercise, neutrophil content remains elevated, and may possibly increase further, for up to 6 h after high-intensity exercise (Nieman, Hensen, et al. 1995; Nieman, Simandle, et al. 1995; Shek et al. 1995). Other studies have shown that neutrophils may have a biphasic response. After exercise, neutrophil concentrations may return to resting levels within 30 min postexercise and increase again 1-2 h postexercise (Hansen, Wilsgard, and Osterud 1991). The magnitude of the second response appears dependent on the duration of the exercise. A longer duration of exercise results in a greater secondary elevation in

©Human Kinetics

neutrophil response (Hansen, Wilsgard, and Osterud 1991).

Acute exercise of moderate intensity (60-85% of $\dot{V}O_2$ max) does not appear to cause any change in monocyte concentrations from resting levels (Foster et al. 1986; Esperson et al. 1990). However, as exercise intensity is increased to maximal effort, elevations in monocyte concentrations may be observed (Gray et al. 1993; Nieman, Hensen, et al. 1995). During prolonged exercise (> 3 h), monocyte concentrations may increase 2.5-fold above resting levels (Gabriel et al. 1994) and may remain elevated between 2-6 h postexercise (Gray et al. 1993; Nieman, Hensen, et al. 1995).

Increases in lymphocyte concentrations are consistently seen during and immediately after moderate- or high-intensity exercise of brief or prolonged duration. However, during prolonged exercise, the increase in lymphocyte concentrations is lower than that of neutrophils (Mackinnon 1999). In addition, lymphocyte concentrations return to baseline levels at a faster rate after exercise and may even decline below resting concentrations (Mackinnon 1999). This decline in lymphocyte concentrations below resting levels is reported after both brief (Gray et al. 1993; Nieman, Hensen, et al. 1995) and pro-

longed exercise (Gabriel et al. 1994; Nieman, Simandle, et al. 1995.

Lymphocyte subsets (T cells, B cells, and NK cells) appear to respond differently to acute exercise. Although all lymphocyte subsets increase during acute exercise, NK cells appear to show the greatest change (Gray et al. 1993; Mackinnon 1999). In addition, the ratio CD4:CD8 also appears to be altered. A greater increase in concentrations of CD8 cells (cytotoxic T cells) is seen relative to changes in CD4 cells (helper and inflammatory T cells). Thus, the ratio CD4:CD8 declines during and immediately after both brief and prolonged exercise (Gabriel, Schwarz, et al. 1992; Gray et al. 1993; Lewicki et al. 1988). Interestingly, the greater relative increase in the CD8 subset is also accompanied by a relatively greater decline during the recovery period. It appears that the CD8 cells are preferentially recruited into the circulation during exercise and removed after exercise (Mackinnon 1999).

Circulating B cells may increase or remain unchanged during or after exercise. Neither brief nor prolonged exercise, either at moderate or high intensity, appears to cause any elevations in B cell numbers (Gabriel, Urhausen, and Kindermann 1992; Iverson, Arvesen, and Benestad 1994; Nielsen et al. 1996; Nieman, Hensen, et al. 1993). However, increases in B cells may occur after repeated maximal exercise (e.g., sprints or resistance training) (Gray et al. 1993; Nieman, Hensen, et al. 1995).

The largest increases in lymphocyte subsets are seen in the NK cells. Increases of up to 200% above resting levels have been reported during both brief (Nielsen et al. 1996) and prolonged (Mackinnon et al. 1988) exercise of high intensity. Increases are seen during and immediately after exercise, and they rapidly return to resting levels. However, after prolonged or more intense exercise, NK concentrations may decline below resting levels for several hours or even days after exercise (Berk et al. 1990; Mackinnon et al. 1988; Shek et al. 1995). Increases in NK cell concentrations appear related to the intensity of exercise. As exercise intensity increases, the NK response is elevated (Nieman, Miller, et al. 1993; Tvede et al. 1993).

Long-Term Training and Leukocyte Response

Cross-sectional comparisons between athletes and nonathletes have generally not reported any significant differences in resting total leukocyte count or any of its subsets (neutrophils, monocytes, and lymphocytes) (Mackinnon 1999). Studies that have examined the effect of prolonged training on resting leukocyte concentrations have also been unable to demonstrate any significant alterations (Baum, Liesen, and Enneper 1994; Hooper et al. 1995; Gleeson et al. 1995; Tvede et al.

1989). Mackinnon (1999), in a thorough review of the literature, has suggested that the magnitude of the leukocyte response to acute exercise remains unaffected by an individual's training status. However, several studies have reported that circulating leukocyte numbers may change during periods of training when volume of training is elevated.

Lehmann and colleagues (1996) examined changes in resting leukocyte numbers in subjects performing prolonged training. When training volume was increased, a decrease in leukocyte count was seen. This decrease corresponded to an elevation in muscle stiffness and fatigue in these subjects. However, when training intensity was increased, no changes in leukocyte numbers were seen. Apparently, the leukocyte response may be more sensitive to changes in training volume than to changes in training intensity. Other studies have also reported decreases in resting leukocyte concentrations after prolonged training (Ferry et al. 1990) or levels that are within the lower range of normal (Keen et al. 1995; Green et al. 1981). Although most studies have not reported any changes in total leukocyte count during prolonged training, those studies that have reported such changes appear to have focused primarily on endurance athletes who are experiencing large increases in training volume.

Benoni et al. (1995), in a study that contrasts with the response of endurance athletes, reported a significant increase in total leukocyte numbers over the duration of a competitive basketball season. In addition, those researchers reported significant increases in neutrophils, monocytes, and lymphocytes. Neutrophil adhesion (an early event in the host defense mechanism) was also shown to be reduced. Thus, the leukocyte response may depend somewhat on the type of training stress (e.g., aerobic versus anaerobic).

In general, leukocyte subsets parallel the response seen in total leukocyte numbers during training. Most studies have not reported any training effects, either in resting counts or in the response to an acute bout of exercise, in neutrophils, monocytes/macrophages, or lymphocytes (including T cells, B cells, and NK cells). However, similar to total leukocyte numbers during periods of prolonged high volume of training, a decrease in the resting concentrations of monocytes (Mackinnon et al. 1997; Ndon et al. 1992) or a reduction in the neutrophil response to exercise (Suzuki et al. 1996) has been reported. Other studies have reported elevations in leukocyte subsets over the duration of a competitive season (Baum, Liesen, and Enneper 1994; Benoni et al. 1995). The physiological significance of these responses is not clear but, as mentioned previously, may be related to the training mode.

Exercise Training and Phagocytic Cell Function

The number of circulating phagocytic cells provides information about the release of neutrophils and monocytes into the vasculature but does not provide any information on the functional capacity of these cells. These functions include migration of the phagocytes in response to chemotactic stimulation, **adherence,** and phagocytic activity.

Neutrophil function appears to be normal in athletes even during periods of intense training (Mackinnon 1999). Acute exercise is generally associated with enhanced neutrophil function (improved phagocytic activity, migration, and expression of complement receptors) (Mackinnon 1999). However, consistent with what we have seen previously, prolonged periods of endurance training may have a deleterious effect on neutrophil function. A 20-30% reduction in neutrophil migration has been reported (Esperson et al. 1991). In addition, several studies have shown a decrease in phagocytic activity in endurance athletes compared with control subjects (Blannin et al. 1996; Hack et al. 1994; Lewicki et al. 1987; Smith et al. 1990). The decreased functional capacity may be related to a decline in granule content, thus reducing the phagocytic ability of the neutrophils. There are decreases observed in

- migrating ability to chemotactic stimulation,
- neutrophil adherence,
- granule content,
- phagocytic activity, and
- sensitivity to stimulation.

The reduction in neutrophil function may result in a greater susceptibility of endurance athletes to infection. However, Smith et al. (1990) have suggested that the lower sensitivity seen in neutrophil function may be a beneficial adaptation in these athletes that limits the inflammatory response to chronic tissue damage caused by continuous exercise training.

Much less research has focused on monocyte/macrophage function as it relates to exercise and training in humans. Lewicki et al. (1987) reported lower monocyte adherence at rest and after exercise in athletes compared with untrained individuals. In addition, reduced monocyte activity in athletes at rest has also been reported (Osterud, Olsen, and Wilsgard 1989). However, in contrast to these reports, improvements between 20 to 60% in the metabolic and phagocytic activity of macrophages in trained athletes after exhaustive exercise have been reported (Fehr, Lotzerich, and Michna 1988, 1989). Whether these changes are related to the athletes' ability to adapt to an increased training stress (overtraining syndrome) is not clear. Further examination of monocyte/macrophage function as it relates to exercise training appears to be needed.

Complement, Acute Phase Proteins, Cytokines, and Exercise

As previously mentioned, complement, acute phase proteins, and cytokines are soluble mediators of innate immunity. They circulate within bodily fluids to inhibit or modulate the immune response to inflammation or infection.

The response of complement to an acute bout of aerobic exercise has shown different results. Several investigators have reported that complement levels are significantly elevated (Castell et al. 1997; Dufaux, Order, and Liesen 1991), whereas others have not seen any change from resting levels (Esperson et al. 1991). These differences may be related to the duration of exercise. Esperson et al. (1991) reported no change in complement levels after a 3-mi (5-K) run by elite athletes. However, in studies examining exercise of longer duration, significant elevations in complement concentration (11-45% increase from resting levels) have been seen (Dufaux, Order, and Liesen 1991; Nieman, Berk, et al. 1989). Complement levels may remain elevated for several hours after exercise and may be responsible for cleaning **proteolytic fragments** released from damaged muscles (Mackinnon 1999).

Lower resting complement levels have been reported in endurance athletes compared with a nonathletic population (Nieman, Berk, et al. 1989). The lowered resting complement levels seen in these athletes also corresponded to a reduced complement response after a session of graded exercise in comparison with the control subjects. The reduced resting response and the attenuated response to exercise seen in the athletic population may be considered a positive adaptation to training. The lower resting and postexercise complement concentrations are thought to reflect long-term adaptation to chronic inflammation from intense daily training (Mackinnon 1999).

The response of acute phase proteins appears dependent on the duration of exercise. C-reactive protein (CRP), the predominant acute phase protein, has been reported to increase when duration of exercise exceeds 2-3 h (Castell et al. 1997; Liesen, Dufaux, and Hollmann 1977; Strahan et al. 1984; Weight et al. 1991). During exercise of shorter durations, CRP levels do not appear to be affected by the exercise stress (Hubinger et al. 1997; Nosaka and Clarkson 1996). Even when the exercise intensity is enough to cause significant muscle cell damage (e.g., maximal eccentric elbow flexion), CRP levels do not

appear to become elevated (Nosaka and Clarkson 1996). The researchers suggested that the lack of acute phase protein response might be related to the lack of a cytokine response seen during this particular exercise protocol.

The effects of training on resting acute phase protein concentrations are not very clear. Mackinnon (1999) has reported that athletes may have CRP levels that are normal, higher, or lower than nonathletes. However, the acute phase protein response to endurance exercise may be reduced after a training program (Liesen, Dufaux, and Hollmann 1977). This may be an adaptation to control the inflammation that could result from daily training sessions and may also be related to the down-regulation of neutrophil function after prolonged training programs.

The cytokine response to exercise is considered difficult to assess. Cytokines, especially the proinflammatory cytokines **IL-1, IL-6** and **TNFα,** appear to be released during and after exercise (Mackinnon 1999). However, because of the rapid clearance of cytokines from the blood and the local production of cytokines at sites of tissue damage, blood levels may be difficult to interpret and may not reflect the true response of the cytokines to exercise. Significant elevations of IL-1, IL-6, IFN, and TNFα in the urine have been reported 3-24 h after endurance exercise (12.5-mi [20-K] run) despite limited changes in the plasma concentration of these cytokines (significant elevation seen in IL-6 only) (Sprenger et al. 1992). In addition, increases in IL-1 have been seen in skeletal muscle for up to 5 days after prolonged eccentric exercise (downhill running for 45 min) (Cannon et al. 1991; Fielding et al. 1993). The elevated IL-1 levels in skeletal muscle after eccentric exercise or after prolonged endurance exercise suggest that the cytokines play some role in the inflammatory and repair process of tissue damage typically seen during this type of exercise.

The effect of training on resting cytokine response is also not very clear. Sprenger and colleagues (1992) have shown that trained endurance athletes have a higher resting concentration of IL-1, IL-6, and TNFα than that seen in control subjects. However, Smith et al. (1992), comparing endurance athletes to nonathletic controls, did not see any significant differences in resting cytokine levels. Although it is difficult to speculate what might have caused these contrasting results, it may be prudent to consider that differences in the training program or training status of the athletes could have played a factor.

Exercise and Immunoglobulins

As previously discussed, immunoglobulins are important mediators of humoral (acquired) immunity. Immunoglobulins demonstrate antibody activity, bind to specific antigens, and stimulate phagocytosis, cytotoxicity, and complement binding. Immunoglobulins are found in both the blood and mucosal secretions, but each may respond differently to exercise.

Immunoglobulin concentrations in the blood appear to remain at resting levels or increase only slightly after exercise in athletes. No changes from resting levels were seen immediately or up to 24 h postexercise in male distance runners after runs of 8 mi (12.8 K) (Hansen and Flaherty 1981) and 13 mi (21 K) (Gmunder et al. 1990). Similar results were seen in cyclists after a 2-h exercise bout (Mackinnon et al. 1989). Immunoglobulin levels also remained at baseline after 30 s of maximal exercise on a cycle ergometer (Wingate Anaerobic Test) when values were corrected for changes in plasma volume (Nieman et al. 1992). Slight increases in serum immunoglobulin levels were seen in overweight females after 45 min of walking. IgG levels remained elevated for 1.5 h postexercise, while IgA and IgM, after showing an increase immediately after exercise, declined to levels below baseline for 5 h postexercise.

Resting blood concentrations of immunoglobulins appear to be similar between athletic and sedentary populations (Nieman, Tan, et al. 1989; Nehlsen-Cannarella et al. 1991). However, several studies have shown reduced levels of immunoglobulins (low end of the clinically normal range) in various athletes (Garagioloa et al. 1995; Gleeson et al. 1995). The reduced levels of immunoglobulins seen in both endurance and nonendurance athletes is likely a function of intense exercise training. The lower immunoglobulin concentration seen in these athletes may result in an increased susceptibility to infection. This will be discussed in further detail later in the chapter.

Salivary IgA levels are generally used to measure mucosal immune states. Several studies have shown reduced resting salivary IgA levels in athletes involved in prolonged high-intensity training programs (Mackinnon and Hooper 1996; Tomasi et al. 1982). Tomasi and colleagues (1982) showed reduced IgA concentrations at rest (50% lower) in elite male and female Nordic skiers compared with age-matched controls. Mackinnon and Hooper (1996) reported lower resting salivary IgA levels in swimmers during a period of high-intensity training, which coincided with feelings of staleness. In that same study, it was also noted that no changes in salivary IgA levels were seen in athletes who apparently adapted well to the changes in training intensity. Tharp and Barnes (1990) and Gleeson et al. (1995) reported significant reductions in resting salivary IgA levels during the course of a swim season. These reductions were still observed even after the athletes began to reduce (taper) their training load. The lower resting IgA concentrations that were maintained during the taper suggest that the cumulative effect of high-intensity training over a prolonged period

(6-7 months) requires a longer period of recovery to return humoral immune function to its normal level. Other studies have also demonstrated the cumulative effect of training on reducing resting salivary IgA levels even during a relatively short duration of training (4-5 days) (Mackinnon and Hooper 1994; Mackinnon et al. 1991). It appears that intense exercise training, regardless of the duration, may suppress resting immunoglobulin concentrations. This immunosuppressive effect may result in a higher incidence of upper respiratory tract infections among athletes experiencing difficulty adapting to changes in the exercise stimulus.

The response of immunoglobulins to acute exercise appears dependent on the duration of exercise. Aerobic exercise exceeding 2 h in duration has been shown to cause reductions in salivary IgA concentrations between 40 to 60% immediately after exercise (Mackinnon et al. 1989; Tomasi et al. 1982). The lowered IgA levels may remain suppressed for up to 24 h postexercise. Salivary IgA levels may also be reduced after exercise of shorter duration and higher intensity (McDowell et al. 1992; Mackinnon et al. 1993; Tharp 1990). However, the decrease in salivary IgA levels will not be of the same magnitude generally seen after prolonged endurance exercise. During exercise of moderate intensity and shorter duration (45-90 min of running), IgA concentrations may not be altered from their resting levels (McDowell et al. 1991; Mackinnon and Hooper 1994). In contrast to more intense and prolonged exercise, moderate exercise does not appear likely to stress the immune system.

Immune Response in Athletes

Exercise training in previously sedentary individuals has been demonstrated to have beneficial effects on reducing the incidence of upper respiratory tract infections (URTI) (Nieman et al. 1990; Nieman, Hensen, et al. 1993). Nieman, Hensen, et al. (1993) reported a 50% reduction in URTI in women exercising 5 days per week compared with sedentary age-matched controls (see figure 5.3). Exercise also enhances the perception of being healthy. A survey of 750 master athletes (ranging in age from 40 to 81 years) showed that these active individuals considered themselves less vulnerable to viral infection than their sedentary counterparts (Shephard et al. 1995). An additional survey reported that 90% of non-elite runners who have been running marathons for an average of 12 years agreed with the statement that they "rarely got sick" (Nieman 2000).

In contrast to the proposed health benefits of training, other reports have suggested that athletes may be at a greater risk for URTI than other populations. Nieman et al. (1990) reported that 12.9% of the runners in a marathon experienced URTI symptoms during the week after the race in comparison with 2.2% of the control runners (runners that did not run the marathon). Similarly, Peters (1990) reported a 28.7% incidence of URTI in runners after a 35-mi (56-K) race compared with a 12.9% incident rate in controls. The greater incidences of URTI appear to occur during the 2 weeks after the event. The elevated risk for URTI after acute athletic events may depend on the distance of the race. During the weeks after races of 3 mi (5 K), 6 mi (10 K), and 13 mi (21 K), Nieman, Johanssen, and Lee (1989) were unable to see any increased prevalence in URTI compared with the week before the race. In that same study, the researchers also noted no differences in the incidence of URTI in runners who ran an average of 26 mi (42 K) per week in comparison with runners averaging 7.5 mi (12 K) per week.

The greater incidence in URTI appears to occur in highly stressful events such as marathons or ultramarathons. However, other investigations have also suggested that intense prolonged training will increase the risk of URTI. Linde (1987) reported that a greater incidence of URTI was seen in elite orienteers (2.5 episodes per subject) compared with age-matched controls (1.7 episodes per subject) during a year of training. Other epidemiological studies have also implied that intense, prolonged training is associated with an increased risk for URTI (Heath, Macera, and Nieman 1992; Nieman et

al. 1994). The increased risk for URTI appears to be limited to endurance athletes. No known studies have reported an increased incidence of URTI in strength or power athletes. However, the volume of research performed on this athletic population is considerably less than that found on endurance athletes.

Evidence does suggest that exercise training is beneficial for lowering the risk of infection. However, intense prolonged training in elite-level athletes performing long-distance endurance events (marathons or ultramarathons) may cause an elevated susceptibility to illness. Athletes in general do not appear to be immunosuppressed. The only illness to which they appear more susceptible is URTI (Mackinnon 1999). In the last few years, many studies referenced in this chapter have indicated that several aspects of the immune system are affected by prolonged periods of intense training (reduced leukocyte counts, lower immunoglobulin concentrations, suppression of antimicrobial activity).

It has been proposed that athletes appear to become more susceptible to infection in the days after intense exercise (Pedersen and Ullum 1994). As seen in figure 5.4, the athlete may begin a training session before full recovery from the previous exercise session (during a period of potential immunosuppression). If this pattern of beginning exercise sessions before complete recovery continues, a cumulative suppression of some aspects of

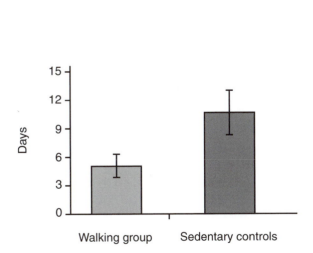

Figure 5.3 Effects of moderate exercise training on URTI days in previously sedentary women.
Data from Nieman et al. 1990.

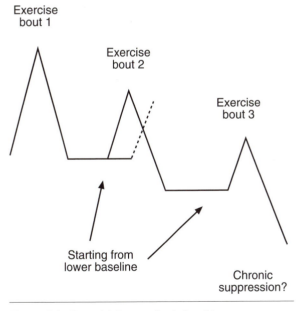

Figure 5.4 Potential for exercise-induced immune suppression.
Adapted, by permission, from B.K. Pedersen and H. Ullum, 1999, "NK cell response to physical activity: possible mechanisms of action," *Exercise and Sport Science Reviews* 26:140-146.

immune function may be seen, causing a further elevation in the risk for infection.

SUMMARY

Exercise may be beneficial for reducing the incidence of infection in previously untrained individuals. However, because exercise intensity remains high during a prolonged period of training or after a bout of prolonged exercise, the athlete may be at a greater risk for URTI. The increased risk for infection may be related to reduced mucosal IgA concentrations, leukocyte counts, and NK cell numbers. The higher incidence of URTI appears confined to endurance athletes. Further research on strength and power athletes appears warranted.

PART II

EXERCISE TRAINING PRINCIPLES AND PRESCRIPTIONS

CHAPTER 6

PRINCIPLES OF TRAINING

©Human Kinetics

For an exercise training program to be properly designed, it is important that the goals and objectives of the program are clear before its development. Because the success of the training program is evaluated by its ability to accomplish these goals, it is imperative that the goals be both reasonable and attainable. Goals that are common to training programs include increasing muscle strength or size, improving sports performance, improving **aerobic capacity**, and improving **body composition**. Through a properly designed exercise program, the individual's training goals can be achieved by both acute and chronic alterations to the structure of the training program. The structure of the training program includes specific component variables that can be manipulated from workout to workout, commonly known as **acute program variables** (Fleck and Kraemer 1997). These variables include the choice of exercise, order of exercise, **intensity** of exercise, **volume** of exercise, **frequency** of training, and length of rest period (between sets or between workouts). Each of these variables may have an important influence on the physiological adaptation to the training stimulus (see chapters 1-5). These acute program variables and their effects on specific training programs are reviewed in the next several chapters.

Unrealistic expectations of a training program can lead to a tremendous amount of frustration on the part of both the athlete and coach. To optimize the exercise program **prescription**, and to set realistic training goals, the basic principles of exercise need to be understood. These principles are the basic tenets of exercise science and are valid whether designing a resistance training program or a running program.

SPECIFICITY PRINCIPLE

According to the specificity principle, adaptations are specific to the muscles trained, the intensity of the exercise performed, the metabolic demands of the exercise, and the **joint angle** trained. For instance, if the goals of the training program were to maximize strength gains, then performing low-intensity, high-volume exercise would not be specific to the objectives of that particular program. Likewise, one would not prepare for a marathon by concentrating solely on running short sprints. **Resistance training** is often part of an athletic conditioning program with the primary objective being to improve sports performance. For strength increases to positively affect sports performance, the training program must have a high **carryover** to the sport. Except for actual practice of the sport, no conditioning program has 100% carryover. To optimize the transfer of strength from the weight room to the field of play, it is important to select exercises that train the specific muscles recruited

during performance. In addition, the exercises selected need to place a similar demand on the neuromuscular coordination of movement that is used during performance (choose exercises that best simulate the actual movement performed on the field of play). For example, a multijoint structural exercise like the push press requires a coordinated movement of both the upper- and lower-body musculature to press a barbell from shoulder height to a position above the head. This exercise is similar to specific movements seen in basketball, such as jumping for a rebound or attempting to score from close to the basket with defenders nearby.

OVERLOAD PRINCIPLE

The basis behind the **overload principle** is that for training adaptations to occur, the muscle or physiological component being trained must be exercised at a level that it is not normally accustomed to. For instance, to maximize muscular strength gains, the muscle needs to be stimulated with a resistance of relatively high intensity (this will be discussed in further detail in chapter 7). If an exercise prescription calls for an individual to perform a 5 **repetition maximum (RM),** and that person uses a resistance that can be lifted for more than 5 repetitions, then that individual may not be overloading the muscle. Subsequently, strength gains may not be maximized. Another example is the endurance athlete who trains for a marathon. If the training goal of the individual is to maximize aerobic capacity in order to run a faster time, then training intensity must be near or at the individual's **anaerobic threshold** (this will be explained in further detail in chapter 9). This can be expressed as a percentage of the individual's maximal heart rate. If the training intensity is not high enough (e.g., heart rate does not reach the required range), then the desired physiological adaptations that can result in an improved aerobic capacity will not be attained.

PROGRESSION PRINCIPLE

During the course of a training program, adaptations occur that change the relative intensity or volume of training. In order to maintain the same **absolute training stimulus** (i.e., intensity or volume of training) the resistance used continually needs to be modified. As an example, if an exercise prescription requires a person to perform 4 sets of 8-10 repetitions of the squat exercise, the objective is for the individual to exercise with a resistance that can be lifted at least 8 times but not more than 10. At the beginning of the training program, the person may be able to squat 135 lb (61 kg) for 10 repetitions in the first set, 9 repetitions in the second set, and 8 repetitions in both the third and fourth sets. After several weeks of training, this individual is now performing 10 repetitions of 135 lb (61 kg) for all 4 sets and feels able to do more than 10 repetitions per set. Obviously, this individual has become stronger. To maximize further strength gains, the resistance needs to be increased for the next exercise session (perhaps to 145 lbs [66 kg]). This process of **progressive overload** is continually being performed throughout the training program. An illustration of the importance of this principle during endurance training can be seen in figure 6.1.

INDIVIDUALITY PRINCIPLE

The **individuality principle** refers to the concept that people respond differently to the same training stimulus. The variability of the training response may be influenced by such factors as **pretraining status,** genetic predisposition, and gender. Many elite bodybuilders publish their training programs, and aspiring bodybuilders attempt to perform the same training regimen with hopes of duplicating the results. Unfortunately, more often than not their results fall far short of their desired outcome. Although

Athlete following the progression principle

Training program for 20 year old cross country runner: Week 10

During his high intensity running days he increased his pace to 6:15 min/mile for the 6-7 miles in order to maintain his training heart rate in the 160-170 beats/min range. His performance times during the cross-country meets continuously improved since the beginning of the season. This is also reflected by a increase in his $\dot{V}O_2$max.

Training program for 20 year old cross country runner: Week 1

During his high intensity running days he runs at a 6:45min/mile pace for 6-7 miles. His heart rate during this run ranges from 160-170 beats/min.

Training program for 20 year old cross country runner: Week 10

During his high intensity running days he continues to run at a 6:45 min/mile pace for 6-7 miles. However, his heart rate during this run now ranges from 150-160 beats/min, and his performance during the cross-country meets has not improved since the beginning of the season. This is also reflected by a lack of change in his $\dot{V}O_2$max.

Athlete NOT following the progression principle

Figure 6.1   Importance of the progression principle during endurance training.

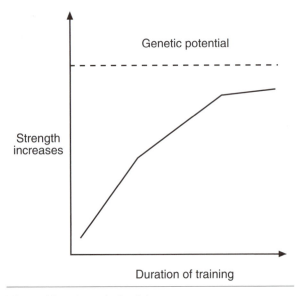

Figure 6.2 Theoretical training curve.

there may be many factors that relate to their disappointment, the primary factor is most likely the large variability between people in response to similar training stimuli.

PRINCIPLE OF DIMINISHING RETURNS

The **principle of diminishing returns** states that performance gains are related to the level of training experience of the individual. Novice weightlifters will experience large strength gains after a relatively short period of time. In contrast, athletes who have strength-trained for several years will make small strength gains over a long period of time. This point is illustrated in a theoretical training curve depicted in figure 6.2. At the onset of a training program, rapid strength gains are made. As training duration continues, the rate of strength improvement begins to slow down. As training continues further, changes in strength and performance are difficult to achieve, and a plateau appears to be reached. This plateau may be considered a **genetic ceiling.** It is at this point that many athletes become frustrated by the lack of performance improvement and may experiment with

anabolic steroids or some other **ergogenic aid** in hopes of pushing past this plateau.

The importance of pretraining status and its effect on strength improvement is reflected in a study performed on elite collegiate basketball players (Hoffman, Maresh, et al. 1991 . The purpose of the study was to examine the effectiveness of an in-season strength training program. The players were placed in two groups for analytical purposes. One group was made up of athletes who had previous strength training experience, and the other group consisted of athletes who had no previous strength training experience and were considered novice lifters. During the course of a season, no strength improvements were observed in the trained group, whereas a significant 4% increase in upper-body strength (1 RM bench press) was noted in the untrained group (see table 6.1). Although both groups participated in identical strength training programs, the difference seen between the groups in strength improvement appeared to be related to the initial strength training experience of the athletes.

The principle of diminishing returns highlights the importance of being able to interpret performance results of the athlete who is training. Hakkinen and colleagues (1987a) examined elite weightlifters for a 1-year period. Small increases in strength were observed, but none of these increases reached statistical significance. Although statistically speaking no changes were seen, practically speaking the athletes and coaches could rate the training program a success. In a group of elite athletes, training improvements are so difficult to achieve that even small improvements can mean the difference between winning or not. In these situations, practical significance takes precedence over statistical significance.

Table 6.1   Effect of an In-season Resistance Training Program.

Variable	Trained	Untrained
	BENCH PRESS	
Pre	101.6 ± 9.6	92.4 ± 26.0
Post	102.9 ± 11.0	96.2 ± 24.2 *
	SQUAT	
Pre	161.4 ± 16.4	131.1 ± 23.3
Post	150.8 ± 13.1	133.7 ± 20.9

* = Significantly different from Pre levels, p< 0.05.

PRINCIPLE OF REVERSIBILITY

When the training stimulus is removed or reduced, the ability of the athlete to maintain performance at a particular level is also reduced, and eventually the gains that were made from the training program will revert back to their original level. This is the basis behind the **reversibility principle.** A common example of the effects of detraining is often seen when an individual is forced to immobilize an injured body part in a cast. When the cast is removed, the muscles that surround the injured limb have undergone a significant reduction in size (referred to as **atrophy**) and strength. This drastic reduction in the size and strength of the muscle occurs because of the lack of activity while the limb is in the cast. However, similar reductions in strength and size, as well as other performance measures, can be seen if the training stimulus that was used to reach that level is removed or even reduced.

When the training stimulus is removed or reduced for an extended duration, the athlete is said to be **detraining.** The removal or reduction of the training stimulus may lead to performance decrements. The extent of performance decrements is related to the length of the detraining or reduced training period and the type of activity. Decreases in aerobic capacity (4-6% reduction in $\dot{V}O_2$max) have been noted after only 2 weeks of inactivity (Coyle, Hemmert, and Coggan 1986; Houston, Bentzen, and Larsen 1979), and longer periods of inactivity will further augment aerobic capacity loss (Simoneau et al. 1987; Drinkwater and Horvath 1972). Strength and power performances also decline during periods of detraining. The magnitude of their decline may depend on the training background, length of the training period before detraining, and specific muscle group

(Fleck and Kraemer 1997). Hakkinen et al. (1989) showed that after a 2-week non-training period, male strength athletes suffered a 3% decline in maximal isometric knee extension strength. In contrast, during the same detraining period, physically active men were able to increase their strength by 2%. In experienced strength-trained individuals, a 24-week strength program was shown to increase maximal isometric-force strength by 27% (Hakkinen, Komi, and Alen 1985). After a 12-week detraining period, maximal isometric force was reduced, but it was still 12% higher than pretraining strength levels. Thus, even after 3 months of detraining, strength levels were still higher than those seen before the 24-week training program.

It does appear that during periods of detraining, muscle and cardiorespiratory endurance changes occur more rapidly than changes in anaerobic forms of activity. This is thought to be partly related to changes in enzymatic activity and stroke volume (Coyle et al. 1984; Coyle, Hemmert, and Coggan 1986). A 12% decrease in stroke volume is evident after 2-4 weeks of detraining (Coyle, Hemmert, and Coggan 1986), and during a similar detraining period, decreases in oxidative enzymes (succinate dehydrogenase and cyctochrome oxidase) are also seen (Wilmore and Costill 1999). Comparing changes in both glycolytic and oxidative enzymes after 3 months of detraining, Coyle and colleagues (1984) reported nearly a 60% decrease in the activities of various oxidative enzymes but no change in the activity of the glycolytic enzymes. The ability to maintain anaerobic exercise performance after relatively short periods of detraining is illustrated in figure 6.3. Performance is

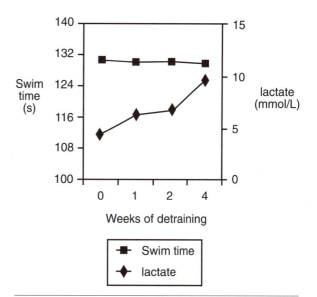

Figure 6.3 Changes in blood lactate and swim performance after 4 weeks of detraining.
Data from Wilmore and Costill 2000

maintained, but the metabolic efficiency has become reduced.

SUMMARY

When designing training programs, it is important to have a clear understanding of the principles of training. This chapter discussed these principles (specificity, overload, progression, individuality, diminishing returns, and reversibility) to help the athlete and coach set realistic goals and develop training programs that will provide the greatest opportunity to achieve performance gains. Detraining was also covered as it related to the principle of reversibility.

CHAPTER 7

RESISTANCE TRAINING

©Terry Wild

The idea of using resistance training to improve sports performance has gained popularity and acceptance only in the last 25-30 years. Resistance exercises were discouraged as a form of training, even for football players, because of the thought that such exercises would limit range of mobility and decrease athletic performance. It was not until the mid- to late-1970s and 1980s that teams began hiring strength coaches to help train their athletes. During this same time period, research began to show the importance of a well-designed strength and conditioning program for helping both competitive and recreational athletes reach their desired goals. This chapter discusses the development of a resistance training program and how manipulation of specific program variables can result in different physiological changes. Discussion also focuses on different modes of resistance training, including the benefits and limitations of each. In addition, the effects that resistance training has on the components of fitness (e.g., speed, vertical jump height) are also explored.

RESISTANCE TRAINING PROGRAM DEVELOPMENT

When developing a resistance training program, several major program design components must be considered (Fleck and Kraemer 1997), including a **needs analysis,** acute program variables, chronic program manipulation, and administrative concerns. Each component contains several different variables that can be manipulated in order to achieve the desired goals. In this section the components will be discussed in detail except for chronic program manipulation. This component, also known as periodization, will be discussed in greater detail in Chapter 11.

Needs Analysis

The needs analysis is the starting point of any training program. It consists of determining the basic needs of the individual in accordance with the target activity. The analysis generally focuses on three primary areas: physiological, biomechanical, and medical.

The **physiological analysis** focuses on determining the primary energy source used during activity. The **duration** and **intensity** of the activity are the primary determinants of energy contribution (see chapter 3). Although a continuum of all three energy sources exists, each source will differ in its magnitude of contribution. As intensity of exercise increases and duration of activity is reduced, a greater reliance on anaerobic metabolism is seen. When intensity of exercise is reduced and duration of activity is elevated, aerobic metabolism appears to be the primary energy source. Additional needs

of the activity or sport are analyzed further by examining the emphasis placed on other major fitness components (**strength, power, speed, agility, and flexibility).** The physiological analysis provides information for acute program variables of resistance training, such as intensity (load) and rest period length.

The **biomechanical analysis** requires the examination of the specific muscles and joint angles used during activity or sport. A correct analysis will help determine the most appropriate exercise. According to Fleck and Kraemer (1997), the following factors are important to consider when developing a biomechanically specific training program:

- The joint around which movement occurs
- The **range of motion** around the joint
- The pattern of **velocity** throughout the range of motion
- The pattern of resistance throughout the range of motion
- The type of muscle action (**concentric, eccentric, or isometric)**

Training adaptations should be specific to the types of exercises chosen to increase the percentage of carryover to the specific activity. Strength training for any activity or sport should be performed throughout the full range of motion of the joint. This not only prevents the loss of flexibility, but it may also improve both joint and muscle flexibility. However, exercises designed for sport-specific training (which may not involve exercising through the full range of motion) should be included to maximize the strength carryover to the respective sport.

One of the advantages of resistance training is its role in the prevention of injury (Fleck and Falkel 1986; Hoffman and Klafeld 1998). Therefore, the **medical analysis** is primarily concerned with locating previous sites of injury in the athlete or understanding the common sites of injury for a respective sport. Specific exercises could be selected that would strengthen specific joints or muscles with the purpose of preventing injuries or at least reducing their severity. This may improve the quality of performance by keeping the better players on the field or court longer.

Acute Program Variables

The acute program variables make up the exercise stimulus, and an infinite number of possible combinations can be used within an exercise session. A proper needs analysis and knowledge of the specific effects that variable manipulations have on training adaptations are crucial to an intelligent exercise prescription.

Exercise Selection

The choice of exercise is related to the specific muscle actions that need to be trained. This decision is based on the biomechanical analysis and injury-site profile (medical analysis) performed during the needs analysis.

Exercises can be classified as structural or body-part. Structural exercises require a coordinated action of many muscle groups. Examples of structural exercises include power cleans, dead lifts, or squats. Exercises that isolate a single joint or muscle group are known as body-part exercises. Examples of such exercises include biceps curls, knee extensions, or triceps pushdowns. Because of the large muscle mass recruited, structural exercises may simulate sport action better and may result in a greater transfer of strength to the actual sport activity. Often, these types of exercises are the core of an athlete's training program. Structural exercises are multijoint exercises that require complex motor control; they should be performed by individuals ready for advanced lifting techniques and be taught by experienced coaches or instructors. If properly taught, these exercises have not been shown to cause any greater risk for injury than isolated body-part exercises. The isolated body-part exercises are often used as assistance exercises and are effective in training specific joints or muscles that are susceptible to injury. In addition, body-part exercises, because of their ability to isolate a specific joint or muscle, are commonly used in rehabilitation from injury.

Order of Exercise

The order of exercise has generally proceeded from large muscle groups to small muscle groups. Because the large muscle-group exercises are generally the structural exercises, reasoning stands that training in this direction (large to small) avoids fatigue before the core lifts. Muscles that are pre-exhausted may not achieve the maximum intensity that can be performed and may cause problems for exercises that require complex motor coordination (e.g., structural lifts). **Pre-exhaustion training,** however, may be used by athletes who do not feel that structural exercises are completely exhausting or fully developing the targeted muscles (Fleck and Kraemer 1997).

The order of exercise may also be specific within the muscle group. For instance, if an individual is training legs, back, and biceps, the order of exercise generally proceeds from large to small between muscle groups. In other words, the individual would begin the exercise program by training the legs, followed by the back, and finally finish the workout by exercising the biceps . Within each set of exercises per muscle group, the order that they are performed also proceeds from large to small. Thus, the order of exercise may proceed as such: squats, leg extensions, leg curls, calf raises, lat pulldowns, seated rows, and dumbbell curls. Notice that the lat pulldowns are performed after the assistance exercises for the legs (i.e., leg extensions, leg curls, and calf raises), even though the lat pulldowns recruit a larger muscle mass. Because the muscles recruited by the leg exercises differ from those recruited by the back exercises, it is doubtful whether performing these leg-assistance exercises would have any detrimental effects on large muscle-mass exercises for the back.

Additional sequences of exercise order may depend on the goals of the training program. Examples of other exercise orders include **super setting** and **compound setting.** Super setting involves using **agonist** and **antagonist** muscle groups in an alternating fashion with little or no rest between exercises (e.g., biceps curls and triceps pushdowns). Compound setting involves performing different exercises for the same muscle group in an alternating fashion with little or no rest between exercises (e.g., incline bench presses and incline dumbbell flys). Depending on the goal or the phase of training, the appropriate exercise order can be determined.

Exercise Intensity

The intensity of exercise, synonymous with training load (the amount of weight lifted per repetition), is probably the most important variable in resistance training. Loading is most likely the major stimulus related to changes in muscle strength or muscle remodeling.

Exercise intensity is generally represented as a percentage of an individual's repetition maximum (RM) for an exercise. The RM refers to the maximum number of repetitions that can be performed with a load. For example, if a person can perform 10 (but not 11) repetitions with 100 lb (45 kg), that individual is said to have a 10 RM of 100 lb (45 kg). Intensity can also be stated relative to an individual's 1 RM. A 1 RM would be the maximum amount of weight that a person can lift for a particular exercise. The exercise prescription can either be written as a relative percentage of an individual's 1 RM (e.g., 80% of 1 RM) or as a range of RM (e.g., 8-10 RM). In the former scenario, if the 1 RM for the bench press was 200 lb (90 kg), then the exercise would require the lifter to perform the required number of sets with 160 lb (73 kg). In the latter scenario, the lifter would select a weight that he or she can lift at least 8 times but not more than 10. If the individual could lift the weight more than 10 times this would no longer be considered a 10 RM. Often the weight used is selected through trial and error.

The major stimulus for either strength or muscle endurance appears to be related to the number of repetitions performed (Anderson and Kearney 1982). Fleck and Kraemer (1997) have referred to this effect as a **repetition maximum continuum** (see figure 7.1). This continuum relates RM loads to the training effects derived from their use. It appears that RM loads of 6 or fewer have the greatest effect on maximal strength or power output. Loads exceeding 20 repetitions have the greatest effect on muscle endurance. This response may also be affected by **pretraining status.** An untrained individual may make significant gains in strength at a 15-20 RM loading, which is likely related to neurological adaptations.

The determination of the appropriate resistance is crucial for optimal performance gains and may be achieved several ways. The most effective method may be through maximal strength (1 RM) testing (Hoffman, Maresh, and Armstrong 1992). As previously mentioned, once a 1 RM is established for an exercise, the training load may be determined through a percentage of the 1 RM. The protocol for determining a 1 RM is depicted in figure 7.2. However, novice lifters may not have developed the skill, balance, and other neurological attributes that would allow for safe and effective 1 RM testing (Wathen 1994). It may then be more appropriate to estimate the 1 RM by using a submaximal weight and asking the individual to perform as many repetitions as possible with that resistance. It should be noted, however, that there could be wide variations in the actual RM (Hoeger et al. 1987, 1990). An estimate of the number of repetitions performed at a percentage of the 1 RM can be seen in table 7.1.

Length of Rest Period

The length of rest between sets is an important variable in exercise design because it determines the amount of recovery of the anaerobic energy sources before the next set. Because the ATP-PC energy source is the most powerful, it is needed for maximal or near-maximal strength performance (1-4 RM). It takes 2.5-3.0 min to replenish the phosphagen stores needed for the next set (Tesch and Larson 1982). Short rest periods (< 1 min) appear to be related to the development of high-intensity muscular endurance and **hypertrophy.** This corresponds to typi-

cal training programs for bodybuilders who are interested in maximizing muscle hypertrophy. The low rest period, combined with a high volume of training (sets × repetitions performed), appears to be a primary stimulus for eliciting the most dramatic anabolic hormone response (see chapter 2). In contrast, longer periods of rest (2-3 min between sets), typical of lifters maximizing their strength potential (e.g., power lifters or Olympic lifters), result in a lower anabolic hormone response (Kraemer, Marchitelli, et al. 1990).

Training Volume

Volume of training can be described as the total amount of weight lifted or repetitions performed in a workout session. Heavy loads used during strength and power training cannot be lifted for many repetitions. Therefore, strength and power programs are generally of low volume. In contrast, training programs that try to maximize muscle growth by using low resistance (for example, 10-15 repetitions per set) typically use a high volume of training. Volume is one of the controlling variables used in periodized training programs, which will be discussed in more detail in chapter 11.

Number of Sets

The optimal number of sets per exercise depends on the specific goal of the program and the training level of the individual. Typically, the number of sets ranges from 3 to 8 per exercise. However, novice lifters may experience strength gains from performing only 1 or 2 sets per exercise. This low number of sets may not provide enough of a stimulus to cause further adaptations in more experienced lifters (Kraemer 1997). Much information concerning the optimal number of sets is derived from empirical evidence. Typical training programs use between 4 to 6 sets for core lifts and slightly lower (3-4 sets) for assistance exercises.

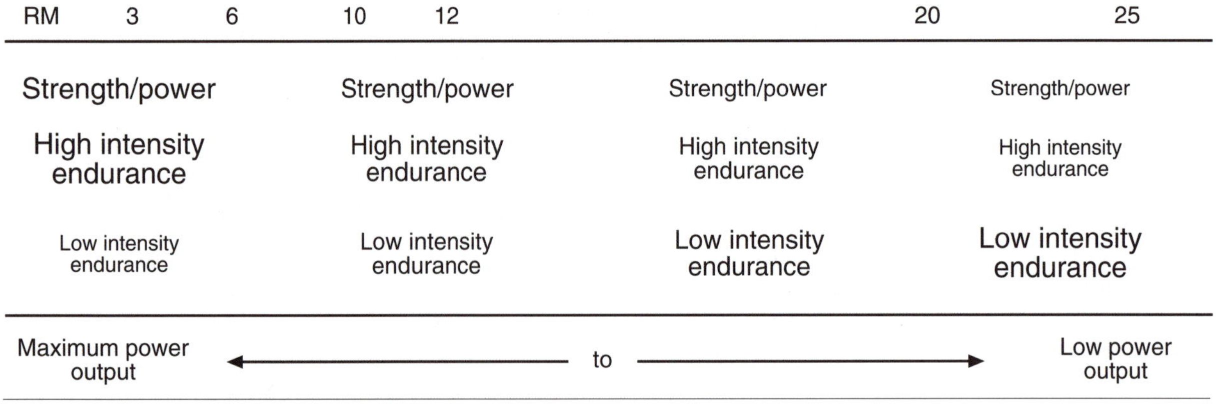

RM	3	6	10	12		20	25
Strength/power			Strength/power		Strength/power		Strength/power
High intensity endurance			High intensity endurance		High intensity endurance		High intensity endurance
Low intensity endurance			Low intensity endurance		Low intensity endurance		Low intensity endurance
Maximum power output	←			to		→	Low power output

Figure 7.1 Theoretical repetition maximum continuum.
Data from Fleck and Kramer 1997.

1. Warm-up set 5-10 repetitions with a resistance equaling approximately 40-60% of the estimated 1 RM.
2. Rest 1-3 minutes
3. A second warm-up set of 3-5 repetitions with a resistance equaling approximately 60-80% of the estimated 1 RM.
4. Rest 3-5 minutes
5. Attempt a 1 RM lift, if successful
6. Rest 3-5 minutes
7. Add resistance and attempt another 1 RM lift, continue until individual is unable to complete a complete repetition.
8. Record the last successful lift as the 1 RM.

Figure 7.2 Protocol for determining 1 RM.

Table 7.1 Estimate of Number of Repetitions Performed at a Percentage of 1 RM

% of 1 RM	Repetitions
100	1
95	2
90	4
85	6
80	8
75	10

Frequency of Training

Frequency of training generally refers to the number of training session per week. The number of weekly training sessions is determined by the training goals, time availability, type of training program, and experience level of the individual. The frequency of training appears to influence the extent of strength improvement. Surprisingly, very few studies have researched the effect that frequency of training has on strength performance. In a study examining the effects of 1-5 days per week of resistance training on nonresistance-trained male high school students, it was reported that the highest frequency of training (5 d · wk) produced the greatest strength gains (Gillam 1981). This study used the same training program (18 sets of 1 RM of the bench press exercise for 9 consecutive weeks) for each training session. Although the training program used is not considered to be a common training style for most resistance training programs, it did lay the groundwork for future studies in this area. That study, as well as others (Hoffman, Maresh, et al. 1991b; Hunter 1985), demonstrated that significant strength improvements can also be made with fewer training days (2-4) per week.

The optimal training frequency appears to be especially important for athletes who already possess a high level of strength. In a study examining strength-trained athletes (Hoffman et al. 1990), it was demonstrated that strength training 3 and 6 times per week did not result in any 1 RM strength improvement in both the bench press or squat exercises. However, significant strength improvements were seen in athletes who trained 4 and 5 days per week. It should be noted that subjects exercising 3 days per week trained their entire body each training session. Subjects training 5 days per week trained their chest and legs 3 times a week and the rest of their body twice per week. Subjects exercising 4 and 6 days per week used a typical split routine and trained each body part twice during the week. It was speculated that training 3 days per week did not provide a sufficient training stimulus (because of a lack of assistance exercises) to improve an already high level of strength, and training 6 days per week may have caused a possible overtraining syndrome that blunted any strength improvement.

Administrative Concerns

The administration of a strength and conditioning program may have a potentially major impact on the design of the optimal training program. Be aware of the ease or difficulty of administration of any training. At times, compromises must be made to accommodate administrative limitations. A training facility may not be able to afford all types of equipment, often because of budgetary constraints. In addition, the training facility may be designed for a population with different training priorities. For example, an athlete may be a member of a health club that is designed primarily for a nonathletic population and lacks much of the equipment typically found in an athletic weight room (e.g., squat racks or lifting platforms). In this situation, the individual may need to make an exercise substitution in the training program.

An additional administrative concern involves time constraints, which may present a problem for completing a workout. This is typically seen in the training of student athletes but may also be an issue for older individuals who are limited by work or family obligations. In such instances, individuals may need to prioritize the exercises in their training programs. Naturally, core exercises take priority over assistance exercises. Other possible solutions may involve designing the training program to include more multi-joint, large muscle-mass lifts or shortening the rest periods between sets. However, depending on the training goals, shortening the rest periods may not be an ideal solution. If left without choice, it may be more advantageous to use shortened rest periods for the assistance exercises and maintain the normal rest interval for the core lifts.

VARIOUS MODES OF RESISTANCE TRAINING

Numerous exercises are available for a resistance training program. However, there are only a few modes of training to consider when developing the program. These modes of training, isometric, **dynamic constant resistance**, **variable resistance**, eccentric, and **isokinetic**, all have certain benefits and limitations in their use. The selection of which mode of training to use is based on the performance needs and goals of the individual and the availability or cost of the equipment. This section addresses the benefits and limitations of each of these modes of exercise.

Isometric Training

Isometrics is also known as **static resistance training.** It refers to a muscle contraction in which no change in the length of the muscle takes place. This type of training can be performed against an immovable object (e.g., a wall) or against a resistance that is greater than the concentric strength of the individual (e.g., a weight-loaded barbell or weight machine). In addition, several commercially produced machines or apparatuses (e.g., dynamometers) are designed to perform isometric contractions.

Increases in static strength have been demonstrated with both submaximal and maximal contractions (Davies and Young 1983b; Fleck and Schutt 1985). However, maximal contractions are superior to submaximal contractions for producing strength gains. In addition, significant increases in both body weight and muscle hypertrophy have been reported after isometric training (Kanehisa and Miyashita 1983; Meyer 1967).

Benefits

Isometric strength increases are joint-angle specific. That is, if isometric contractions are performed at specific joints, strength increases will be seen at the specific joint angle trained. This strength increase will have a carryover of ±20° from the joint angle trained (Knapik, Mawdsley, and Ramos 1983). The benefit of such training may be related to the specific strength improvement observed at the **sticking point** during dynamic exercise. The sticking point refers to the joint-angle position of the contraction in which the muscle is at its weakest. Examining a maximum repetition during an arm-curl exercise, the muscle appears to be capable of moving a resistance throughout its full range of motion except at a particular angle in which the strength of the muscle is not enough to move the resistance. At this point the muscle will be unable to complete the repetition. By using isometric contractions at this specific sticking point, the strength of the muscle at that angle will increase, eventually allowing the individual to complete the lift with that resistance.

Limitations

Joint-angle specificity, which can be a great benefit of isometric training, is also its biggest limitation. Because strength is developed only at a specific angle and strength carryover is minimal, dynamic strength and power increases are not observed. For such increases to occur (strength increases throughout the range of motion of a joint), isometric training must be performed at several joint angles. This is not an efficient method of training.

An additional limitation to isometric training is its inability to increase motor-performance ability (Fleck and Schutt 1985). Many individuals exercise to improve sports performance (e.g., increase vertical jump height or sprint speed). Isometric strength training lacks the sport specificity to positively affect motor performance, thus negating its use as a primary training modality for athletes.

Dynamic Constant Resistance Training

The term dynamic constant resistance refers to a muscular contraction in which the muscle exerts a constant tension against a set resistance. This type of contraction has been generally referred to as an **isotonic** contraction. However, several problems arise with the use of this term when training with free weights or weight machines. Although a muscle contraction is considered isotonic during such exercises, in reality the tension is not constant but varies with the mechanical advantage of the joint involved in the movement (Fleck and Kraemer 1997). A more appropriate definition of isotonic would be a resistance exercise where external resistance does not vary (Fleck and Kraemer 1997). Because of this confusion, it has become more popular to refer to these exercises as dynamic constant resistance training.

Benefits

The primary benefit of dynamic constant resistance exercise is its ability to simulate sport movement and recruit a large muscle mass. This offers an advantage over other training methods by offering a greater possibility of strength carryover to actual sport performance. The effect of strength training on motor-performance improvements (e.g., vertical jump height, speed, and agility) will be discussed later in the chapter.

Limitations

The primary limitation of training with dynamic constant resistance exercises is the necessity of proper supervision of training. Sound safety protocols prohibit individuals from training alone (especially with free weights). Furthermore, adequate facility space is a must to ensure a safe lifting environment.

Variable Resistance Training

Variable resistance training equipment operates through a lever arm, cam, or pulley arrangement. Its purpose is to alter the resistance throughout the range of motion of an exercise in an attempt to match the increases and decreases in strength throughout the entire range of motion. The goal of such a design is to force the muscle to contract maximally throughout its range of motion. However, because of differences in limb length, muscle/tendon attachment to the bone, and body size, it is very unlikely that a particular machine could match the mechanical arrangement of all individuals.

Benefits

Similar to dynamic resistance exercise, exercising on variable resistance machines can lead to significant strength gains. Safety issues (e.g., spotting or weights falling down) do not present themselves as a major concern because of the controlled nature of training.

Limitations

The controlled training environment, which offers a safety benefit, may create a hindrance to motor-performance improvement by limiting the number of multijoint exercises that recruit a large muscle mass and simulate actual sport movements. Furthermore, the cost of variable resistance machines may strain many budgets.

Eccentric Training

Eccentric training, which consists of a contraction in which the muscle lengthens, is frequently referred to in many gyms as "negatives." This type of contraction is commonly seen in downhill running or the lowering phase into a semi-squat position prior to jumping. The use of dynamic constant resistance exercises requires an eccentric component. Whenever a resistance is lowered, the muscle contracts eccentrically (lengthens); as the resistance is raised to complete the repetition, the muscle contracts concentrically (shortens).

The **force capability** during an eccentric contraction is much greater than the force capability during either concentric or isometric contractions (see figure 7.3). For this reason, the use of eccentric training has been suggested to produce greater increases in strength than the other modes of training (Atha 1981). The optimal resistance for eccentric training appears to be 120% of the 1 RM for a particular exercise (Johnson et al. 1976). This resistance will allow the lifter to lower the resistance slowly and stop at will.

Benefits

The primary benefit of eccentric training is its strength-development capability. Specifically, this type of training may be appropriate for those individuals interested in improving bench press and squat strength (e.g., power lifters).

Limitations

Eccentric contractions are the primary stimulus for postexercise muscle soreness (Clarkson 1997). Soreness appears to peak 48 h after a training session. The effect that postexercise soreness has on performance is not completely understood. However, it stands to reason that the muscle damage and decreased flexibility associated with postexercise muscle soreness may negatively affect performance. Therefore, it would be prudent for individuals to avoid heavy eccentric exercise before important competitions. Needless to say, any sudden change in training intensity, even with dynamic constant resistance training, may cause elevations in muscle soreness. Thus, careful consideration must be given to ensure that all

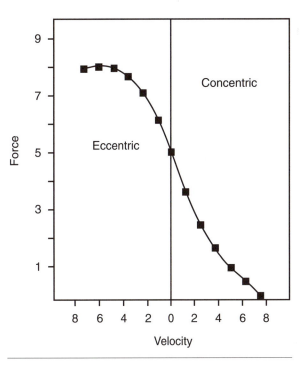

Figure 7.3 Force-velocity curve.
Data from Knuttgen and Kramer 1987.

intensity changes occur at appropriate time points in the training program. An additional limitation posed by eccentric training is the requirement of a partner to prevent injuries from the supramaximal weight lifted.

Isokinetic Training

Isokinetics involves a muscular contraction performed at a constant angular limb speed. In contrast to other types of strength training, there is no set resistance to overcome. Maximal force is applied to a controlled velocity throughout the full range of motion of the muscle.

Benefits

Proponents of isokinetic exercise propose that such training allows for maximal strength development throughout the full range of motion because of the constant maximal force applied by the muscle during a repetition. In addition, muscle contractions can be performed at velocities that are encountered during actual sport performance. The isokinetic training device trains the subject in a controlled environment with isolation of a particular joint and, if needed, possible limitations on joint range of motion. Furthermore, because of the concentric/concentric nature of each repetition (most companies now offer an eccentric option), muscle and joint soreness is minimal. These benefits make isokinetic training an ideal tool for rehabilitation.

Limitations

An isokinetic machine is the most expensive equipment of any of the modes of training discussed. It is not suitable for large groups of individuals training at the same time.

RESISTANCE TRAINING EFFECTS ON THE COMPONENTS OF FITNESS

Besides increases in strength and improvements in muscle hypertrophy and body composition, resistance training programs can also improve various components of fitness. The extent of much of the motor-performance improvements is related to the design of the training program. Similar to what has been previously discussed concerning training experience and strength and size improvements, the ability of a resistance training program to affect performance variables, such as jump height or sprint speed, is also related to the pretraining status of the individual. High-level athletes who begin a training program have a limited capacity for further improvements (remember, the closer the individual is to his or her genetic potential or performance ceiling, the more difficult it is to achieve performance improvements). Nevertheless, the importance of resistance training for these individuals is not diminished by their pretraining status. It is

©Human Kinetics

only necessary to adjust the pretraining expectations and to set realistic training goals. In this section, the effect that resistance training has on strength improvement and its influence on the components of fitness is discussed.

Strength

Strength is the maximal force a muscle or muscle group can generate at a specified velocity (Knuttgen and Kraemer 1987). If strength improvement is the measuring stick of program success, then several factors that can dramatically affect strength improvement need to be acknowledged (e.g., pretraining status, program duration, and training frequency). Table 7.2 shows the improvement in strength from various resistance training programs. Although these differences may be partially explained by variations in training duration and frequency, the initial strength level (pretraining status) of the subjects (not shown in table) may have had a greater effect on the relative strength gains induced by training. As mentioned earlier, novice lifters may experience relative

Table 7.2 Strength Improvements to Various Resistance Training Programs

Reference	Sex	Duration of training (wks)	Days per week	Sets and repetitions	Exercise	% improvement
Berger 1962	M	12	3	3 × 6	BP	30
Berger 1963	M	9	3	6 × 2	BP	17
				3 × 6	BP	21
				3 × 10	BP	20
Fahey and Brown 1973	M	9	3	5 × 5	BP	12
Brown and Wilmore 1974	F	24	3	8 wk = 1 × 10, 8, 7, 6, 5, 4 16 wk = 1 × 10, 6, 5, 4, 3	BP	38
Hunter 1985	M	7	3	3 × 7–10	BP	12
	F	7	3	3 × 7–10	BP	20
	M	7	4	2 × 7–10	BP	17
	F	7	4	2 × 7–10	BP	33
Hoffman et al. 1990	M	10	3	4 wk = 4 × 8 4 wk = 5 × 6 2 wk = 1 × 10, 8, 6, 4, 2	BP SQT	2 5
			4*		BP SQT	4 7
			5*		BP SQT	3 8
			6*		BP SQT	4 7
Fry et al. 1991	F	12	4*	3 × 12–15 3 × 8–10 3 × 3–5	BP	10
Stone and Coulter 1994	F	9	3	3 × 6–8	BP SQT	19 33
				2 × 15–20	BP SQT	17 31
				1 × 30–40	BP SQT	12 25
Hoffman and Klafeld 1998	F	10	3 wk = 4* 7 wk = 1	3 wk = 4 × 10–12 7 wk = 4 × 8–10	BP SQT	23 27

* = split routine training program, generally each exercise was trained twice per week. M = males, F = females, BP = bench press, SQT = Squat

strength gains that are much greater than experienced strength-trained individuals.

Very few published reports exist concerning the improvements of strength during an athletic career. Hunter, Hilyer, and Forster (1993) reported a 24% and 32% increase in bench press and squat strength, respectively, in men over a 4-year intercollegiate basketball career. Most of these improvements appeared to occur within the athlete's freshman year of training (8% and 15% improvement in the bench press and squat strength, respectively). Another study examining female intercollegiate basketball players reported 20-25% improvements in strength measures in these athletes from their freshman to senior years (Petko and Hunter 1997). Unfortunately, no studies have been published concerning strength changes in athletes over their careers in the more traditional strength/power sports (e.g., football).

Anaerobic Power

Improvement in power may be the most important element for enhancing athletic performance. Power (P) can be expressed as

$$P = \text{force (strength)} \times \text{velocity.}$$

Explosive muscle power is needed for throwing, jumping, and striking. In addition, power is needed to rapidly change direction or to accelerate from a run to a maximal sprint.

Maximal strength is attained at a very low shortening velocity. Although this strength expression is important for several sports (e.g., power lifting and certain positions in football), most activities require strength at faster velocities. The strength capability of the muscle differs across its velocity of contraction. As seen in figure 7.3, the highest force outputs occur at the slowest velocities of concentric muscle contraction, while the lowest force outputs are produced at the fastest velocities of concentric muscle contraction. The maximal power output appears to occur at approximately 30% of the maximal shortening velocity (Knuttgen and Kraemer 1987). Typical heavy-resistance training programs have the greatest effect on strength improvement at the slow velocity/high force portion of the power-velocity curve seen in figure 7.4.

Improvement in 1 RM strength, which is coveted by both athletes and coaches, is not without some merit for power development. It is true that slow-velocity strength capability has less impact on the ability of the muscle to produce force at rapid shortening velocities (Kanehisa and Miyashita 1983; Kaneko et al. 1983). However, all explosive movements begin at zero or slow velocities, and it is during these phases of the movement that slow-velocity strength may contribute to power development (Newton and Kraemer 1994).

Resistance Training and Jump Performance

Strength in both isokinetic and dynamic constant resistance exercise measures has been shown to have a significant, positive correlation to vertical jump height (Bosco,

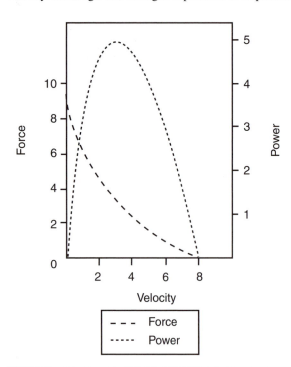
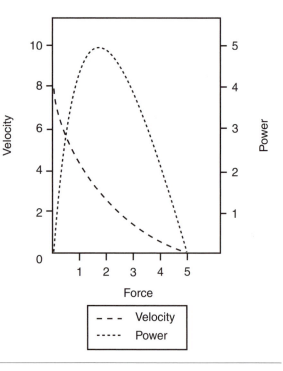

Figure 7.4 Power-velocity curve.
Data from Knuttgen and Kraemer 1987.

Mognoni, and Luhtanen 1983; Podolsky et al. 1990). The relationship between jump height and isokinetic strength is seen when velocities of muscle contraction exceed 180° · s^{-1}. In addition, the power clean, an exercise with a high power output and fast velocity of movement, has been shown to be a significant factor in predicting jumping ability (Mayhew et al. 1987). This relationship between jumping ability and strength in exercises with high speeds of joint movement is consistent with the angular velocity of the knee joint during the vertical jump (Eckert 1968).

Squats, cleans, snatches, and push presses are the primary exercises that have been suggested to improve jumping ability (Garhammer and Gregor 1992; Young 1993). The overall explosive nature of these exercises, and their ability to integrate strength, explosive power, and neuromuscular coordination among several muscle groups, suggests that they may be effective for improving jump performance. Increases in vertical jump height are generally seen when improvements in lower-body strength are reported in resistance training programs. However, the relative improvement in vertical jump performance is of a much lower magnitude than that seen in leg-strength improvement.

Acute program training variables (e.g., frequency, intensity, and volume of training) appear to be important factors influencing vertical jump performance. Resistance training programs of 5-6 days per week have been demonstrated to elicit greater vertical jump improvements (2.3-4.3%) than programs of 3-4 days per week (0.0-1.2%) in resistance-trained Division I-AA college football players (Hoffman et al. 1990). Although these differences were not statistically different, they may have more practical significance for coaches of elite athletes. The potential benefit of a higher frequency of training may be the inclusion of a greater number of assistance exercises and their subsequent effect on sport performance changes (Hoffman et al. 1990).

The effect of training intensity (relative load of weight lifted) on vertical jump improvements is unclear. Several studies (Hakkinen and Komi 1985b; Wilson et al. 1993) have shown significant improvements in jump height using light loads (< 60% of 1 RM), which supports the theory of high-velocity training and consequent improvements in rate of force development. However, other reports suggest that increases in vertical jump height can also be achieved using higher intensities (> 80% of 1 RM) of training (Young 1993).

Resistance training programs using multiple sets per exercise have been shown to be superior for improving vertical jump performance than single-set training programs (23.1% versus 6.9%, respectively) after 24 weeks of training (Kraemer 1997). In a 12-week study examining a periodized resistance training program (both linear and nonlinear) versus a traditional, nonperiodized resistance training program, no significant benefit was observed in any training method for improving vertical jump performance (Baker, Wilson, and Carlyon 1994). In addition, no significant differences in 1 RM squat strength between the training programs were reported in that study. Studies comparing periodized versus nonperiodized training programs of longer duration have not been examined.

Many athletes are required to perform their off-season resistance training programs concurrently with a sport-specific conditioning program. These programs generally consist of agility, endurance, flexibility, speed, and **plyometric training.** The inclusion of these additional training variables to the athletes' training regimen may have an effect on sport performance gains. The effect may be more related to the specific training program design. The addition of endurance training to a resistance training program may (Hennessy and Watson 1994; Hunter, Demment, and Miller 1987) or may not (McCarthy et al. 1995) reduce the relative improvement in vertical jump performance. However, the inclusion of plyometric training may provide a greater stimulus for improving vertical jump performance than resistance training alone (Komi et al. 1982; Newton and McEvoy 1994). This may be related to the increased neural stimulation (disinhibition of the golgi tendon organ reflex) and a more efficient recruitment of agonist and synergistic muscle groups reported after plyometric training. Plyometric training will be discussed further in chapter 12.

Resistance Training and Sprint Performance

Strength is also related to sprint performance (Alexander 1989; Anderson et al. 1991) and appears to be a better indicator of speed when strength testing is performed at velocities greater than 180° · s^{-1} (Perrine and Edgerton 1978). Substantial strength of the hamstring and gluteal muscles is necessary to pull the body over the supporting foot during the ground contact phase of the sprint, while strength in the hip flexors (quadriceps and iliopsoas) allows for the rapid acceleration of the thighs. Stepwise multiple regression analysis of isokinetic data has demonstrated that concentric knee-extension strength at 230° · s^{-1} (correlation coefficient [r] = 0.74) and eccentric hip-flexor strength at 180° · s^{-1} (r = 0.82) are the best predictors of speed performance in male sprinters (Alexander 1989). Less impressive but significant relationships (r = 0.57) have also been demonstrated between a slower speed of contraction (60° · s^{-1} of concentric knee-flexion strength) and sprint performance (Anderson et al. 1991). Absolute maximal strength in multijoint structural exercises (e.g., squats and power cleans) do not appear to be significantly related to sprint speed (Baker and Nance 1999). However, relative strength values in these exercises have been shown to have significant negative correlations (r = -0.66 and -0.72 in the squat and power clean, respectively) to sprint speed (Baker and Nance 1999).

The benefit of traditional high-intensity (70-100% of 1 RM) resistance training programs on speed improvement is not well understood. Statistically insignificant improvements (< 1% decrease in sprint time) for 30-yd (27-m) or 40-yd (37-m) sprints have been observed in collegiate athletes after off-season periodized resistance training programs (Fry et al. 1991; Hoffman et al. 1990). These very modest decreases in sprint speed may be more reflective of the pretraining speed ability of these athletes than any failure of the exercise prescription. The inclusion of high-velocity movements (stretch-shortening exercises) in the athletes' resistance training program has been suggested to be important for achieving improvements in sprint speed (Delecluse et al. 1995). However, only slight (p > 0.05) changes in sprint speed have been shown when a high-velocity stretch-shortening program is added to a periodized resistance training program (Delecluse et al. 1995). The only notable training effect reported from the inclusion of high-velocity training was a significant improvement in initial sprint acceleration. Interestingly, speedskaters appear to have a higher peak torque output at $30° \cdot s^{-1}$ than land-based speed athletes (Smith and Roberts 1991). This may be the result of speedskating technique, which requires isometric contractions of the knee extensors during the glide phase. Thus, the use of traditional resistance training programs for athletes involved in ice sports (e.g., hockey and speedskating) may be potentially more beneficial for speed improvement in those athletes than in land-based athletes.

Resistance Training and Agility Performance

Strength is an important factor in an athlete's ability to stop and change direction rapidly (Anderson et al. 1991; Hoffman et al. 1992). The quick acceleration and deceleration of the body during these movements suggest that eccentric actions contribute to the ability to rapidly change direction. A significant relationship (r = 0.58) has been reported between peak eccentric hamstring force at $90° \cdot s^{-1}$ and agility run-time and thus may be an important indicator of successful agility run-time (Anderson et al. 1991).

Very few studies have reported on the effects of resistance training programs on agility measures. In the few studies that have reported on changes in agility performance after off-season resistance training programs, either no change (Hoffman et al. 1991) or an increase in time (Fry et al. 1991) for the T-drill have been observed. Part of this problem may be related to a greater emphasis on agility training during the preseason training period in contrast to the off-season training program.

Resistance Training and Swimming, Kicking, and Throwing

The importance of strength in swimming, kicking, and throwing has implications on the resistance training pro-

grams developed for swimmers, soccer players, and baseball players. In a study on the importance of strength to swim performance, it was reported that as the distance of the swim increased, the importance of strength decreased (Sharp, Troup, and Costill 1982). Thus, a greater emphasis on strength training appears warranted for swimmers competing in races of short duration.

For soccer players, the importance of strength in the kicking limb is reflected by a significant correlation (r = 0.82) between soccer ball velocity and isokinetic knee-extensor strength (Poulmedis et al. 1988). The strength of the dominant limb in these athletes may contribute to the bilateral deficits reported in soccer players (Mangine et al. 1990). Therefore, resistance exercises that emphasize bilateral movements should be incorporated into their training programs.

The relationship between strength and throwing speed is complex because of the interaction of various muscle groups involved in the throwing action. However, significant correlations have been reported between wrist extensors (r = 0.71) and elbow extensors (r = 0.52) and throwing speed (Pedegna et al. 1982). These results clearly imply the importance of strength to throwing speed. Several studies have demonstrated significant increases (2.0-4.1%) in throwing velocities of baseball players (Lachowetz, Evon, and Pastiglione 1998; McEvoy and Newton 1998; Newton and McEvoy 1994). These increases in throwing velocity were seen after 8-10 weeks of both traditional (Lachowetz, Evon, and Pastiglione 1998; Newton and McEvoy 1994) and ballistic (McEvoy and Newton 1998) resistance training programs. Whether these results can be reproduced in an elite athletic population needs to be examined.

Resistance Training and Cardiovascular Fitness

Resistance training is not considered a primary means of improving aerobic capacity as measured by $\dot{V}O_2max$. However, some forms of resistance training, specifically circuit training, have been shown to improve $\dot{V}O_2max$ (Gettman and Pollock 1981). Participation in a circuit resistance training program for up to 20 weeks may result in 5-8% improvements in aerobic capacity, although the improvements are moderate when compared with increases in aerobic capacity after endurance-based exercise programs.

In a high-volume, Olympic weightlifting program (i.e., exercises for the clean and jerk and the snatch), an 8% increase in aerobic capacity was observed after 8 weeks of training (Stone, Wilson, et al. 1983). Although resistance training is highly anaerobic, it appears that a greater aerobic effect is experienced when a high volume of work is performed. In addition, the use of large muscle-mass exercises (e.g., power cleans, squats, and high pulls) and shorter rest periods between sets may also contribute to improving aerobic capacity after resistance training programs.

Aerobic capacity improvements in weightlifters may be related to the oxygen consumption during resistance training. Depending on the type of training program, oxygen consumption may range from 38% to 60% of peak $\dot{V}O_2$ (Stone et al. 1991). The upper range is within the training threshold reported for aerobic training of untrained individuals. Additional mechanisms may involve the interaction of central (heart rate and stroke volume) and peripheral (arteriovenous oxygen difference) factors. Because resistance training causes cellular adaptations that are in direct contrast to those considered to promote aerobic adaptations (e.g., increase in mitochondrial and capillary densities), peripheral factors are less likely to be involved in improvements in aerobic capacity after resistance training. It is more probable that such improvements are the result of a central mechanism.

Resistance Training and Body Composition

Positive alterations in body composition have been demonstrated after short-term resistance training. Decreases in the percentage of body fat are achieved through increases in lean body mass as well as decreases in fat content (Kraemer, Deschenes, and Fleck 1988; Stone et al. 1991). Increases in lean body mass may result from increases in muscle tissue and possibly bone density. In addition, positive changes in body composition may be associated with increases in body weight. The volume of training may be the key factor in these body compositional changes.

Resistance Training and Flexibility

Flexibility is defined as the range of motion around a joint. It is related to both injury prevention and performance, and it has a large genetic component. Athletes with greater flexibility appear to be more resistant to muscle strains and pulls than athletes with less flexibility. A concern for many individuals is the loss of flexibility as a result of a heavy resistance training program. However, Olympic weightlifters are second only to gymnasts in flexibility tests (Fisher and Jensen 1990). If resistance training exercises are used in a manner that permits full range of motion, then flexibility is greatly enhanced. In addition, to achieve maximal joint flexibility, exercises for both the agonist and antagonist muscle groups should be performed.

WOMEN AND RESISTANCE TRAINING

The presence of women in the weight room has grown rapidly in the last few years, leading to questions about trainability and the women's physiological adaptation to resistance training. Large differences between the genders are seen when examining strength performance. Women's mean total body strength is reported to be 63.5% of men's total body strength (Laubach 1976). However, there is a great deal of variation between body parts. Large absolute strength differences between the genders is seen in the upper body (women are 55.8% as strong as men), while the magnitude of difference is much less in the lower body (women are 71.9% as strong as men). When males and females are compared relative to body mass, the wide gap in strength performance narrows. Wilmore (1974) showed that upper-body strength in women, when compared with men relative to body mass, was 46% as strong. In the lower body, relative to body mass, women were 92% as strong. When these strength comparisons were made relative to lean body mass, Wilmore showed that men were still almost twice as strong as females in the upper body, but females may be stronger than males in the lower body (106% as strong). These differences are for the average male and female and do not include trained athletes or account for body-size differences. These results were confirmed by later studies (Hoffman, Stauffer, and Jackson 1979).

EXAMPLES OF RESISTANCE TRAINING PROGRAMS

The following are examples of different types of resistance training programs. Depending on the resistance training experience of the individual, time available for training, and the type of equipment available in the training facility, adjustments to these training programs can be made. If a substitution must be made for a particular exercise, it is important to choose an exercise that recruits the same muscle groups. In addition, the exercise selected should not be too advanced for the training level of the individual.

It is recommended that the training programs be performed in the order listed. Exercises are generally written from larger muscle mass to smaller muscle mass, but at times the program may have the exerciser complete a muscle group before proceeding to the next body section. The numbers next to each exercise represent the volume and intensity of exercise. For example, if the program is written as squat 1×10, $3 \times 6\text{-}8$ RM, it should be interpreted as 1 warm-up set of 10 repetitions and 3 work sets of a 6-8 repetition maximum. As mentioned earlier in the chapter, the individual would select a weight that he or she can lift at least 6 times but not more than 8 times. If the weight cannot be lifted for 6 repetitions, then it is too heavy and should be reduced. Likewise, if more than 8 repetitions can be performed, the resistance is too light and weight should be added.

Training frequency: 4 d · wk^{-1}

Rest period between sets: 2-3 min

Rest period between workouts: 72 h between body parts. That is, if the legs are trained on Monday, then the chest should be trained at the next session, and the leg exercises should not be repeated until Thursday.

DAYS 1 AND 3		DAYS 2 AND 4	
Exercise	**Sets × Repetitions**	**Exercise**	**Sets × Repetitions**
Squat	1 × 10, 3 × 6-8 RM	Bench press	1 × 10, 3 × 6-8 RM
Leg extension	3 × 6-8 RM	Incline bench press	3 × 6-8 RM
Leg curl	3 × 6-8 RM	Shoulder press	1 × 10, 3 × 6-8 RM
Calf raise	3 × 6-8 RM	Upright row	3 × 6-8 RM
Lat pull-down	4 × 6-8 RM	Triceps push-down	3 × 6-8 RM
Seated row	4 × 6-8 RM	Sit-up	3 × 20
Biceps curl	4 × 6-8 RM		
Bent-knee sit-up	3 × 20		

TRAINING PROGRAM 2: Example of a Bodybuilding Program

Training frequency: 4 d · wk^{-1}

Rest period between sets: 1 min

Rest period between workouts: 72 h between body parts

DAYS 1 AND 3		DAYS 2 AND 4	
Exercise	**Sets × Repetitions**	**Exercise**	**Sets × Repetitions**
Squat	1 × 10, 3 × 10-12 RM	Bench press	1 × 10, 3 × 10-12 RM
Leg extension	3 × 10-12 RM	Incline bench press	3 × 10-12 RM
Leg curl	3 × 10-12 RM	Incline fly	3 × 10-12 RM
Calf raise	3 × 10-12 RM	Shoulder press	1 × 10, 3 × 10-12 RM
Lat pull-down	3 × 10-12 RM	Upright row	3 × 10-12 RM
Seated row	3 × 10-12 RM	Lateral raise	3 × 10-12 RM
Standing barbell biceps curl	3 × 10-12 RM	Triceps push-down	3 × 10-12 RM
Seated dumbbell biceps curl	3 × 10-12 RM	Dip	3 × 12-15
Bent-knee sit-up	3 × 20	Sit-up	3 × 20

TRAINING PROGRAM 3: Example of a Muscle-Toning and Strength Program

Training frequency: 3 d · wk^{-1}
Rest period between sets: 1-2 min
Rest period between workouts: 48 h between workouts

DAY 1		DAY 2		DAY 3	
Exercise	**Sets × Repetitions**	**Exercise**	**Sets × Repetitions**	**Exercise**	**Sets × Repetitions**
Leg press	3 × 8-10 RM	Dumbbell lung	3 × 8-10 RM	Leg press	3 × 8-10 RM
Bench press	3 × 8-10 RM	Incline bench press	3 × 8-10 RM	Bench press	3 × 8-10 RM
Shoulder press	3 × 8-10 RM	Upright rows press	3 × 8-10 RM	Shoulder press	3 × 8-10 RM
Lat pull-down	3 × 8-10 RM	Seated row	3 × 8-10 RM	Lat pull-down	3 × 8-10 RM
Leg curl	3 × 8-10 RM	Leg extension	3 × 8-10 RM	Leg curl	3 × 8-10 RM
Triceps push-down	3 × 8-10 RM	Triceps extension	3 × 8-10 RM	Triceps push-down	3 × 8-10 RM
Biceps curl	3 × 8-10 RM	Biceps curl	3 × 8-10 RM	Biceps curl	3 × 8-10 RM
Sit-up	3 × 20	Situp	3 × 20	Sit-up	3 × 20

TRAINING PROGRAM 4: Example of a Circuit Training Program

Training frequency: 2-3 d · wk^{-1}
Rest period between sets: 1-2 min
Rest period between workouts: 48 h between workouts
Instructions: The circuit is performed continuously from exercise to exercise without rest. When the complete circuit is performed, a 5 min rest is given before proceeding to the next circuit.
Number of circuits: 2-3

Exercise	Sets × Repetitions
Leg press	1 × 12-15 RM
Bench press	1 × 12-15 RM
Leg curl	1 × 12-15 RM
Shoulder press	1 × 12-15 RM
Leg extension	1 × 12-15 RM
Lat pull-down	1 × 12-15 RM
Triceps push-down	1 × 12-15 RM
Biceps curl	1 × 12-15 RM
Sit-up	1 × 12-15
Cycle ergometer	2 min

SUMMARY

This chapter discussed how changes in acute program variables such as intensity, volume, and rest can result in significantly different physiological adaptations. The ability to properly assess the needs of the athlete is important for designing an appropriate resistance training program. There are several different modes of resistance training; this chapter discussed the benefits and limitations of each. Finally, a resistance training program can not only increase the size or strength of an individual, it can also improve components of fitness (e.g., speed and jump height) that affect the performance potential of the athlete.

CHAPTER 8

ANAEROBIC CONDITIONING AND THE DEVELOPMENT OF SPEED AND AGILITY

©Human Kinetics

The importance of both **speed** and **agility** for determining the success of an athlete or team is well acknowledged among coaches of **anaerobic** sports such as football, basketball, or hockey. Several studies have demonstrated that both speed and agility performance can differentiate between starters and nonstarters in NCAA Division I football (Fry and Kraemer 1991; Black and Roundy 1994). In addition, speed and agility have been reported to have a significant positive relationship to playing time in NCAA Division I basketball players (Hoffman et al. 1996). However, the importance of these variables to the athlete is not only related to how fast or how quick he or she is, but whether the athlete can maintain speed and agility performance at maximum levels throughout the duration of a competitive contest. This chapter discusses the conditioning of athletes participating in anaerobic sports, and the ability of athletes to develop both speed and agility.

ANAEROBIC CONDITIONING

Part I of this book was devoted to the physiological adaptations to various training programs, including anaerobic training. Table 8.1 lists the significant physiological adaptations that are generally seen in athletes who train anaerobically. These adaptations help the athlete perform high-intensity activity with rapid recovery between each exercise session. This enables the athlete to perform repeated bouts of exercise with minimal reductions in performance. To develop the most effective program, it is important that the coach understands the physiological demands the athlete experiences during competition. However, very little is known about the physical demands and physiological responses of many anaerobic athletes playing team sports such as basketball or football. This is in contrast to coaches training individual-sport athletes, such as sprinters, who generally perform an isolated event requiring maximal effort. The physiological demands placed on those athletes are easier to understand; consequently, the design of their training program is more straightforward. It should not be misconstrued that it is easy to develop a training program for the individual-sport athletes. However, the **needs analysis** might be clearer and the subsequent **exercise prescription** could be made more specific to the demands of the event.

In team sports such as football, basketball, or hockey, high-intensity activity (short sprints with various changes in direction) is performed repeatedly for the duration of the game. However, the type of movement patterns and the duration of activity could be variable. In addition, there are several positions on each team that may have different demands and require separate training programs. The goal for coaches developing training programs is to bring each athlete to a level of conditioning that allows for maximum performance to be maintained during each spurt of activity throughout the game. A vital issue, however, is proper assessment of the activity requirements of each position in each sport. By having a thorough understanding of the activity demands of the sport, the coach is able to maximize the effectiveness of the training program through greater **specificity** in the types of exercises and in the **work/rest ratio.**

As mentioned previously, limited data is available concerning the physiological demands placed on athletes competing in anaerobic sports such as basketball or football. However, several studies have been published that provide at least a starting base of information to assist in program development. The types of movements, the **intensity** of exercise of these movements, the number of consecutive plays, the number of groupings of plays, and the length of rest between plays are all variables that could provide for a more specific exercise prescription. The following section is an example of how such an analysis may be applied to two popular anaerobic team sports (basketball and football). This examination will be used as a basis for the anaerobic exercise prescription.

Basketball

The game of basketball has a continual flow of play, with a smooth transition from offense to defense. All players have shared responsibilities (e.g., rebounding and shooting) that require them to perform similar movements on the basketball court. These movements range from jumps to runs (from a jog to a sprint) and shuffles (backward and side) at various degrees of intensity. A study by McInnes et al. (1995) was the first to categorize the movement patterns of a basketball game. In that study, the researchers separated the movement patterns into eight

Table 8.1   Potential Physiological Adaptations to High Intensity Anaerobic Training

Increase in the transformation of type II fibers to a more glycolytic subtype.

Significant elevations in glycolytic enzymes (phosphofructokinase, phosphorylase, lactate dehydrogenase).

Increase in maximum blood lactate concentrations.

Reduced blood lactate concentrations during submaximal exercise.

Improved buffering capacity.

different categories (stand/walk, jog, run, stride/sprint, low-, medium-, or high-shuffle, and jump). Their results illustrated the intermittent nature of basketball by demonstrating 997 ± 183 changes in movement during a 48-min basketball game. This equated to a change of movement every 2 s (players averaged 36.3 min played per game). Shuffle movements (all intensities) were seen in 34.6% of the activity patterns of a basketball game, and running (intensities varying from a jog to a sprint) was observed in 31.2% of all movements. Jumps were reported to occur in 4.6% of all movements, whereas standing or walking was observed during 29.6% of the playing time. Movements characterized as high intensity were recorded once every 21 s of play. When considering both high-intensity shuffles and jumps, only 15% of the actual playing time was spent engaged in activity that could be categorized as high intensity. However, the mean blood lactate concentration during the game was 6.8 ± 2.8 mmol, and heart rates were reported to be at 85% of peak for more than 75% of the actual play and at 95% for 15% of the game. Although the majority of activity in the basketball game appeared to be performed at intensities considered aerobic in nature, some of the physiological measures did indicate the high anaerobic characteristics of the sport. It should also be noted that the subjects in that study were Australian National League players; their style of play may be different from that seen in other leagues (e.g., NCAA, NBA). Hoffman and colleagues (1996) analyzed 4 years of basketball at a NCAA Division I institution and reported that speed and other anaerobic performance variables were positively correlated to playing time, whereas aerobic capacity had a significant negative correlation to playing time. Thus, it does appear that to train for basketball, the athlete's training program should place a large emphasis on improving anaerobic conditioning.

An anaerobic exercise prescription specifically for basketball players can be developed. In general, the training of the anaerobic energy system is not initiated until the preseason training program begins (Hoffman and Maresh 2000). Until this time, the athlete's off-season conditioning program should focus primarily on resistance training, sport-specific skill development (including agility and speed), and playing basketball (scrimmages or summer leagues). Anaerobic conditioning is avoided until the preseason phase to prevent the possibility of **overtraining syndrome** during the season (see chapter 21). During the preseason (approximately 6 weeks long at the collegiate level), the goal is to bring the athlete close to, but not necessarily at, peak condition. Once official basketball practice starts, the team begins practicing with the coaching staff. This period before the competitive season is also referred to as the preseason; however, for the athlete, this will be preseason two. Depending on the specific situation involved, many basketball coaches de-

vote much of this time to conditioning drills. During this phase of training, it may be wise for the strength and conditioning specialist to still provide room for improvement. A problem of **fatigue** or **overreaching** may develop if the athlete peaks too soon.

An example of a preseason anaerobic conditioning program can be seen in table 8.2. This program is performed 4 days per week with a progression in both **volume** and intensity. It is also important to note how the intensity of the workouts is manipulated by changes in the work/rest ratio during the sprint training. The aim in reducing the work/rest ratio is to improve the recovery time between bouts of high-intensity activity typically seen during a basketball game. In addition, as described previously, there is a great deal of variability in the movement patterns and intensity levels of these movements. Thus, to provide for a greater similarity to the game, it may be wise to design the training program with exercises that simulate such changes. For example, interval or fartlek training may become an integral part of the training program. Considering the variation in intensity and length of sprints that these drills can employ, they provide a better opportunity to simulate the changes that may occur in a game of basketball. Specific descriptions of these types of training exercises will be presented later.

Football

Football is primarily a game of repeated maximal-intensity bouts of exercise. Every player on the football field has specific responsibilities that may vary considerably (e.g., lineman versus wide receiver). Subsequently, the physical demands experienced by each player will also be different. However, each player must perform these responsibilities at their maximum ability at all times. It has been suggested that the anaerobic energy system is the principal energy system used by the body during a game of football (Gleim, Witman, and Nicholas 1984; Kraemer and Gotshalk 2000). Up to 90% of the energy production during a football game is provided by the **phosphocreatine system,** and the remaining energy production is the result of **glycolysis.**

A football game can be separated into a series of plays. Figure 8.1, a and b, shows the sequence of series in a NCAA Division III college football game. In this particular game, each team had 15 series on offense with an average of 4.5 plays per series (range 1-12). Table 8.3 shows an average number of series and plays seen in 9 games (entire season) of Division III football. During each game, there was an average of 14.4 offensive series per team with an average of 4.6 plays per series. This is slightly more than the average number of series reported (12.3 and 13.3) for the two Super Bowl teams during the 1991-1992 NFL playoff season (Plisk and Gambetta

Table 8.2 Example of an Anaerobic Conditioning Program for Basketball

	Day 1	Day 2	Day 3	Day 4
WEEKS 1–2	INTERVALS	SPRINTS (DISTANCE × REPETITIONS)	INTERVALS	SPRINTS (DISTANCE × REPETITIONS)
	3–4 laps	400-m × 1 100-m × 2 30-m × 8 Work/rest ratio = 1:4	3–4 laps	200-m × 4-5 Work/rest ratio = 1:4
WEEKS 3–4	INTERVALS	SPRINTS (DISTANCE × REPETITIONS)	INTERVALS	SPRINTS (DISTANCE × REPETITIONS)
	4–5 laps	400-m × 1 100-m × 3-4 30-m × 8-10 Work/rest ratio = 1:4 Work/rest ratio = 1:4	4-5 laps	200-m × 5-6
WEEKS 5–6	INTERVALS	SPRINTS (DISTANCE × REPETITIONS)	INTERVALS	SPRINTS (DISTANCE × REPETITIONS)
	5–6 laps	400-m × 2 100-m × 4-5 30-m × 10-12 Work/rest ratio = 1:3	5-6 laps	200-m × 6-7 Work/rest ratio = 1:3

1997). However, the NFL teams ran approximately one more play per series than reported for the college football teams (between 5.3 to 5.6 plays per series). Each play was reported to last for an average of 5.49 s (ranging from 1.87 to 12.88 s) in college football (Kraemer and Gotshalk 2000), whereas the average NFL play was reported to be 5.0 s in duration (Plisk and Gambetta 1997). Between each play, each team has a maximum of 25 s. However, this time clock does not begin until the referee has set the ball. Thus, the rest interval between plays generally exceeds 25 s in duration. In limited reports, the average time between plays in a college football game is 32.7 s (Kraemer and Gotshalk 2000), whereas in the NFL, the average rest interval between plays has been reported to range between 26.9 to 36.4 s (Plisk and Gambetta 1997). The average time per play and rest time between plays allows for a more precise development of the work/rest ratio needed in the development of the anaerobic exercise prescription. Thus, according to the data, it appears that a work/rest ratio of 1:5 could be used in the off-season conditioning programs for football players performing short-duration sprints that simulate the movement patterns of an actual football game.

An anaerobic exercise prescription specifically for football players can also be developed. During the off-season conditioning program, the primary emphasis should be placed on resistance training. In addition, specific speed and agility exercises are often incorporated into the normal winter-conditioning programs (Hoffman et al. 1990). However, these drills are not designed to bring the athletes' anaerobic conditioning to in-season levels. This is reserved for the preseason conditioning program, which emphasizes anaerobic conditioning to maximize the athletes' readiness to compete. This conditioning program begins 6 to 10 weeks before training camp for reasons similar to those discussed in preparing the basketball player. The type of drills and progression of volume and intensity are also similar to what is seen in table 8.2. However, specific adaptations for the football player can be made. For example, for the college athlete it appears that there are 4-5 plays per series that last approximately 5 s (see table 8.3). Considering that there are 3-4 series per quarter, a conditioning program can be developed that would simulate a quarter of a football game with realistic work/rest ratios. In addition, incorporation of a range of sprinting distances can simulate the varied runs that are frequently seen in a game. Plisk and Gambetta (1997) have suggested that this type of training program can also incorporate both "successful" (sets of 10 plays or more) and "unsuccessful" (set of 4-5 plays) series.

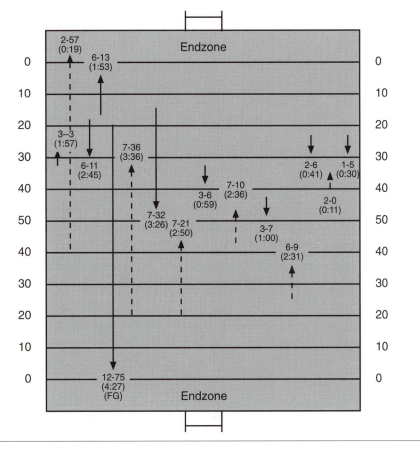

Figure 8.1a Drive chart for first half.

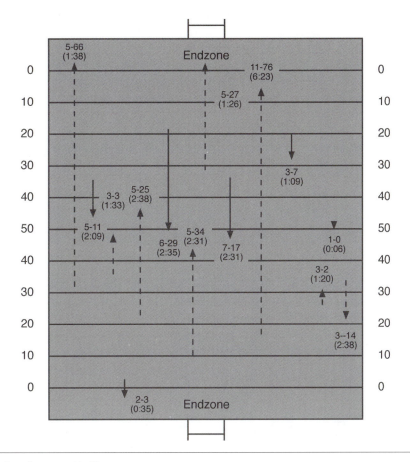

Figure 8.1b Drive chart for second half.

©Michael Zito

Table 8.3   Average Number of Series and Plays in NCAA Division III Football (9 games observed)

Observations	Total number
Plays observed	1193
Series observed	259
Series per game	14.4
Plays per series	4.6
Percentage of series 6 plays or greater	31.2%
Percentage of series 10 plays or greater	8.1%

Athletes in Anaerobic Individual Sports

In contrast to the conditioning programs for team sports such as football or basketball, the exercise prescription for athletes participating in an individual event such as sprinting is easier to develop. Unlike basketball or football players, where the movements and intensity of action may be varied, athletes such as sprinters have a more straightforward needs analysis. A sprinter is required to run a single sprint at maximum ability during a competition. The sprinter may compete in several different races but the requirements are similar. Sprinting is considered any race category that includes distances up to 440 yd (400 m) (Carr 1999); distances greater than 440 yd (400 m) are considered middle-distance events. The training program for the sprinter is focused primarily on developing power, improving running technique and speed, and increasing speed endurance. An example of a training program for sprinters can be seen in table 8.4.

The training program for the sprinter is unlike that of the basketball or football player, whose primary emphasis during anaerobic conditioning is to prepare for repeated bouts of high-intensity activity with limited rest intervals. In contrast, the sprinter's training program is less concerned with improving fatigue rate and more concerned with the quality of each sprint. Table 8.5 depicts the rest intervals recommended for enhancing speed endurance for the 400-m sprinter. Notice the long time intervals between sprints. Obviously, the goal for the 400-m sprinter is maximizing the quality of each sprint.

Table 8.4　Example of a Training Program for a 100-m or 200-m Sprinter

Day 1	Day 2	Day 3	Day 4	Day 5	Day 6
		PRESEASON			
Sprint drills Speed development Turn-around 40-m or 50-m sprints Falling acceleration starts	Interval training	Sprint drills Running technique Block starts and starts to 30-m Stick drill running	15-min jog	Sprint drills Running technique Rotary running Speed development Stick drill Rollover starts	2 × 300-m @ 80-85% of max 4–6 × 100-mm uphill runs
		EARLY SEASON			
Sprint drills Speed development Turn-around 40-m or 50-m sprints	Sprint drills Speed stick running 4 × 60 block starts 4 × 30 flying starts bounding with a weighted vest	Technique work 4 × 100-m relaxation strides	Sprint drills 4 × 120-m 6 × 50-m weighted sled pulls	Sprint drills Rotary running 3-4 × 20-m starts 2 × 100-m smooth striding	Early competition
		LATE SEASON			
100-m athletes 5 × 200-m @ 90% of max – 3-min between sprints 200 – m athletes 4 × 300-m @ 85–90% of max. 10–12 min of recovery between sprints	Sprint drills Speed stick drill 2 × 100-m acceleration drill 2 × 120-m strides	6 × 100-m relaxation strides	Sprint drills 2–3 × 30-m block starts 1 × 150-m @ 100% of max	Warm-up only	Most important competition

Data from USA Track and Field 2000.

Table 8.5   Speed-Endurance Training for a 400-m Sprinter

Number of sprints	Distance of each sprint (m)	Recovery time between sprints (min)
10	100	5–10
6	150	5–10
5	200	10
4	300	10
3	350	10
2	450	10

The distance of each sprint per workout can vary. However, the total distance run per workout should be approximately 2.5 times the distance of the athlete's event. Thus, if the athlete were a 400-m sprinter then 1000-m in sprints per workout would be run. The length of the rest period is long to ensure complete recovery.

Data from USA Track and Field 2000.

ANAEROBIC CONDITIONING EXERCISES

A number of exercises can be used as part of a conditioning program designed to prepare an anaerobic athlete for competition. Often these types of drills are described as enhancing **speed-endurance** and are traditionally used to enhance or maintain speed during long-duration sprint events. These drills have also been described as metabolic conditioning, which is a broader term for anaerobic conditioning or anaerobic endurance.

Interval Sprints

Interval sprints are an excellent way to develop anaerobic endurance. This drill can be performed on a 400-m track. Typically, the athlete sprints the straightaways and jogs or walks the turns (see figure 8.2). This combination of a 100-m sprint followed by a 100-m jog is continued for the length of the workout. The length of the workout and the rest period (jog or walk) depends on both the conditioning and performance level of the athlete.

Fartlek

Fartlek can be performed on either a track or a cross-country course of 2-3 mi (3-5 K). The athlete alternates short bursts of sprinting with jogging. The length of the sprint can be alternated between short and long distances with appropriate adjustments to the rest interval between each sprint. Generally, the same relative work/rest ratio can be maintained for both long and short sprints.

Repetition Sprints

Repetition sprints require the athlete to perform maximum sprints for a given distance. The distance can be either short (e.g., 22-44 yd [20-40 m]) or long (e.g., 110-440 yd [100-400 m]). After a passive rest, the athlete repeats the sprint. The number of repetitions and the work/rest ratio depend on the conditioning level of the athlete.

Repetition Sprints From Flying Starts

Although similar to the repetition sprint drill, **repetition sprints from flying starts** have one difference. The athlete begins each sprint from a running start and accelerates over 22 yd (20 m) before maintaining the sprint for the required distance. The number of repetitions and the work/rest ratio are again dependent on the conditioning level of the athlete.

Repetitive Relays

Repetitive relays use a group of athletes who form a relay team (see figure 8.3a). Athlete A sprints to and tags athlete B, who races to athlete C. Athlete C sprints to athlete D and this process continues for the length of

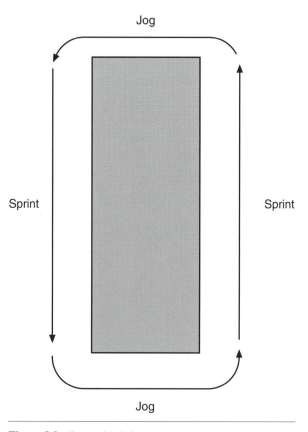

Figure 8.2   Interval training.

a

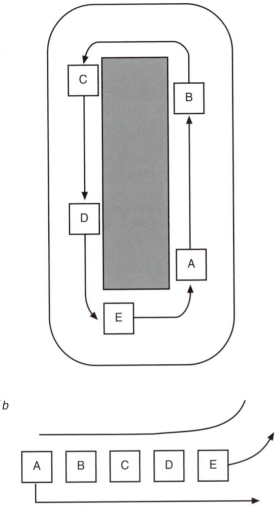

b

Figure 8.3 Repetitive relays *(a)* and rolling sprints *(b)*.

the track. The athlete remains in the position of the runner that he or she replaced. It is possible to make this drill competitive by having relay teams compete against each other. The number of repetitions depends on the conditioning level of the athletes. The work/rest ratio is controlled by the number of members on the relay team. For instance, if you assume that each member of the relay team is similar in speed, then having five relay members would result in approximately a 1:4 work/rest ratio.

Rolling Sprints

Rolling sprints are performed with at least four or more athletes jogging or running slowly one after the other around the track (see figure 8.3b). On the coach's signal, the last athlete sprints to the front of

the line. As the athlete gets to the front, the coach signals again and the athlete that is now last sprints to the front. This continues for the duration of the run. To increase the intensity of the run, the coach can reduce the time between signals or add more runners to the group.

SPEED DEVELOPMENT

Speed is the ability to move a certain distance in a particular time. If a runner can sprint 110 yd (100) m in 11 s, the actual time measured represents the runner's average speed over 110 yd (100 m). If the runner was examined for 11-yd (10-m) splits, he or she gained speed at the start but began to fatigue and slow down toward the end of the run. So the 11 s that was timed is the average sprint speed for that distance. It does not provide any information about **acceleration** from the start of the sprint nor does it give any information about the athlete's speed-endurance, which is the ability to maintain speed without fatiguing.

The ability to accelerate from the start may vary considerably among athletes. Some Olympic-caliber athletes can continue to accelerate through the 70-m mark in a 100-m sprint. Although the ability to accelerate is important, the rate at which this acceleration occurs may have even greater importance. This is especially true of the basketball or football player who needs to reach peak velocity as quickly as possible. This is a component of sprinting that the resistance training program (see chapter 7) is attempting to improve. As long as the runner can still accelerate, he or she will gain speed. Only if the runner begins to decelerate will he or she begin to slow down. To maintain running speed, the athlete focuses on improving speed-endurance. This can be accomplished in part by some of the training programs outlined in table 8.5.

Running speed is the interaction of **stride rate** and **stride length.** The stride contains two steps, or footstrikes. Each step, or footstrike, can be defined as the point of contact of one foot with the ground. Stride rate is the number of steps taken with each leg during the distance of a run. For example, if a sprinter in a 110-yd (100-m) run took 24 steps with his right leg and 23 steps with his left leg, it equates to 47 strides. If the sprinter ran the race in 11.5 s, he would have a stride frequency of 4.1 strides $\cdot$ s^{-1}. Elite sprinters have a stride rate of approximately 5 strides $\cdot$ s^{-1} (Mero, Komi, and Gregor 1992). As the stride rate increases, the amount of time spent on the ground (**support phase**) decreases, while the time spent in the **flight phase** increases. If stride rate increases and the stride length remains constant, running speed will increase. Similarly, if stride rate remains constant but stride length increases, then running speed will also increase. Figure 8.4 shows the effect of changing stride length and stride rate on running velocity.

As can be seen in figure 8.4, both stride rate and stride length increase as running velocity increases. However, stride length appears to increase only up to velocities of about 8 m · s^{-1} (Enoka 1994). The contribution of both stride rate and stride length to sprint speed changes at different running velocities. Initial changes in speed are the result of an increase in the runner's stride. As running velocity increases further, increases in both stride length and stride rate contribute to the higher running velocities. However, as speed is increased even further, a slight decrease in stride length is seen along with a sustained increase in stride rate. Thus, stride rate appears to be more important than stride length in determining the runner's maximum velocity (Mero, Komi, and Gregor 1992).

Both stride rate and stride length appear to be variable among individuals. Stride length is dependent on the athlete's height and leg length (Mero, Komi, and Gregor 1992; Plisk 2000). The taller the athlete or greater the leg length, the longer the stride. Stride rate is also variable, with large differences seen between trained and untrained individuals (Mero, Komi, and Gregor 1992; Plisk 2000). Trained sprinters are able to achieve a greater stride rate than untrained runners and, from a static start, are capable of increasing stride rate for about 27 yd (25 m). Untrained runners appear to reach their maximum stride rate at about 11-16 yd (10-15 m) (Plisk 2000). In addition, elite runners are able to reach their maximum velocity much earlier than untrained runners. Stride rate appears to be very trainable (Plisk 2000). By improving power performance, the athlete can increase acceleration ability by decreasing the ground-contact time of each stride and increasing the impulse production on takeoff (Delecluse 1997; Mero, Komi, and Gregor 1992; Plisk 2000). Improvement in running technique may also result in improvement in running speed, especially for individuals with technical flaws in their sprint style. Table 8.6 and figure 8.5 review the mechanics of sprinting.

AGILITY DEVELOPMENT

Agility is the ability of an individual to react to changes in direction without loss of speed or accuracy. The expression that someone can "stop on a dime" describes the ability of an athlete to sprint at maximal velocity and rapidly change direction (either in response to a coach's signal in a practice drill or during an actual competition) without any reduction in speed. In addition, agility is often used to describe the ability of an athlete to change from one type of movement to another. For example, a defensive back who goes from running backward (backpedaling) to a forward sprint or a linebacker who goes from a side shuffle to a forward sprint are examples of common changes in movement and direction during competition. It requires a combination of strength, power, balance, and coordination to change from a movement performed at a maximal velocity, decelerate as quickly as possible, and accelerate in the new direction, possibly with a new movement. The ability to accomplish this at a high level of precision is an important determinant of an athlete's success, especially in sports such as basketball, football, hockey, and soccer. However, much of this is based

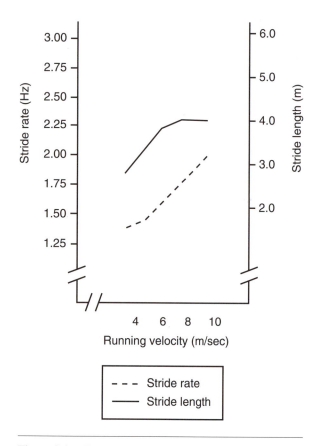

Figure 8.4 Change in stride length and stride rate with running velocity.
Data from Luhtanen and Komi 1978.

Table 8.6   Running Mechanics

Start phase

Athlete is positioned in the blocks with body weight resting equally on the back leg and hands.

Arms are shoulder-width apart and the hands are just rear of the starting line with the fingers and thumb forming a 'V'.

The shoulders are rotated forward approximately 3-4 inches in front of the hands.

The athlete's dominant or stronger leg is usually in the front block and 1 - 2 foot lengths behind the starting line. The rear foot is approximately 1 foot lengths behind the front foot.

In the set position, the athlete is like a coiled spring; athlete's rear end is raised with the angle of the front leg approximately $80° - 90°$ at the knee and $110° - 130°$ at the knee in the rear leg.

The athlete's back and head form a straight line and vision is forward and toward the ground.

Upon the sound of the starting gun the athlete forward leg is extended while still in the blocks and the knee of the back leg is driven forward. The arm action is a counterbalance of the legs.

During the start the athlete's body is inclined formed for the first 5-6 m of the sprint. Beyond this distance the athlete assumes a more upright position.

Acceleration and stride phase

Acceleration is achieved by driving or pushing with the drive or strike leg. This requires a forward lean from the legs through the waist and upper body. The angle of the lean will be directly proportional to the acceleration.

The free leg will drive low and fast.

Footstrike will occur with the foot coming down under the body and possible behind the body's center of gravity, which may cause the athlete to stumble if he or she does not accelerate properly.

Heel recovery of the drive leg will be very low coming out of the blocks and during acceleration in order to get the foot down as fast as possible to drive again and overcome inertia.

As the athlete reaches their maximum velocity the forward lean of their body will be lessened and the athlete's heel will rise higher as they get into a steady sprint stride.

During the acceleration phase the arms will be driven very high and forward relative to the athletes upper body. The arms play an important role in maintaining balance, rhythm and relaxation.

During the stride the foot that is striking the ground should be moving backwards as it lands and strike the ground as close as possible to a point directly under the body's center of mass.

The ankle of the swing leg should cross the support leg above the knee.

The arms should be bent at the elbows (less than $90°$ on the upswing and greater than $90°$ on the downswing). The arms swing backward as if they are reaching for the hip pockets. The forward swing of the arm should be as high as the chin but not cross the midline of the body.

The shoulders, neck, jaw and face should be relaxed and the ankle should be dorsally flexed (toes pointed up) just prior to the footstrike.

The head remains erect and eyes focused on the finish line.

Finish phase

The athlete maintains good sprint posture and normal stride action through the finish line.

The athlete should maintain the same stride length that was used in the middle of the race and should not overstride which sometimes occurs as the athlete nears the finish line.

The athlete needs to maintain their stride rate (speed-endurance) throughout the sprint and as they near the finish line may lean forward at the tape. This is effective only is performed during the last stride at the finish line.

Data from USA Track and Field 2000.

Figure 8.5 Sprinting mechanics.

on empirical evidence (i.e., not scientifically-based) since there is limited scientific data available that has actually examined the relationship between agility and athletic performance. Hoffman and colleagues (1996) showed that agility performance was significantly related to playing time in Division I college basketball players. The quicker or more agile the player (as determined by a T-test), the more playing time he received. In addition, agility has also been suggested to be an important component of successful performance in football (Kraemer and Gotshalk 2000) and soccer (Kirkendall 2000).

If agility is to be one of the variables measured as part of a comprehensive testing program for an athlete, the particular agility test used must be part of the athlete's agility program. Agility is often trained as part of the off-season conditioning program. It is important that the agility drills selected simulate the movements and actions performed during actual competition. There are a number of drills that can be used to improve agility. Many coaches also incorporate specific movements that are common to their respective sports and use these movements as drills to enhance their players' agility. Following are several examples

of agility drills that are common to several different sports.

An interesting question is whether speed and agility can be improved. There are many unsubstantiated claims made by private trainers that they have been able to reduce 40-yd sprint time by 0.3 s or more. However, most studies published have seen only limited improvements in both sprint speed and agility. It is likely that these large speed gains are related to the conditioning level of the athlete measured. If the athlete is not in peak condition, or has not been performing sprints regularly (which may be the case during the middle of an off-season conditioning program), normal times for a 40-yd or 100-yd sprint may increase remarkably. This may be due to flaws in technique or lack of speed-endurance. As the athlete begins to train specifically for speed, sprint times will improve but likely only to the point of peak condition of the previous season. Evaluation of a training program should be based on the peak sprint time of the previous season and not from the point at which the training program began (a state of detraining). That's more a function of anaerobic conditioning than speed improvement.

Side Shuffle

Two cones are placed 16-20 feet apart. The athlete starts in the center of the two cones with the feet slightly wider than shoulder width and with a slight forward lean. At the start, the athlete side shuffles to either cone and touches it with his or her closest hand. The athlete then side shuffles to the other cone. This can proceed for 10 touches or for 10 s depending on the coach's discretion. It is important to ensure that the athlete does not cross his or her feet and shuffles as quickly as possible between each cone (see figure 8.6a).

T-Drill

Cones are set in a T formation. The athlete sprints from the starting cone to second cone 30 ft away. The athlete then side shuffles to a third cone 15 ft away, turns, and proceeds to shuffle past cone 2 to a fourth cone that is 30 ft away. The athlete then side shuffle back to the second cone and does a backward sprint to the startign cone (see figure 8.6b).

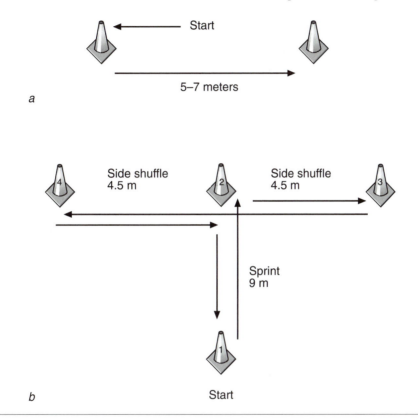

Figure 8.6   *(a)* Side shuffle agility drill and *(b)* T agility drill.

Zig-Zag Drill

Cones are set up in a zig-zag formation. The athlete starts at the first cone and sprints to each diagonally placed cone. The athlete can perform this drill using more than one method. The athlete can run directly to each cone and then cut to the next cone. The athlete can run around each cone, staying as close to the cone as possible (see figure 8.6c).

Four Corner Drill

Cones are placed in a square about 16-20 ft apart. The athlete begins the exercise by performing a backward run to the second cone; with a tight transition around the cone, the athlete then performs a side shufffle from cone 2 to cone 3; as the athlete makes another tight transition around cone 3, he or she then performs a karioka exercise (shuffle over, shuffle under) to cone 4; the athlete then sprints to cone 1 (see figure 8.6d).

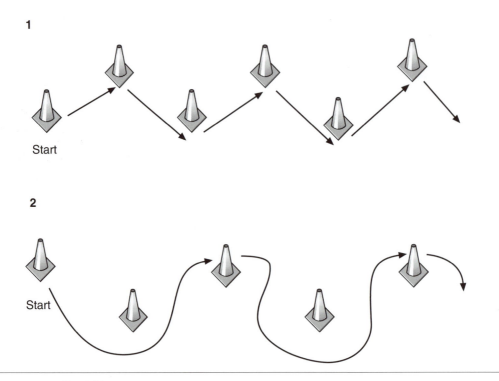

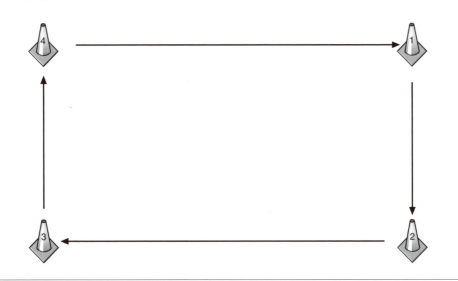

Figure 8.6c   Zig-zag agility drill.

Figure 8.6d   Four corner agility drill.

Quick Feet

Using a line, the athlete stands with feet together and takes small choppy steps back and forth across the line. The object of this drill is to stay as close as possible to the line without touching and to be as quick as possible (see figure 8.6e).

L-Drill

Three cones are lined up in a L formation about 30 ft apart. The athlete lines up either standing or in a 3-point stance. At the start, the athlete sprints from the first to the second cone, makes a tight transition at the second cone, and runs a figure eitht around cone 3 and back to 2. The athlete moves tightly back around the second cone and then sprints back to the starting cone.

Plisk (2000) has suggested that the ability to decelerate from a given velocity may be the most important factor that determines the athlete's ability to rapidly change direction. He suggested using a progressive, or tiered, approach to enhance agility. At first, the athlete performs the drill at half speed. On hearing a signal (whistle), the athlete begins to decelerate and, on hearing a second whistle, stops within three steps. The athlete then progresses to running at three-quarters speed and stopping within five steps. Once the athlete is successful at this stage, she then performs the drill at full speed using a seven-step braking action. A similar approach can be used for all movements (backward or lateral), with the velocity of movement and braking distance dependent on the athlete's ability (see figure 8.6f).

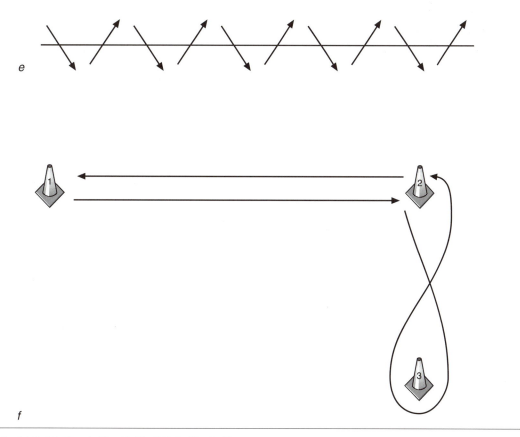

Figure 8.6   *(e)* Quick Feet Agility Drill; (f) L Agility Drill.

Hoffman and colleagues (1990, 1991a, 1991b) at the University of Connecticut examined the ability of off-season conditioning programs to improve sprint speed and agility in basketball and football players. After 6 months of off-season conditioning, the basketball players showed no improvement in either sprint speed (27-m sprint) or agility (T-test) (see figure 8.7). In a 10-week winter conditioning program for football players, no significant changes in 40-yd sprint speed were observed, despite the incorporation of sprint technique and plyometric training exercises performed twice a week. Considering that these athletes were Division I players, the inability to improve speed or agility during the off-season conditioning programs may be related to the high pretraining level. Other studies have suggested that speed can be improved if high-velocity movements such as

stretch-shortening (plyometric) drills are part of the sprint and resistance training program (Delecluse et al. 1995; Rimmer and Sleivert 2000). However, a significant training effect appeared to occur in these studies only when using subjects with average speed (Rimmer and Sleivert 2000).

There have not been enough published studies examining the ability to improve sprint speed or agility. Of the studies that are available, it appears that if the athlete has a high pretraining level of speed, the chances of becoming faster may be limited. However, if the athlete has only average speed or is deficient in a specific component of speed (e.g., explosive strength or running technique), the ability to improve sprint speed is much greater. Similar to strength (chapter 7), speed may also have a ceiling that limits progress;

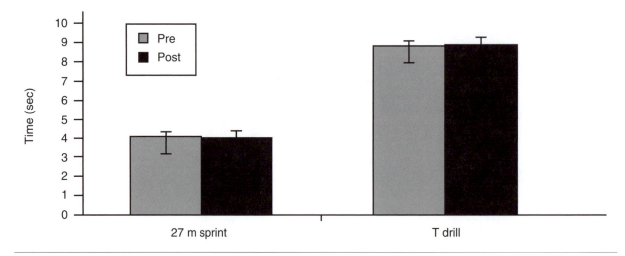

Figure 8.7   Speed and agility changes in basketball players after a 6-month off-season conditioning program. Adapted from Hoffman et al. 1991.

much of this is determined by genetic potential. As discussed in chapter 1, to run fast it would be advantageous to have predominantly **fast-twitch** muscle fibers. However, if an individual was not born with such a beneficial fiber-type composition, it would not be possible through training to cause any fiber-type transformation. Thus, an individual who is considered slow may become faster but may not be able to become fast.

An important issue, however, is the difference between changes that are statistically significant and practically significant. Statistically, examining sprint data may not prove the efficacy of a particular training program, especially when training elite-level athletes. However, for this caliber of athlete, an improvement of perhaps 0.05 s may be the difference between a medal or 4th place. For example, 0.04 s was the difference between Dwain Chambers (10.08 s), who finished 4th, and Obadele Thompson (10.04 s), who received the bronze medal in the 100-m sprint at the 2000 Sydney Olympics. Practically, it holds much significance even though it may not represent any statistical significance. Therefore, when interpreting research on the efficacy of training programs, the data, including subject population, must be carefully examined.

SUMMARY

Several studies have demonstrated the importance of both speed and agility to performance success. The ability to improve these variables during off-season conditioning programs does not appear to be very clear. Athletes who are fast and agile may not be able to significantly improve sprint time or agility during these training programs. However, athletes with an inadequacy in any particular component variable of speed or agility (e.g., technique or strength) may have a larger window of adaptation, possibly resulting in significant performance changes. There are training differences between anaerobic athletes who play team sports such as football or basketball and anaerobic athletes who are training for a specific event such as a 400-m sprint. In the former situation, emphasis is placed on preparing the athlete to play a complete game that involves repeated bouts of high-intensity activity with limited rest periods. For the sprinter, emphasis is placed on speed-endurance so that minimal fatigue is seen toward the latter stages of the sprint. One of the primary differences between the training program of the sprinter versus the football or basketball player is the emphasis of the program. The sprinter is interested in the quality of each individual sprint, whereas basketball and football players are primarily focused on the quantity of high-intensity activity common to their sport.

CHAPTER 9

ENDURANCE TRAINING

Endurance training enhances the ability to perform prolonged exercise. Endurance may be specific to either muscular endurance or cardiorespiratory endurance. Muscular endurance involves the ability of a specific muscle or group of muscle fibers to sustain high-intensity, repetitive, or static exercise (Wilmore and Costill). Athletes involved in anaerobic sports such as wrestling or boxing generally train to improve muscle endurance, and it is often accomplished through a specific resistance training program (see chapter 7). Cardiorespiratory endurance refers to the ability to sustain long-duration exercise. Enhancing cardiovascular endurance is the primary goal of distance runners, cyclists, long-distance swimmers, and triathletes. This chapter focuses on the conditioning programs of athletes who are training to improve or maintain cardiorespiratory endurance. In addition, many athletes who participate in anaerobic sports such as football, basketball, or hockey often incorporate cardiorespiratory endurance training into their conditioning programs. This chapter discusses the efficacy of such training for these athletes. For reasons of simplicity, the remainder of this chapter refers to *cardiorespiratory endurance* as simply *endurance*.

PHYSIOLOGICAL ADAPTATIONS TO ENDURANCE TRAINING

Endurance training results in physiological adaptations that are different from those seen through either resistance or anaerobic training. The effects of endurance training have been described in detail in chapters 1 through 5 and are summarized in the following section. The physiological changes common to endurance training programs positively affect the ability of the body to supply **ATP** to fuel muscular exercise aerobically. The degree to which these adaptations occur is primarily dependent on the training status of the individual and his or her genetic makeup. Individuals who are deconditioned or new to endurance training programs experience physiological adaptations that reflect a relatively large improvement in endurance performance. However, there are limitations to the extent that these adaptations can occur. Similar to what was described in chapters 7 and 8 concerning resistance-trained and anaerobically-trained athletes, the physiological adaptation and performance gains of endurance athletes who are highly experienced and in good condition will not match the degree of that seen in the novice or deconditioned individual. The extent of such improvements and the potential for success in endurance sports, as well as in other sports, are largely dependent on the genetic makeup of the individual.

Cardiovascular

Increase in **cardiac output**

Increase in **stroke volume**

Increase in blood volume and hemoglobin concentration

Increase in blood flow to exercising muscles

Decrease in resting heart rate and blood pressure

Metabolic and Musculoskeletal

Increase in mitochondrial size and number

Increase in **oxidative enzymes**

Increase in **capillary density**

Increase in reliance on stored fat as an energy source

Possible increase in myoglobin content

FACTORS RELATING TO ENDURANCE PERFORMANCE

A number of factors determine the success of an endurance athlete. The extent that many of these factors can be improved is related to the genetic potential of the individual (e.g., **maximal aerobic capacity** and **fiber type**), whereas other factors (e.g., **lactate threshold** and **exercise economy**) are limited more by the training status of the individual. There are still other factors that influence endurance performance that may be more acute in nature (e.g., nutritional and hydration status, rest, and psychological well-being). These acute factors are not within the scope of this chapter but will be covered elsewhere in the book.

Maximal Aerobic Capacity

Aerobic capacity, or $\dot{V}O_2max$, is widely considered the most objective measure of endurance capacity. $\dot{V}O_2max$ is defined as the highest rate of oxygen consumption achieved during maximal exercise. As the **intensity** of aerobic exercise increases, oxygen consumption also rises. As exercise intensity is further elevated, a point is reached where this elevation is not matched by any further increase in oxygen consumption. The point at which this plateau occurs is known as the maximal aerobic capacity, or $\dot{V}O_2max$ (see figure 9.1).

Table 9.1 provides a range of $\dot{V}O_2max$ values seen in both young (20-29 years) sedentary males and females of average fitness and elite-level male and female distance runners. The differences between the sedentary individuals and the elite-level distance runners can be attributed to both training and genetics. However, it is interesting to note that the $\dot{V}O_2max$ values seen in the young sedentary population are similar to those seen in highly trained elite-level basketball players (Hoffman and Maresh 2000). This not only highlights the importance of understanding the **specificity principle** of training, but also indicates that care must be taken when interpreting maximal aerobic capacity as an absolute indicator of fitness.

When individuals begin an endurance training program, their $\dot{V}O_2$max typically increases. This increase in aerobic capacity can justifiably be interpreted as an increase in fitness level. $\dot{V}O_2$max values have been reported to increase 15-30% over the first 3 months of an endurance training program and may rise as much as 50% within 2 years of training (McArdle, Katch, and Katch 1996). By beginning an endurance training program, a previously sedentary individual could increase his maximal aerobic capacity and reach a level of conditioning that enables him to compete in a marathon. However, the ability of that individual to actually win the race may be limited by factors that he is unable to control. As shown in table 9.1, elite male distance runners typically have $\dot{V}O_2$max values exceeding 70 ml · kg · min⁻¹ (ranging between 70 to 90 ml · kg · min⁻¹). As an example, consider a sedentary male with an initial $\dot{V}O_2$max value of 50 ml · kg · min⁻¹. After 2 years of training, a 50% improvement in maximal aerobic capacity can be hypothetically assumed, resulting in a $\dot{V}O_2$max value of 75 ml · kg · min⁻¹. Even with this remarkable increase in aerobic capacity, he still will not be competitive against elite distance runners. With all other factors considered equal (e.g., lactate threshold, exercise economy, nutritional and hydration state, psychological well-being), $\dot{V}O_2$max is the single most important variable for determining success in distance activities. If the lactate threshold of two people is at 85% of $\dot{V}O_2$max, the one with the higher $\dot{V}O_2$max would have a greater potential for success. The rest of this section discusses the additional factors that are critical for the success of endurance athletes.

Muscle Fiber Type

The importance of **slow-twitch (type I) fibers** for endurance performance has been more extensively reviewed in chapter 1 and summarized earlier. Endurance athletes such as long-distance runners have been reported to possess a higher percentage of slow-twitch fibers (approximately 70% type I fibers) than either middle-distance runners or sprinters (Costill, Fink, and Pollock 1976). This is illustrated in figure 9.2. For the endurance athlete, the advantage of a high percentage of slow-twitch fibers is related to the metabolic mechanisms that these fibers possess for providing aerobic energy to the exercising muscle. The greater capillary density, **mitochondrial content,** and oxidative enzymes common in type I fibers compared with **type II (fast-twitch) fibers** are more conducive for the metabolic breakdown of stored fat and carbohydrates used for energy during prolonged activity. It would be an advantage if an individual training aerobically were able to convert existing type II fibers to the more beneficial type I fibers. However, there are significant limitations on the ability of an individual to cause any fiber-type transformations, regardless of the training program.

There does not appear to be any convincing evidence that endurance training causes an increase in the percentage of slow-twitch fibers. Although several studies have suggested that endurance training may be able to transform a percentage of type II fibers to the more oxidative type I fibers (Howald et al. 1985; Simoneau et al. 1985),

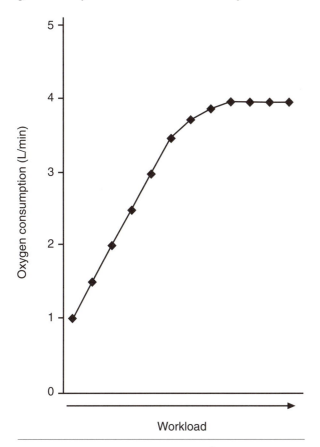

Figure 9.1 Maximal aerobic capacity ($\dot{V}O_2$max).

Table 9.1 $\dot{V}O_2$max Values for Young (20-29 years) Sedentary Individuals and Elite Level Endurance Athletes

Test groups	VO₂max (ml · kg · min⁻¹)	
	Males	Females
Sedentary individuals with average fitness	44–51	35–43
Elite level distance runners	71–90	60–75

Adapted from Martin and Coe 1997.

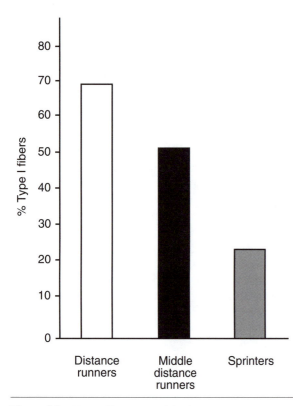

Figure 9.2   Type I fiber composition.

the majority of studies have been unable to see any transformation between fiber types. However, it is likely that a transformation of fiber subtypes does occur. Thus, a muscle fiber may become either more oxidative or more glycolytic depending on the type of training program being used (see chapter 1).

Despite limitations on fiber-type transformations, significant alterations in the metabolic mechanism can still be accomplished. Changes in capillary density and mitochondrial content are important adaptations to endurance training. Although capillary density is increased with training and a positive relationship ($p < 0.05$) has been seen between capillary density and $\dot{V}O_2$max (Saltin et al. 1977), the primary importance behind increasing capillary density is to prolong the time that oxygen spends within the muscle. That is, the longer that oxygen remains in transit within the muscle (because of the greater capillary network), the greater the opportunity to extract it, even at a high rate of muscle blood flow (Bassett and Howley 2000).

As previously mentioned, the high percentage of slow-twitch fibers seen in the endurance athlete is also accompanied by a high concentration of oxidative enzymes in the mitochondria of the cell. Costill, Fink, and Pollock (1976) showed a significantly greater concentration of **succinate dehydrogenase (SDH)** in long-distance runners compared with middle-distance runners and sprint-

ers. In addition, they reported a significant relationship between muscle SDH concentration and $\dot{V}O_2$max ($r = 0.79, p < 0.05$). The higher concentration of SDH within the muscle appears to be indicative of the greater oxidative potential of the muscle. It is known that endurance training can increase oxidative enzyme concentrations almost twofold from untrained levels (Gollnick et al. 1972; Holloszy et al. 1970) as a result of increasing the number and size of the mitochondria within the cell. Although increases in mitochondrial content and capillary density result in a higher $\dot{V}O_2$max, the magnitude of the increase in maximal aerobic capacity after endurance training is not in proportion to the increases in either mitochondrial number or capillary density.

Lactate Threshold

Blood lactate concentration represents the balance between its rate of production and its rate of removal. During light to moderate exercise, the energy demands of the muscles are met sufficiently through adequate oxygen availability. As exercise intensity increases, the muscles are unable to maintain the balance between energy production and energy demand through aerobic metabolism. It is at this point that blood lactate concentrations begin to accumulate. The point at which aerobic metabolism is unable to meet the demands of exercise and the muscles must rely on anaerobic metabolism for energy supply is termed the **onset of blood lactate accumulation (OBLA).** It is also known as the **anaerobic threshold,** or lactate threshold (Farrell et al. 1979), and is typically reported as a percentage of $\dot{V}O_2$max.

The pattern of lactate threshold is similar for both endurance-trained or untrained individuals, the only difference being the percentage of $\dot{V}O_2$max at which lactate threshold occurs. In untrained individuals, blood lactate begins to accumulate at exercise intensities exceeding 55% of their maximal aerobic capacity (Davis et al. 1979). Trained endurance athletes may perform at exercise intensities between 80 to 90% of their $\dot{V}O_2$max (Martin and Coe 1997). Figure 9.3 compares blood lactate responses at different intensities of exercise in trained and untrained endurance subjects. The differences seen in the lactate thresholds can be attributed to several possible mechanisms relating either to genetic differences or to physiological adaptations. These include a greater percentage of slow-twitch fibers, mitochondrial content, oxidative enzymes, and capillary density in the trained endurance athlete (Gollnick and Saltin 1982; Holloszy and Coyle 1984).

The lactate threshold is trainable and is frequently recognized as an indicator of endurance performance. It is often used as part of the **exercise prescription**. Often, exercise intensity is set at a person's lactate threshold or slightly above it. This is continually adjusted as the indi-

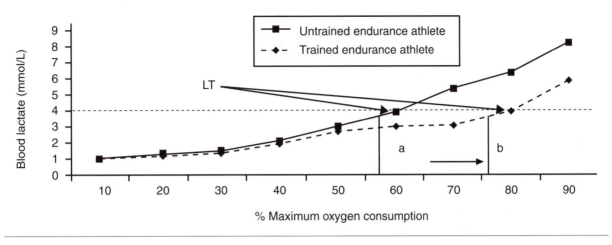

Figure 9.3 Blood lactate response at different intensities of exercise in trained and untrained endurance athletes.

vidual becomes better conditioned. It is relatively easy to determine the lactate threshold. Typically, blood lactate is plotted against exercise intensity (running speed). The running speed at which a blood lactate level of 4 mmol · L⁻¹ is reached is considered to be the lactate threshold (Heck et al. 1985) and is typically recommended as the training intensity.

Exercise Economy

The term *exercise economy* is used to describe the oxygen consumption needed to run at a given velocity. It has been shown that exercise economy may explain some of the variability in distance running performance seen in subjects with similar $\dot{V}O_2$max levels (Bassett and Howley 1997). In addition, it has also been shown to be an important factor for predicting performance in athletes running a 10K race (Conley and Krahenbuhl 1980). Figure 9.4 illustrates an example of a comparison in the running economy between two subjects with a similar $\dot{V}O_2$max. Differences in exercise economy have also been seen in cyclists (McCole et al. 1990) and swimmers (Van Handel et al. 1988). Often, differences in running economy can be attributed to differences in body position or in performance technique. As technique improves, the energy demand for a given exercise velocity is reduced, resulting in a more economical performance. Runners who increase their **stride length** from optimum levels experience a greater energy demand during exercise (Cavanagh and Williams 1982). Changes in stride length and **stride rate** appear to have a significant impact on exercise economy and subsequently on endurance performance. Besides differences in running, cycling, or swimming technique, other factors such as increases in body temperature, wind resistance, and weight (e.g., weight of shoes) also contribute to differences in the economy of exercise between individuals by causing elevations in $\dot{V}O_2$max (Daniels 1985).

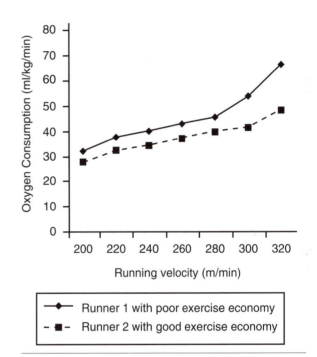

Figure 9.4 Comparison of running economy between two subjects with similar aerobic capacities.

ENDURANCE EXERCISE PRESCRIPTION

To design the exercise prescription for the endurance athlete, three primary training variables need to be considered: **training frequency, exercise duration,** and training intensity. The mode of exercise depends on the type of endurance sport (e.g., cycling, running, or swimming) that the individual performs. Although **cross-training** (performing a different mode of exercise) does provide some benefit in the training programs of athletes, the specifics of this type of training will be discussed later in the chapter.

©Human Kinetics

Exercise Duration

The length of time for an exercise session depends on the exercise intensity. If exercise intensity is above the lactate threshold, then the duration of exercise will be relatively short because of **fatigue**. However, if the intensity of exercise is mild to moderate, training could be of much longer duration. An appropriate duration of exercise has not been determined. For deconditioned or sedentary individuals, 5 min a day of low-intensity exercise could elicit a training effect. In contrast, NCAA endurance runners average workouts of 8-10 miles (13-16 K) a day depending on the phase of training during the season (Kurz et al. 2000). The training session for those athletes would obviously be of much longer duration. For practical purposes, to produce a training effect it is recommended that most individuals exercise 20-30 min per session at an exercise intensity of at least 70% of $\dot{V}O_2$max (ACSM 2000a).

One of the concerns for the coach and athlete is the risk-to-benefit relationship that exists when increasing the duration of training. Increasing the duration of training places the athlete at a higher risk of overtraining (see chapter 21). Because of the high level of conditioning common among distance athletes, coaches may increase the length of each training session to elicit further physiological adaptations. However, increasing the duration of training may still not bring about the desired performance outcomes. Costill and colleagues (1991) failed to see any differences in swimming power, endurance, or performance in collegiate swimmers when training duration was doubled from 1.5 h a day to two daily 1.5-h training sessions.

Training Intensity

Intensity of exercise may be the single most important variable to be consider when writing the exercise prescription. All of the other acute program variables depend on training intensity. As training intensity increases, the duration of exercise becomes shorter; likewise, as exercise intensity is reduced, the duration of exercise can be prolonged. As mentioned in the discussion on training frequency, if intensity of exercise is high, the frequency of training needed to elicit a training effect may be reduced. However, if training intensity is low, a greater frequency of training may be needed to elicit the desired physiological adaptations and enhance endurance performance. The proper adjustment of exercise intensity is important for achieving the desired training goals. To maximize aerobic capacity, training intensity must create an overload on the physiological processes of the body, resulting in the desired adaptation. However, if exercise intensity is too high, fatigue may occur

Frequency of Training

How often an athlete should train for endurance is debatable. Research has shown that training twice per week appears to produce similar changes in $\dot{V}O_2$max as training 5 days per week (Fox et al. 1973; McArdle, Katch, and Katch 1996). However, this may be related to the intensity of training because as training intensity is reduced, a greater frequency of training may be needed to increase aerobic capacity (McArdle, Katch, and Katch 1996). In addition, the frequency of training may also depend on the training status of the individual. Fewer days of endurance training per week may be needed to maintain aerobic fitness than was needed to reach that level of performance (Potteiger 2000). The frequency of training, however, is also related to the goals of the training program. If the primary goal of exercise is to reduce body fat, then a higher frequency of training will be more beneficial to the individual. The greater the frequency of training, the greater the caloric expenditure. It is for this reason that daily exercise is recommended for individuals interested in losing weight (ACSM 2000a).

prematurely, and the training stimulus may be insufficient.

Training intensity can be expressed in several ways. It may be expressed as a relative intensity of a person's $\dot{V}O_2$max or as a percentage of **maximal heart rate.** In addition, a **rating of perceived exertion (RPE)** has also been used to prescribe training intensity (see table 9.2). The ideal method for determining exercise intensity is to find the velocity of exercise (whether it is running, cycling, or swimming) that corresponds to the athlete's lactate threshold. As described earlier, the relative percentage of $\dot{V}O_2$max that elicits the lactate threshold may be varied and depends on the athlete's training status. If the velocity of exercise needed to reach the lactate threshold can not be calculated, then the other methods may be used to prescribe exercise intensity.

Heart Rate

The use of **heart rate** to prescribe aerobic exercise is well accepted by the exercise science community because of the close relationship between heart rate and oxygen consumption (see figure 9.5). Regardless of age, conditioning level, or gender, the relationship between percent $\dot{V}O_2$max and percent maximal heart rate is maintained (ACSM 2000a). This relationship is also maintained during different modes of exercise. However, it is important to note that during different forms of exercise, or during exercise that uses a smaller muscle mass (e.g., arm versus leg), the maximal heart rate may be different. For example, the maximal heart rate during swimming is approximately 13 beats · min^{-1} lower than that seen during running in trained and untrained men and women (Magel et al. 1975; McArdle et al. 1978). The reason for the lower maximal heart rate during swimming is likely related to the horizontal body position, the cooling effect of water, and the reduced neural stimulation from the smaller active musculature of the upper body (McArdle, Katch, and Katch 1996).

When writing an endurance exercise prescription for healthy males or females, it is generally acknowledged that exercise should be performed at about 70% of the maximal heart rate to elicit improvements in aerobic capacity. For example, a 20-year-old male whose maximal heart rate is 200 (the formula for predicting maximal heart rate is 220 – age) should run at a pace that elicits a heart rate of 140 beats · min^{-1}. The use of a percentage of the maximal heart rate provides a relatively simple method for determining exercise intensity. An alternate method is to calculate the **heart rate reserve (HRR),** which provides the functional capacity of the heart rate. This formula, also known as the Karvonen method, is as effective as using a percentage of the maximal heart rate and is also used with regularity. The HRR can be calculated as follows:

$$HRR = \text{maximal heart rate} - \text{resting heart rate}$$

Once the HRR is calculated, the target heart rate can be determined. Target heart rate is given as a range of the functional capacity of the heart plus the resting heart rate:

$$\text{Target heart rate} = (HRR \times \text{exercise intensity}) + \text{resting heart rate}$$

For example, an exercise prescription for a 20-year-old college student with a resting heart rate of 60 beats · min^{-1} calls for exercise to be performed at 60-70% of the functional capacity. After determining the age-predicted maximal heart rate, the HRR is then calculated.

Table 9.2   Ratings of Perceived Exertion

RPE Scale (6-19)

6	No exertion at all
7	Extremely light
9	Very light
11	Light
13	Somewhat hard
15	Hard (heavy)
17	Very hard
19	Extremely hard

Borg RPE scale © Gunnar Borg, 1970, 1985, 1994, 1998.
Adapted, by permission, from G. Borg, 1988, *Borg's Perceived Exertion and Pain Scales,* (Champaign, IL: Human Kinetics), 47.

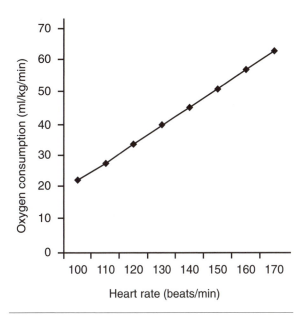

Figure 9.5 Relationship between heart rate and oxygen consumption.

$$HRR = \text{maximal heart rate} - \text{resting heart rate}$$

$$HRR = 200 \text{ beats} \cdot \text{min}^{-1} - 60 \text{ beats} \cdot \text{min}^{-1}$$

$$HRR = 140 \text{ beats} \cdot \text{min}^{-1}$$

With a heart rate reserve calculated as 140 beats · min[-1], the next step is to calculate the target heart rate:

$$\text{Target heart rate (THR)} = (HRR \times \text{exercise intensity}) + \text{resting heart rate}$$

$$THR = (140 \times 0.60) + 60 = 144 \text{ beats} \cdot \text{min}^{-1}$$

$$THR = (140 \times 0.70) + 60 = 158 \text{ beats} \cdot \text{min}^{-1}$$

Thus, the exercise intensity for this person should range between 144 to 158 beats · min[-1]. Endurance training causes physiological adaptations that are expected to decrease the heart rate for an absolute exercise load. To maintain the same exercise stimulus, adjustments must be made to the exercise intensity.

Ratings of Perceived Exertion

Ratings of perceived exertion (RPE) can also be used to prescribe endurance training programs. Table 9.2 shows Borg's (1982) 15-point rating scale of perceived exertion. The exerciser rates how he or she feels in relation to the exercise stress. As energy expenditure and physiological strain increase, the perceived strain on the exerciser also increases. An RPE of 13 or 14, which corresponds to a verbal anchor of "somewhat hard," coincides with a heart rate of about 70% of the maximal heart rate (McArdle, Katch, and Katch 1996). However, recent research suggests that the use of the RPE may not be valid for exercise durations of 20 min or longer (Kang et al. 2001).

ENDURANCE TRAINING PROGRAMS

How athletes train has changed considerably through the years as the scientific knowledge base regarding the best way for athletes to reach peak performance has expanded. This is no different for endurance athletes. Historically speaking, Kurz and colleagues (2000) reported that many coaches believed that the longer the race, the more important aerobic training became. In support of this, previous research has demonstrated that **training volume** (total number of miles run per week) is a significant predictor of marathon run-time (Dotan et al. 1983). However, recent thought has refocused training tactics from high volume, low intensity to low volume, high intensity. Still, one accepted theory of training does not exist and, as might be expected, a number of methods are used to train endurance athletes.

Kurz and colleague (2000) published a survey on training methods of NCAA Division I distance runners. The survey, which included 14 Division I programs, reported the training methods employed and the number of weekly sessions during a particular training cycle. These results were adapted into figure 9.6, and the various training methods are listed and described in table 9.3. In this study, the training year was separated into three different phases. The transition phase included the period between the months of May to August, the competition phase ranged from August to October, and the peaking phase was the month of November. In general, the transition phase for an endurance athlete is designed as a period of development in which the volume of training (weekly mileage) is gradually increased as the athlete enters the competitive phase of the season. During the competitive season, a greater emphasis is placed on improving the lactate threshold, with training directed toward the final competitions of the peaking phase. During the peaking phase, the athlete focuses all workouts on preparation for the final race. Training volume is tapered, and a greater emphasis is placed on speed work. Typical training volumes for each phase of training can be seen in figure 9.7.

Cross-Training

Athletes often use cross-training (performing exercise different from their usual mode of training) to maintain general fitness. Cross-training is beneficial during periods of injury if the athlete is unable to perform his or her normal exercise routine. For example, a runner suffers an injury that prevents her from performing her normal training regimen. She can perform an alternative method of training such as swimming or cycling to maintain fitness levels during the time that she is unable to run. Cross-training also prevents boredom and monotony by providing variety to the athlete's training program. In addition, the use of different modes of exercise may reduce the likelihood of overtraining because it distributes the stress of training among different muscle groups (Potteiger 2000). Cross-training is also critical for athletes who compete in multiple events. For example, triathletes need to swim, cycle, and run in preparation for their competitions. Although cross-training appears able to maintain aerobic capacity in other modes of training, it is unable to improve aerobic capacity to the same degree as mode-specific training (Foster et al. 1995; Gergley et al. 1984; Magel et al. 1975).

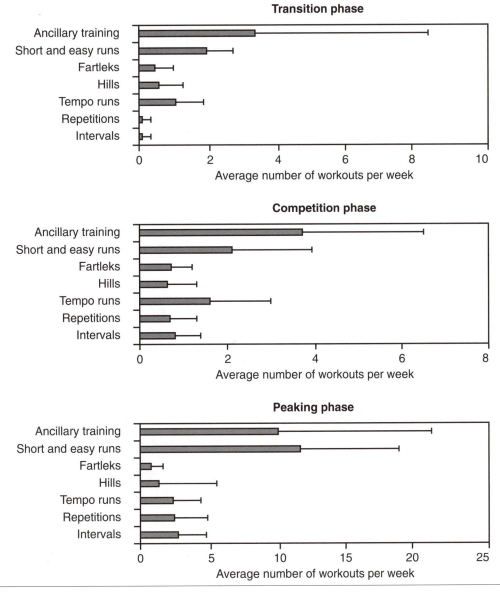

Figure 9.6 Training methods of Division I cross-country runners.
Adapted from Kurz et al. 2000.

Table 9.3 Training Methods for Distance Runners

Training method	Description
Intervals	Repeated bouts of high intensity runs at or exceeding the runners' face pace, interspersed with a recovery period lasting no longer than the time of the high intensity run.
Repetitions	A series of runs of at least 400 m in distance performed faster than a race pace with a complete recovery period between each run.
Tempo runs	These are training runs performed at a pace 20–30 seconds slower than the normal race pace.
Hills	Runs repeatedly performed on a graded hill at 85–90% effort for a time interval between 30 s and 5 min. The recovery period would be the jog back down the hill.
Fartleks	These are runs performed at various intensities over various distances.
Short and easy running	As the name implies, this is a run performed at an easy pace over a relatively short distance.
Ancillary training	This includes all supplementary training including resistance training, plyometrics, and form drills.

Adapted from Kurz et al. 2000.

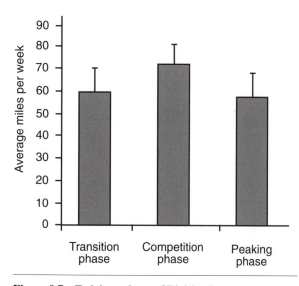

Figure 9.7   Training volume of Division I cross-country runners.

Adapted from Kurz et al. 2000.

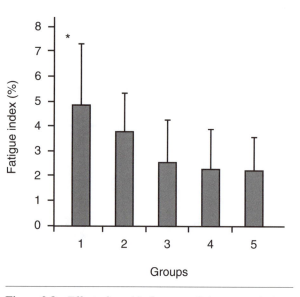

Figure 9.8   Effect of aerobic fitness on fatigue rates during anaerobic exercise. * = Significant change.

Adapted, by permission, from J.R. Hoffman, 1997, "The relationship between aerobic fitness and recovery from high-intensity exercise in infantry soldiers," *Military Medicine* 162:484-488. © Military Medicine: International Journal of AMSUS.

Endurance Training for Anaerobic Athletes

Athletes who play **anaerobic** sports such as football or basketball are often required to perform conditioning drills designed to enhance their aerobic capacity (e.g., long jogs). This may appear strange, considering that the movement patterns of these sports are performed at predominantly maximal intensities. In addition, aerobic capacity has been reported to have a significant negative correlation to playing time of basketball players (Hoffman et al. 1996). That is, the greater the athletes' aerobic capacity, the less time they actually played. Thus, direct performance benefits from increasing aerobic capacity in these athletes may be questionable and would therefore have important implications for their exercise prescriptions. However, considering the relationship between oxygen supply and skeletal muscle **recovery** (Idstrom et al. 1985), aerobic fitness may be critical for the recovery processes during athletic performance involving repeated high-intensity activities.

Few studies have examined the ability of aerobic capacity to enhance recovery from anaerobic activity. In one study, Koziris et al. (1996) examined recreationally-trained males and females but were unable to establish any relationship between oxidative metabolism and fatigue during the latter stage of a Wingate Anaerobic Test (30 s of high-intensity cycle ergometry). Similarly, Hoffman, Epstein, Einbinder, et al. (1999) were unable to find any relationship between aerobic capacity and

recovery after high-intensity exercise in elite-level basketball players. However, in another study performed on a large group of subjects with diverse aerobic fitness levels, a relationship between aerobic fitness and recovery during repeated bouts of high-intensity exercise was indeed reported (Hoffman 1997). However, this relationship appears to be limited. A reduction in the rate of fatigue between bouts of high-intensity exercise was observed as aerobic fitness improved (reduced time for a 2,000-m run) to values that approached the population mean. In subjects whose level of aerobic fitness neared or exceeded the population mean, no further benefit in fatigue rate was seen (see figure 9.8). Although aerobic conditioning appears to provide some advantages for the anaerobic athlete, it is likely that once an average level of aerobic fitness is achieved, there would be minimal benefits from attempting to increase aerobic capacity further. In addition, once an aerobic base is achieved, sport-specific team practices and games are sufficient to maintain aerobic fitness (Hoffman, Fry, et al. 1991).

SUMMARY

This chapter reviewed the specific adaptations seen after endurance training. Although every athlete who performs endurance exercise experiences these adaptations, the ability of the athlete to be competitive may be limited by genetic factors. There does appear to be an upper limit, or ceiling, to improvement of aerobic capacity. In addition, endurance athletes must exercise at or slightly above their lactate threshold to maximize their performance gains. For those individuals who are unable to calculate lactate threshold, exercise intensity can be given as a percentage of maximal heart rate or ratings of perceived exertion. Finally, in regard to endurance training for the anaerobic athlete, it appears that once a baseline aerobic fitness level is reached, further concentration on improving aerobic capacity may not be justified.

CHAPTER 10

CONCURRENT TRAINING

©Human Kinetics

For many sports, the reliance on more than one energy system dictates the inclusion of various modes of exercise in the training regimen of the athlete. However, training multiple energy systems and performing various types of training simultaneously, referred to as **concurrent training,** has important implications for the physiological adaptations of such training programs. The preceding chapters demonstrated that the adaptive responses of the body are specific to the type of training program used. **Endurance training** results in physiological adaptations (e.g., increases in oxidative enzyme activity, capillary density, and mitochondrial content) that are conducive to improving and maintaining prolonged aerobic activities. **Resistance training** produces changes that are often in direct contrast to those seen during endurance training. These adaptations often include increases in muscle mass that may parallel decreases in mitochondrial volume density (MacDougall et al. 1979). Such contrasting adaptations from performing either endurance or resistance training have created a hesitance on the part of both endurance and strength athletes to engage in the opposite form of training for fear that it may compromise desired training adaptations. This chapter focuses on the compatibility of these forms of training and the effect that concurrent training has on development of **maximal aerobic capacity** and muscle **strength,** growth, and adaptation. In addition, the use of concurrent training for individuals interested in general fitness and reducing body fat is also discussed.

EFFECT OF CONCURRENT STRENGTH AND ENDURANCE TRAINING ON V̇O$_2$MAX

Hickson, Rosenkoetter, and Brown (1980) were one of the first groups to examine the effects of resistance training programs on aerobic power and short-term endurance. In their study, college-aged males who were active in recreational sports initiated a 5-day per week resistance training program. After 10 weeks of training, the subjects averaged a 38% increase in lower-body strength (1 RM squat), but their aerobic capacity went unchanged when expressed relative to body weight. Interestingly, their time to exhaustion in both cycle and treadmill exercise increased significantly by 47% and 12%, respectively. This was one of the first studies to provide evidence not only contrary to the belief that resistance training may be detrimental to endurance performance but that resistance training may in fact improve endurance performance. Several mechanisms were thought to underlie the improved endurance performance in these subjects. The authors suggested that an improved **glycolytic enzyme capacity** (a potential training adaptation) would

provide a greater capacity to resynthesize **ATP** as an immediate source of energy. In addition, the neurological adaptations commonly observed during the initial stages of a resistance training program may have resulted in an alteration in motor unit recruitment patterns, providing for an improved **exercise economy.**

Later studies combining endurance and resistance training demonstrated similar effects on aerobic capacity and endurance performance. These studies, using primarily untrained individuals, have consistently reported that the combination of resistance and endurance training does not compromise the ability of these subjects to improve their maximal aerobic capacity (Dudley and Djamil 1985; Hunter, Demment, and Miller 1987; Kraemer et al. 1999; McCarthy et al. 1995). The studies lasted 6-12 weeks and typically used a 3-day per week resistance and endurance training program. Several of these studies included both resistance and endurance training on the same day (McCarthy et al. 1995; Hunter, Demment, and Miller 1987; Kraemer et al. 1999). However, in some studies, resistance and endurance training were performed on alternate days for a total of 6 consecutive days of training (Dudley and Djamil 1985). In either training scenario, the inclusion of a resistance training program did not impede the ability to improve maximal aerobic capacity (see figure 10.1).

Combined endurance and resistance training has produced similar results for untrained subjects and physically active and trained subjects (Bishop et al. 1999; Hennessy and Watson 1994; Hickson et al. 1988). These studies have ranged in duration from 8 to 12 weeks, and the frequency of training has varied from study to study. Bishop and colleagues (1999) examined the effects of adding a 2-day per week resistance training program to the off-season regimen of endurance-trained female athletes. After 12 weeks of concurrent training, no significant differences in endurance performance, **lactate threshold,** or maximal aerobic capacity were reported between subjects that performed resistance training and those that did not. Hennessy and Watson (1994) investigated an 8-week, 5-day per week combined strength (3 days per week) and endurance training (4 days per week) program for anaerobic athletes. When compared with other athletes similar in training background who performed only endurance training 4 days per week, no differences were seen in aerobic capacity improvements. Other investigators have shown that when a resistance training program is added to the normal training regimen of experienced endurance-trained individuals, not only was aerobic capacity maintained, but short-term endurance was also improved (Hickson et al. 1988). Improvements in endurance performance **(time to fatigue)** were seen in both cycling (11% improvement) and treadmill exercise (13% improvement). Although evidence

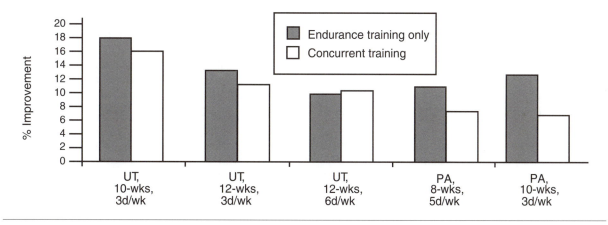

Figure 10.1   Effect of concurrent training on maximal aerobic capacity. UT = intrained subjects; PA = physically active subjects. Adapted from McCarthy et al.1995; Kraemer et al. 1999; Hunter et al. 1987; Hennessy and Watson 1994; Dolezal and Potteiger 1998.

consistently reflects the ability of concurrent training programs to improve maximal aerobic capacity, the benefits of resistance training in individuals with endurance training experience may not always be realized. Some researchers have reported that the addition of a resistance training program improves aerobic capacity in physically active subjects, but the magnitude of this improvement may be reduced when compared with subjects with similar training backgrounds that are performing only endurance training (Dolezal and Potteiger 1998). To summarize the research on concurrent training and aerobic capacity, it appears that the addition of resistance training to the aerobic conditioning programs of either previously sedentary individuals or endurance-trained athletes does not compromise improvements in aerobic capacity, and benefits to short-term endurance may also be realized. Both resistance and endurance training appear compatible as long as the frequency of endurance training is not reduced. If resistance training was added at the expense of aerobic exercise, the ability to maintain or improve aerobic capacity may indeed become compromised.

EFFECT OF CONCURRENT STRENGTH AND ENDURANCE TRAINING ON MAXIMAL STRENGTH

Results from studies examining the effects of concurrently training for both endurance and strength improvements have been inconclusive. Several investigations have suggested that combining endurance and resistance training may compromise the potential for strength gains. However, a number of studies have shown no detrimental effects on strength improvement after combining both modes of training. These contrasting results are likely related to differences in the training status of the sub-

jects and specific differences in the training program design.

Dudley and Djamil (1985) reported that the combination of endurance and resistance training in previously untrained individuals compromised the magnitude of maximal torque (force acting at a distance from the axis of rotation) increases after 7 weeks of training. However, this effect was seen primarily at fast speeds of contraction. At slower contraction velocities, strength gains were still observed. In addition, the subjects performed strength and endurance training on alternate days, resulting in a 6-day per week training program. Such a high frequency of training may impede **recovery** and possibly result in an **overtraining syndrome** (see chapter 21). If an individual is in a chronic state of fatigue, the potential for strength gain could be affected. However, because the stimulus that causes overtraining syndrome is individualistic, training 6 days per week may not always result in fatigue. This was demonstrated in a study by Hunter, Demment, and Miller (1987), who also investigated an untrained subject population performing a 6-day per week concurrent training program. They reported no detrimental effects in strength gains when compared with a strength-trained only group. When concurrent training programs were performed on the same day (endurance training preceded by resistance training or the reverse), the number of exercising days per week was reduced. When frequency of training is reduced, whether it is in subjects with limited or no resistance training experience, strength gains do not appear to be compromised (Bell et al. 1991; Gravelle and Blessing 2000; McCarthy et al. 1995).

The effect of adding endurance training to the exercise programs of subjects with experience in strength training does appear to impede the ability to maximize strength performance. Investigations on experienced resistance-trained subjects have shown that upper- (1 RM bench press) and lower- (1 RM squat) body strength may be compromised when endurance training is performed concurrently with strength training (Dolezal and Potteiger 1998; Hennessy and Watson 1994). In both of these studies, subjects performed endurance and resistance training on alternate days, resulting in a total frequency of training of 6 days per week. Thus, the reduction of strength improvement in these subjects may be related to possible chronic fatigue from the high frequency of training. Figure 10.2, a and b, shows the effect of concurrent training on the percent improvement of strength in previously untrained and trained subjects, respectively.

The cumulative effect of fatigue appears to be a large contributing factor to diminished strength improvements in subjects with some resistance training experience. In studies that examined novice weightlifters, even a 6-day per week concurrent training program did not compromise strength gains (Hunter, Demment, and Miller 1987). In addition, initial increases in strength are generally the result of neurological adaptations occurring within the first 2 months of training. It is possible that these neurological adaptations are not impeded in the novice lifter even at a higher frequency of training. In subjects with some resistance training experience, the physiological adaptations that contribute to further increases in strength involve muscle structural changes and may be more sensitive to additional modes of training and fatigue.

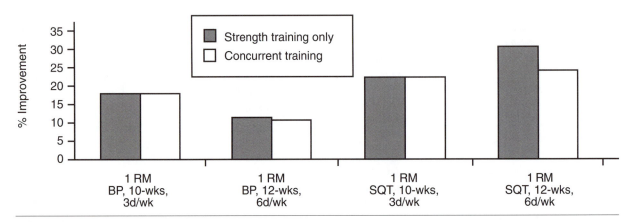

Figure 10.2a   Effects of concurrent training on maximal strength in untrained subjects. BP = bench press; SQT = squat. Adapted from McCarthy et al. 1995; Hunter 1987.

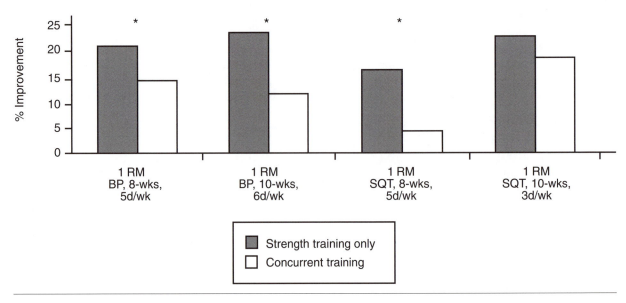

Figure 10.2b   Effect of concurrent training on maximal strength in trained subjects. BP = bench press; SQT = squat; * = significant difference.
Adapted from Hennessy and Watson 1994; Delezal and Potteiger 1998.

Leveritt and Abernethy (1999) studied the acute effects of high-intensity endurance exercise on a resistance training session in recreationally trained subjects. It was shown that endurance exercise performed 30 min before resistance training reduced strength performance (see figure 10.3). The number of repetitions performed during 3 sets of squat exercises was reduced 13-36% when preceded by a high-intensity endurance workout in comparison with the number of repetitions performed when no endurance exercise was performed before the training session. In addition, isokinetic testing reduced peak torque 10-19% over a range of testing velocities when endurance training preceded the resistance exercise. These results suggest that if endurance training is performed before resistance training, the training stimulus to the muscle may be reduced, possibly limiting the extent of the physiological adaptations. However, if the strength training session occurs before endurance exercise, then the resulting strength improvements may not be diminished.

EFFECT OF SEQUENCE OF TRAINING ON ENDURANCE AND STRENGTH IMPROVEMENTS

The number of studies examining the effect of the **sequence of training** on both strength and aerobic capacity improvements is limited. In a study by Collins and Snow (1993) on previously untrained subjects, the sequence of training (whether endurance preceded resistance training or the reverse) did not appear to have any effect on either aerobic capacity or strength improvements. Thus, it appears that in an untrained population, the sequence of training may not be very important. However, in this study, the subjects were required to perform only 2 sets of 3-12 repetitions per exercise using 50-90% of their 1 RM. Although such an exercise regimen can produce significant strength gains in this subject population, this would not be a typical resistance training workout in a more experienced group of resistance-trained subjects (see chapter 7). Further study on the effect of sequence of training using a more experienced subject population would be welcome.

Another concern related to sequence of training involves whether concurrent training should be performed on alternate days or whether resistance and endurance training should be performed on the same day. By training on alternate days, the subject may possibly extend the weekly training sessions to 6 days per week. In the untrained subject, this does not appear to impede strength or aerobic capacity improvements (Hunter, Demment, and Miller 1987). However, in more experienced subjects, a high frequency of training (e.g., 6 days per week) may not provide a sufficient time for recovery between exercise sessions, which compromises strength improvements (Dolezal and Potteiger 1998; Hennessy and Watson 1994). This may be related to a higher sensitivity in these individuals for making further physiological adaptations. In contrast, when endurance and strength training are performed on the same day, there do not appear to be any detrimental effects on the ability of the subjects to increase their strength or aerobic capacity (Bell et al. 1991; Gravelle and Blessing 2000; McCarthy et al. 1995). Although the daily training volume may be greater in reduced weekly training frequencies, the rest between each exercise session may be sufficient to allow complete recovery.

Sale and colleagues (1990a) looked specifically at the question of same-day versus alternate-day training in

Figure 10.3 Acute effect of endurance training on strength performance. * = significant difference.
Adapted from Leveritt and Abernethy 1999.

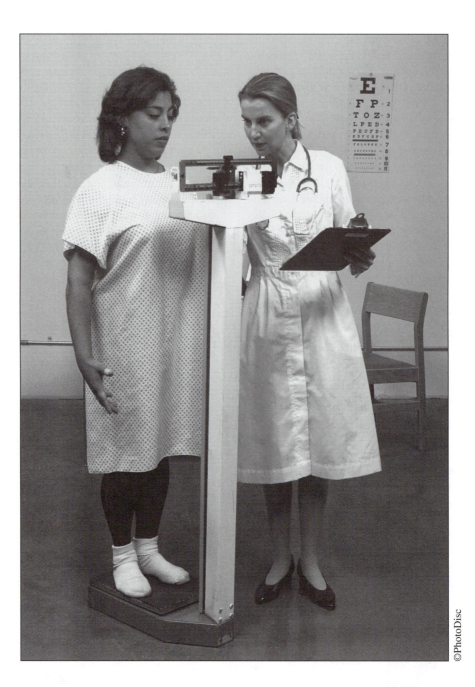

©PhotoDisc

previously untrained college-aged male subjects. Resistance and endurance training were performed twice per week. Both groups made significant improvements in aerobic capacity with no significant difference observed between the groups. However, the group that trained 4 days per week (alternate-day training group) made significantly greater strength gains than the group that trained only twice per week (same-day training group) (24% versus 13%, respectively). Perhaps by training on alternate days, but limiting the total number of weekly training sessions, the subjects have a more complete recovery, resulting in significantly greater strength gains. The subjects in this group may have had a better-quality

workout because it was not preceded by an endurance training session. Thus, when adequate rest is provided, significantly greater strength performance gains may be realized, even when the frequency of training is high.

EFFECT OF CONCURRENT TRAINING ON MUSCLE GROWTH AND MUSCLE FIBER CHARACTERISTICS

Many endurance athletes are concerned that the inclusion of a resistance training program would cause physiological changes to the muscle that would be detrimental to endurance performance (e.g., muscle **hypertrophy,** decrease in mitochondrial volume and capillary density). In studies examining this particular question, no significant alterations in muscle size or muscle fiber composition were reported in endurance-trained subjects performing resistance exercise for the first time (Bishop et al. 1999; Hickson et al. 1988). It should be noted, however, that these studies were 10-12 weeks in duration. It is generally understood that muscle hypertrophy occurs approximately 6-8 weeks after the initiation of a resistance training program in previously untrained individuals (see chapter 1 for a review). It is possible that the time course for muscle adaptations in individuals with limited resistance training experience may be longer when concurrently performing another mode of training. Based on available evidence, it appears that 10-12 weeks of resistance training added to the exercise program of endurance-trained subjects do not cause any significant changes to muscle fiber composition or to fiber cross-sectional area. The effect of training durations exceeding 3 months is not known. However, by manipulating acute program variables, the coach and athlete can specifically focus on the physiological adaptations that are beneficial to the endurance athlete.

The effect of concurrent training on muscle adaptations in other population groups may be different. In previously untrained subjects, 6 weeks of resistance training resulted in a significant increase in the muscle fiber area of both **type I** and **type II** fibers in a group that performed only strength training (Bell, Syrotuik, et al. 2000). These muscles continued to hypertrophy even after 12 weeks of training. In comparison, a group of subjects performing both endurance and resistance training (using the same resistance training program) showed no significant changes in muscle fiber area after 6 weeks of training. However, after 12 weeks of training, significant increases were observed in the muscle fiber area of type II fibers only. This study, although showing significant increases in muscle fiber area, did suggest that combining both endurance and strength training might suppress some of the adaptations seen when performing only strength training. Kraemer, Patton, and colleagues (1995) showed similar results in physically active subjects. In their study (see figure 10.4), significant increases in both type I and type II fibers were seen in both strength- and combined strength- and endurance-trained groups. A significant reduction in the type I fiber area was seen in a group of subjects performing only endurance training. This study demonstrated an apparent benefit of resistance training for the endurance-trained individual. Not only did resistance training not compromise aerobic capacity, but the inclusion of this mode of training appeared to prevent the muscle fiber **atrophy** that may accompany endurance training. As will be seen later in this chapter, such adaptations may have important implications for individuals interested in increasing their daily energy expenditure as part of weight-reduction programs.

Kraemer, Patton, and colleagues (1995) also demonstrated fiber subtype transformations that were specific

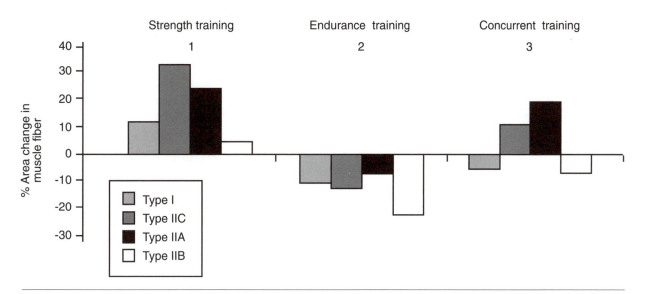

Figure 10.4   Effect of concurrent training on muscle fiber alterations.
Adapted, by permission, from W.J. Kraemer et al., 1995, "Responses of IGF-1 to endogenous increases in growth hormone after heavy-resistance exercise," *Journal of Applied Physiology* 79:1310-1315.

to the training program employed. Subjects who performed only strength training (4-day split routine using 2 heavy days [5 RM] and 2 moderate days [10 RM]) or performed both strength and endurance training (4 days per week using 2 days of prolonged runs [45 min at 80-85% $\dot{V}O_2$max] and 2 days of interval training) saw similar type II fiber subtype transformations (IIa ←IIb) after training. In contrast, subjects who performed only endurance training had a type II muscle fiber subtype transformation from fibers that were more glycolytic to fiber subtypes that were more oxidative in nature (IIc ←IIa ←IIb). In other words, the inclusion of a high-intensity resistance training program appears to reduce the magnitude of the type II fiber transformation to more oxidative fibers.

EFFECT OF CONCURRENT TRAINING ON HORMONAL ADAPTATIONS

Combining endurance and resistance training in previously untrained subjects does not compromise performance gains in either variable. However, in subjects with resistance training experience, concurrent training may potentially hinder the ability of these individuals to maximize their strength gains, likely related to a reduced adaptive ability of the muscle. As mentioned earlier, the neurological adaptations that lead to initial strength increases are not likely to be affected, considering that strength improvements in the novice lifter do not appear to be diminished by concurrent training. Because **anabolic and catabolic hormones** are intimately involved with both exercise recovery and muscle growth, it would appear that any difference in the way hormones respond to concurrent training might play a key role in regulating muscle adaptations.

In one study (Kraemer et al. 1999) examining the effect of concurrent training on hormonal responses, previously sedentary subjects were placed in one of three training groups: a strength only, an endurance only, or a combination of both strength and endurance training. Each of these groups exercised 3 days per week. After 12 weeks of training, no significant differences in the resting hormonal concentrations of either **testosterone** or **cortisol** were seen between the groups.

In studies examining physically active or trained subjects, the hormonal response to these various modes of training appears to be specific to the exercise regimen employed. Kraemer, Patton, and colleagues (1995) examined the compatibility of endurance and strength training in physically active men for 12 weeks. In the group of subjects that performed only resistance training, no changes in testosterone concentrations were observed

during the course of the training program. However, cortisol concentrations were reduced by the 8th week of training, suggesting that a greater anabolic environment existed for muscle growth (reflected by a greater testosterone/cortisol ratio). In the group that performed only endurance training, no significant changes in testosterone were seen, but significant elevations in cortisol concentrations were reported. In the group of subjects that performed both endurance and resistance training, significant increases in both testosterone and cortisol were observed at the end of the training period. However, the relative increase in cortisol far exceeded that of testosterone, reflecting a greater exercise stress from the combination of both endurance and resistance training. These results were confirmed by other studies showing a greater catabolic response to combined strength and endurance training versus endurance training only in previously trained individuals (Bell et al. 1997).

The importance of the hormonal environment for muscle growth was described in chapter 2. Briefly, changes in the anabolic and catabolic hormonal profile influence cellular changes relating to **protein synthesis** and muscle fiber adaptations. If the catabolic hormone response is elevated, increases in protein degradation may be seen. This was demonstrated by Kraemer, Patton, et al. (1995), who showed a decrease in fiber size that corresponded with an increase in cortisol concentrations in subjects performing endurance exercise. Kraemer, Patton, and colleagues (1995) further showed that the addition of resistance training to an endurance training program may preserve muscle mass by increasing the anabolic response. This may not only prevent the catabolic effects of the elevated cortisol concentrations but may also provide enough of a stimulus to cause slight increases in type IIa muscle fibers. Figure 10.5 shows a diagram of the hormonal response to various modes of training, including concurrent training, and the possible effects on muscle fiber.

EFFECT OF CONCURRENT TRAINING ON BASAL METABOLIC RATE AND WEIGHT LOSS

To reduce body weight and alter **body composition,** it is generally accepted that prolonged endurance training is more advantageous for increasing energy output than any other form of training. Endurance training is also thought to play a role in potentiating both **basal metabolic rate (BMR)** and **resting metabolic rate (RMR)** (Ballor and Poehlman 1992). However, other studies investigating the ability of endurance training to increase BMR or RMR have been ambiguous. Some investigations have shown either no change in BMR (Sjodin et al. 1996) or a de-

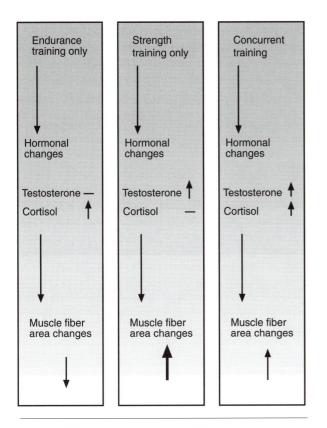

Figure 10.5   Hormonal and muscle fiber response to concurrent training.

crease in RMR (Thompson, Manore, and Thomas 1996) after endurance training. In contrast, resistance training is recognized as the primary mode of training to increase muscular strength, and it also positively alters body composition, primarily by increasing lean tissue. An increase in lean body mass increases the basal metabolic rate and results in an increase in total energy expenditure. Thus, for individuals interested in weight reduction and body composition changes, the combination of both endurance and resistance training appears to provide a more optimum advantage in eliciting a training effect.

Very few studies have actually examined the effect of concurrent training on basal metabolic rate and weight loss. Dolezal and Potteiger (1998) were two of the first investigators to study the effects of combining endurance and resistance training on basal metabolic rate and body composition changes compared with resistance training or endurance training alone. They showed that after 10 weeks of concurrent training, subjects were able to significantly increase their BMR and reduce their body fat from preexercise values. Nevertheless, the magnitude for each change was still lower than that seen when subjects performed each mode of training by itself. However, it was clear that by combining both modes of training, the subjects were able to achieve all the benefits that are typi-

cally derived when performing each mode of exercise alone. The benefits of adding resistance training to an endurance training program or to a diet and endurance training program appear to reside in the ability to prevent a decline in lean tissue resulting from both a low caloric diet and the catabolic effects of endurance training (Kraemer et al. 1999).

EFFECTS OF COMBINED SPRINT AND RESISTANCE TRAINING

The exercise science literature has generally considered only the combination of endurance and resistance training as the typical modes of exercise when discussing concurrent training. However, considering that both resistance and anaerobic endurance training (e.g., sprint and interval training) are two separate forms of training and are often performed concurrently in the training programs of many anaerobic athletes, it seems appropriate to make some mention of it here. Both modes of training appear to complement each other. However, the major concern would be increasing the risk of overtraining. Typically, training programs are developed that alter both the intensity and volume of training for each type of exercise in order to minimize the risk of fatigue in the athlete (see chapter 11). However, during the competitive season, the emphasis on anaerobic training appears to compromise the ability of these athletes to improve their strength (Hoffman, Maresh, et al. 1991). Thus, the goals of the training program are adjusted to accommodate the realistic expectations of training at specific times of the year. Further discussion of the combination of both anaerobic and resistance training appears in other chapters in this textbook.

SUMMARY

Concurrent training does not appear to compromise the ability of either trained or untrained individuals to realize gains in aerobic capacity. In addition, the inclusion of resistance training in the training programs of endurance athletes may improve their short-term endurance. However, the addition of an endurance training program to the resistance training regimen of physically active or experienced resistance-trained individuals appears to negate the potential for improvement in maximal strength. This hindrance of strength gains does not appear to occur in previously untrained individuals. Further benefits derived from combining both resistance and endurance training are the ability to maintain or improve lean muscle mass and decrease fat mass, suggesting that this is an ideal method of training for individuals whose primary goal is to reduce body-fat percentage.

CHAPTER 11

PERIODIZATION

©Human Kinetics

Coaches manipulate training programs at regular time intervals to help athletes achieve optimal performance gains and meet their training objectives. Training goals vary according to the type of sport being played. An athlete who plays football, or any other sport that places emphasis on performance over a complete season, aims to achieve peak condition at the onset of the season. The athlete is then concerned with maintaining that level of conditioning throughout the season. In contrast, other athletes (e.g., gymnasts, swimmers) may concentrate on a major competition that occurs toward the end of the competitive year. As such, peak condition needs to be attained at that finite point in the competitive season. To achieve these varying goals, coaches manipulate training variables (e.g., **intensity, volume**). This chapter provides a background on **periodization** and discusses the efficacy of periodization and different models of periodized training programs. Much of the recently published literature has focused on the effect of periodization on strength/power training. However, an attempt will be made to discuss periodization in relation to all sport disciplines.

PERIODIZATION FOR ALL DISCIPLINES

Periodized training (development of an annual training program) has been reported as far back as the ancient Olympic Games and has been employed by coaches throughout the 20th century (Bompa 1999). However, it was not until the latter half of the 20th century, when sport scientists began to publish their work, that periodized training received much attention (Bompa 1999). Most of this work was published by scientists from the Eastern bloc countries such as Russia, Romania, and East Germany.

In 1965, Russian sport scientist Dr. Leonid Matveyev published a model of his periodized training program that divided the training year into several different phases and cycles. Most of this program was related to the **General Adaptation Syndrome (GAS)** developed by Dr. Hans Selye (Stone, O'Bryant, and Garhammer 1981). The GAS (see figure 11.1) suggests that there are three response phases to stressful demands placed on the body. The first phase, referred to as the **alarm** phase, is the initial response to the stimulus (i.e., exercise) and consists of both shock and soreness. An exercise stimulus or a change in the **exercise prescription** frequently results in a reduction in performance. The second phase is an **adaptation** to this new stimulus. The body adapts to the training stimulus or change in exercise prescription and an improvement in performance is observed. The third phase is one of **exhaustion.** The body is unable to make any further adaptation to the training stimulus. Unless the stimulus is reduced, a situation leading to chronic fatigue **(overtraining)** may occur. If sufficient recovery is allowed, the body can then make further adaptations and performance may increase further. The goal of periodization is to avoid or minimize periods of exhaustion and to maintain an effective exercise stimulus, which leads to maximizing athletic potential.

The basic principle of periodization is a shift from an emphasis on high volume and low intensity to an emphasis on low volume and high intensity. This can be seen in a modified version of Matveyev's periodization model (see figure 11.2). Matveyev divided his training year into three distinct phases: **preparatory, competitive, and transitional.** Each phase of the training program relates to a change in the volume and intensity of training. The preparatory phase of training may comprise two subphases known as **general preparation** and **specific preparation.** During the general phase of preparatory training, volume is high and the intensity of exercise is low. The primary purpose of this phase is to prepare the athlete for more intense and sport-specific training in later phases. The second part of this phase of training is a more specific preparation in which intensity of exercise increases and the volume of training is reduced. The competitive phase, which consists of all the competitions, may also be divided into subphases. The exhibition contests may be considered part of the **precompetition phase** and the primary or most important competitions are considered part of the **main competition phase.** The essential difference between these two competition phases is how the volume and intensity of exercise are manipulated. During the precompetition phase, volume of training is reduced as intensity of training steadily increases.

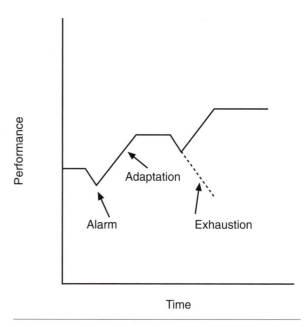

Figure 11.1 Selye's General Adaptation Syndrome.
Adapted, from H. Selye, 1956, *The stress of life* (New York: McGraw-Hill Companies). By permission of McGraw-Hill Companies.

As the athlete gets closer to the main competition, training intensity reaches its highest level and volume of training falls to its lowest point of the competitive year. The final phase of the year is the transitional phase, when the athlete enters a period of **active rest** in which both volume and intensity of training are substantially reduced.

In the latter half of the 20th century, Western scientific literature contained a number of studies examining different combinations of sets and repetitions to produce maximal strength gains (Berger 1962, 1963; O'Shea 1966). However, it was not until the complete domination of the Eastern European countries in the weightlifting events at the Olympic Games of the 1960s and 1970s that the optimal way to develop a training program was reexamined. Stone, O'Bryant, and Garhammer (1981) published a strength-training model that adapted the work of

Matveyev. Their training program was divided into four different phases, or **mesocycles,** (see figure 11.3a). Each mesocycle may last 2-3 months, depending on the athlete. The initial mesocycle is typically called the **preparatory,** or **hypertrophy,** phase. Similar to Matveyev's model, this cycle consists of high-volume and low-intensity training. It is designed primarily to increase muscle mass and endurance. Its objective is also to help prepare the athlete for the more advanced or intense training seen in the later cycles of the training program. In the next two mesocycles **(strength** and **strength/power),** intensity of training is elevated and volume of training is reduced. As the names of these two mesocycles indicate, they are primarily concerned with strength and power development. The final mesocycle of the training year is the **peaking** phase. During this cycle, the athlete prepares for a single

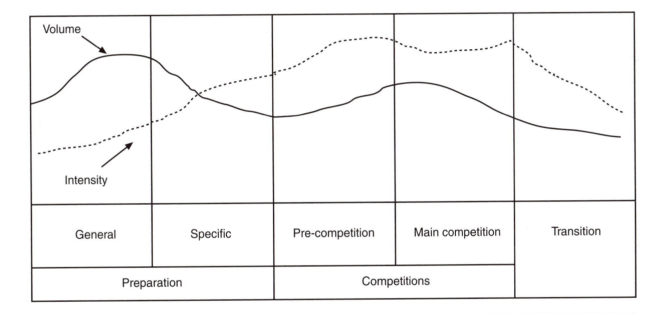

Figure 11.2 Matveyev's Model of Periodization.
Adapted from Bompa, 1999.

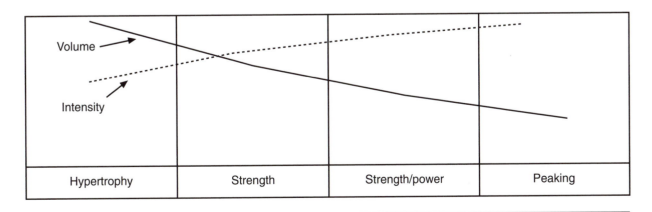

Figure 11.3a Periodization model for a strength/power athlete training for a single competition.

contest by reducing volume and increasing intensity in similar fashion to Matveyev's main competition phase. Table 11.1 compares volume and intensity across the various mesocycles in a strength/power athlete.

Athletes who participate in a sport that places importance on an entire season must achieve peak condition by the onset of the competitive year. Thus, after the peaking phase, which may last for several weeks before the preseason, the athlete performs a **maintenance** phase to maintain the strength gains made during the off-season (see figure 11.3b). During the maintenance phase, the intensity of exercise is reduced to a level similar to what might have been used during the strength phase (6-8 RM). The volume of training is also lowered by reducing the number of assistance exercises performed. These athletes may also have a short peaking phase that precedes a training camp or the start of the season. It is also not uncommon to have short training cycles called **microcycles** that ease the transition from mesocycle to mesocycle.

Efficacy of Periodization

Up to this point, periodization has been discussed as it relates to maximizing athletic potential. However, ath-

letes are unable to maintain peak condition for a prolonged period of time without fatiguing. Manipulating both training volume and intensity allows the athlete to make the necessary physiological adaptations while reaching peak condition at the appropriate time. In addition, the athlete minimizes the risk of developing an overtraining syndrome by altering volume and intensity of training in conjunction with short, appropriately timed **unloading phases** (tapering) after increases in training work.

Relatively few studies have examined whether a periodized training program is more effective than a nonperiodized training program. Recent studies have focused primarily on the effects of a periodized resistance training program on maximal strength or on the enhancement of components of athletic performance, such as vertical jump height, speed, or agility (Baker, Wilson, and Carlyon 1994; Kraemer 1997; McGee et al. 1992; O'Bryant, Byrd, and Stone 1988; Stowers et al. 1983; Willoughby 1992, 1993). Table 11.2 reviews the results of these studies.

Significant increases in strength have been shown in both periodized and nonperiodized resistance training programs. Improvements in upper-body strength (as

Table 11.1   Volume and Intensity in a Periodized Strength Training Program

	Volume		Intensity
MESOCYCLE	SETS	REPETITIONS	1 RM
Hypertrophy	3–5	8–12	60–75%
Strength	3–5	6–8	80–85%
Strength/power	3–5	4–6	85–90%
Peaking	3–5	2–4	>90%

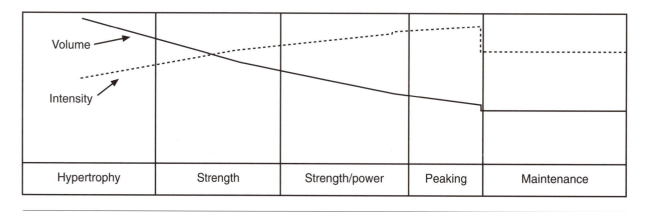

Figure 11.3b   Periodization model for a strength/power athlete training for an entire season.

Table 11.2 Examination of Periodized Resistance Training Studies

Reference	Duration of training	Training frequency	Sets × repetitions[1] of core exercises	Performance tests	% increase
Baker et al. 1994	12	3	5 × 6	Bench press	12*
				Squat	26*
				Vertical jump	9*
			4 wk 5 × 10	Bench press	11*
			4 wk 5 × 5	Squat	18*
			4 wk 3 × 3	Vertical jump	4*
			2 wk 5 × 10	Bench press	16*
			2 wk 5 × 6	Squat	28*
			2 wk 5 × 4	Vertical jump	10*
			2 wk 5 × 6		
			2 wk 4 × 3		
Kraemer 1997	14	3	1 × 8–10 forced reps	Bench press	3*
				Hang clean	4*
				Vertical jump	3*
			3 wk 2–3 × 8–10	Bench press	11*,#
			2 wk 3–4 × 6	Hang clean	19*,#
			50% of 1 RM,	Vertical jump	17*,#
			70–85% of 1 RM,		
			85–95% of 1 RM		
			Repeat all weeks		
	24	3	1 × 8–10, forced reps	Bench press	13*
			Strength session:	Leg press	6*
		4	2–4 × 12–15, 8–10, 3–5	Bench press	29*,#
			Hypertrophy session:	Leg press	20*,#
			2–4 × 8–10		
McGee et al. 1992	7	3	1 × 8–12	Squat reps to exhaustion	46
			3 × 10	Squat reps to exhaustion	71*
			2 wk 3 × 10	Squat reps to	74*
			3 wk 3 × 5	exhaustion	
			2 wk 3 × 3		
O'Bryant et al. 1988	11	3	3 × 6		
			81–97* of	Squat	32*
			pretraining	Cycle power	6*
			1 RM	Squat	38*,#
			4 wk 5 × 10	Cycle power	17*,#
			4 wk 3 × 5,1 × 10		
			3 wk 3 × 2,1 × 10		
			70–117% of pretraining		
			1 RM		

(continued)

Table 11.2   *(continued)*

Reference	Duration of training	Training frequency	Sets × repetitions[1] of core exercises	Performance tests	% increase
Stowers et al. 1983	7	3	1 × 10	Bench press	7*
				Squat	14*
				Vertical jump	0
			3 × 10	Bench press	9*
				Squat	20*
				Vertical jump	1*
			2 wk 5 × 10	Bench press	9*
			3 wk 3 × 5	Squat	27*,#
			2 wk 2 × 3	Vertical jump	10*
Willoughby et al. 1992	12	2	3 × 10	Bench press	8*
				Squat	13*
			3 × 6–8	Bench press	17*,**
				Squat	26*,**
			4 wk 5 × 8–10	Bench press	28*, #
			4 wk 4 × 5–7	Squat	48*,#
			4 wk 3 × 3–5		
Willoughby et al. 1993	16	3	5 × 10	Bench press	8*
				Squat	14*
			6 × 8	Bench press	10*
				Squat	22*,#
			4 wk 5 × 10	Bench press	23*,#
			4 wk 4 × 8	Squat	34*,#
			4 wk 3 × 6		
			4 wk 3 × 4		

[1] = all repetitions were performed as a RM (repetition maximum)

* = Significant increase from pre- to post-testing

= Significant difference from nonperiodized group

** = Significantly different from other non-periodized group

Adapted from Fleck 1999.

determined primarily by a 1 RM bench press) ranged from 9-17% from pretest values in individuals performing a nonperiodized resistance training program. Lower-body strength increases (as determined primarily by a 1 RM squat) ranged in those same individuals from 20-32%. These increases, reported from nonperiodized, multiple-set training programs, reached statistical significance in all studies. Significant improvements were also noted pre- to posttraining in studies that examined single-set nonperiodized training (3-13% and 6-14% increases in upper- and lower-body strength, respectively). However, the magnitude of improvement was significantly less than that seen from the multiple-set nonperiodized training programs. When comparing multiple-set nonperiodized resistance training programs with periodized resistance training programs, most studies reported significantly greater strength gains when subjects performed a periodized training program (see table 11.2). Strength improvements ranged from 9 to 29% in upper-body strength and from 27 to 48% in lower-body strength in subjects performing periodized resistance training programs. In addition to greater strength development, subjects performing periodized resistance training also showed a greater improvement in vertical jump power (Stone, O'Bryant, and Garhammer 1981) and vertical jump height (Stowers et al. 1983) than subjects in nonperiodized training programs. The results of these studies appear to indicate that periodized resistance training is more effective than nonperiodized training in eliciting strength and motor-performance improvements.

In a critical review of the literature, Fleck (1999) reported that although there is convincing evidence concerning the effectiveness of periodized resistance training, the limited number of studies examining this form of training leaves many questions still unanswered. For example, of the few studies that have examined periodized training, only a small number have examined motor performance, body composition, or short-term endurance changes. Of these studies, the significant improvements reported in strength and motor performance do suggest that periodized training is more effective than nonperiodized training. However, Fleck (1999) suggests that this may be largely dependent on the training status of the individual. The magnitude and the rate of strength increases are much greater in untrained individuals than in trained individuals (see chapter 7). Thus, in consideration of the rapid strength increases seen in novice lifters, periodized training may not be necessary until a certain strength base has been established.

Models of Periodization

The periodization model that has been the focus of discussion until now consists of uniform changes in training intensity and volume that remain relatively constant throughout each mesocycle. This **linear** model of periodized training is the classic form for designing most periodized training programs. However, **nonlinear,** or **undulating,** periodization models of resistance training are also becoming popular (Fleck and Kraemer 1997). This model of periodization varies the volume and intensity of training from workout to workout. An example of this model of periodization can be seen in table 11.3. Light, moderate, and heavy intensities can be alternated during each week of training. This model of training has been shown to be as effective as the traditional or linear model of periodization (Baker, Wilson, and Carlyon 1994; Poliquin 1988) and may be appropriate as a training program for sports that place importance on a season of competitions (Fleck and Kraemer 1997). For instance, athletes who play a varied schedule, such as basketball, hockey, baseball, or soccer players, may have several games or competitions in a given week, or their schedule of competitions and travel do not permit a regularly scheduled in-season strength maintenance program. It may be preferable to some coaches and athletes to use relatively light intensities of training preceding or on days of competition. In the undulating periodized program, the athlete can still train at a high intensity but at a more appropriate time of the week.

The following are examples of periodized training programs for a strength/power athlete playing a team sport, a strength/power athlete whose primary goal is an isolated competition (e.g., national meet) during the training year, and an endurance athlete. Each program should be considered as only an example and a possible guideline for developing a periodized training program to meet the needs of an athlete. It is important to remember that the entire athletic conditioning program must be considered during the development of the periodization program and not just the resistance training component of the program.

Periodized Training Program for a Strength/Power Athlete in a Team Sport

The training program of an athlete preparing for a season of competition involves bringing the athlete to peak condition for a specific time period of the year. A periodized training program that has only one peaking phase is referred to as a **monocycle** (Bompa 1999). An example of an annual periodized training program for a strength/power athlete playing a team sport (e.g., football) can be seen in figure 11.4. After the competitive season, the athlete generally goes through a transitional period. During this time the athlete undergoes an active rest in which there are no formal workouts, and the only activity that the athlete may participate in is low-intensity and low-volume recreational sports such as jogging, swimming, basketball, or any of the racket sports. For the collegiate or high school football player, this period of time generally coincides with exams and winter vacation. Once the athlete returns to school, the off-season conditioning program usually begins. As seen in figure 11.4, the initial mesocycle is the preparatory period, a resistance training program that prepares the athlete for the more strenuous training of the subsequent phases. This mesocycle may last 6-8 weeks. In addition, the intensity and volume of training during this phase is beneficial to those athletes needing to add additional muscle mass and is also referred to as the hypertrophy phase. During this training phase, the athlete also performs some form of conditioning activity to maintain aerobic capacity. This may include normal endurance-type activities such as jogging, cycling, and swimming, or the athlete may play recreational basketball, volleyball, or some other sport that lasts 20-30 min. This conditioning activity should be performed two to three times each week. At the conclusion of this mesocycle, there may be an

Table 11.3 Example of Nonlinear Periodized Training

	Sets	Repetitions	Rest between sets
Day 1	3–4	8–10 RM	2 min
Day 2	4–5	3–5 RM	3–4 min
Day 3	3–4	12–15 RM	1 min

Adapted from Fleck and Kraemer 1997.

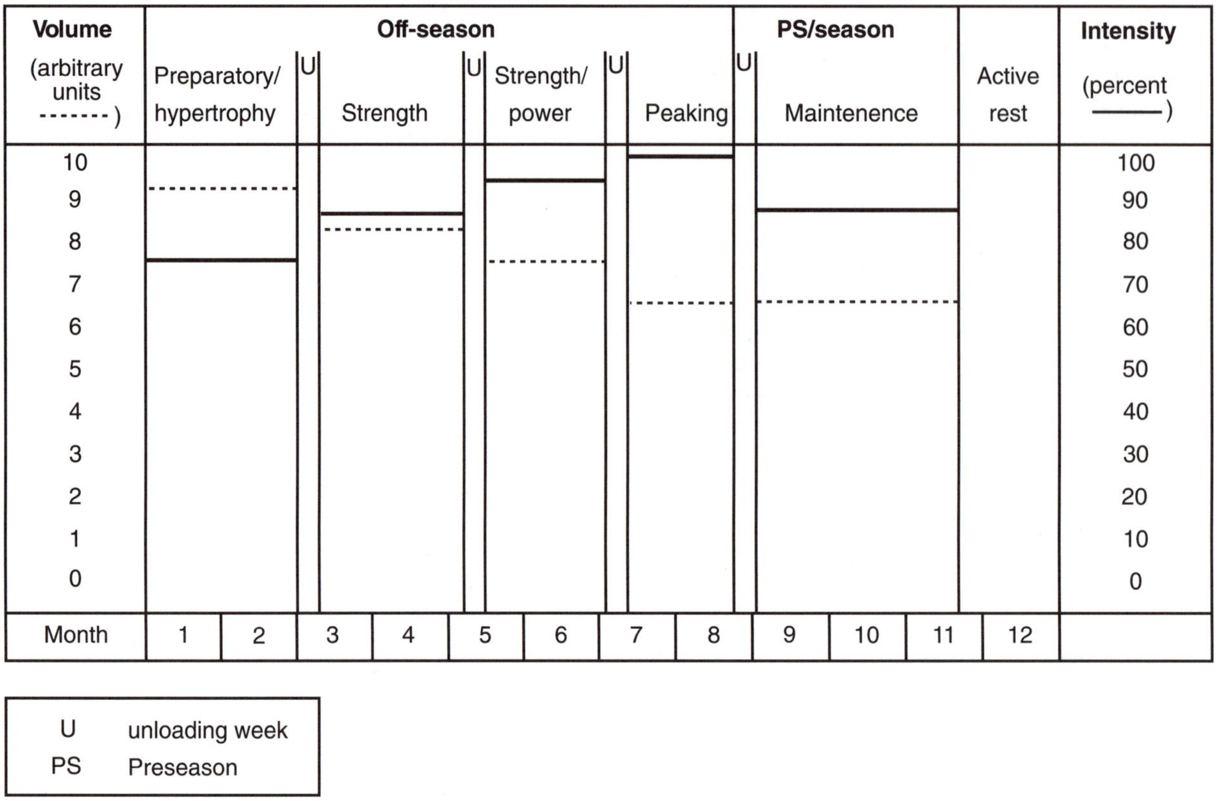

Volume (arbitrary units ------)	Off-season					PS/season		Active rest	Intensity (percent ——)
	Preparatory/ hypertrophy	Strength	Strength/ power	Peaking		Maintenence			

Volume	Preparatory/hypertrophy	U	Strength	U	Strength/power	U	Peaking	U	U	Maintenence		Active rest	Intensity
10													100
9	-------------				————								90
8	————		-------------						————				80
7							-------------						70
6								-------------	-------------				60
5													50
4													40
3													30
2													20
1													10
0													0
Month	1	2	3	4	5	6	7	8	9	10	11	12	

U	unloading week
PS	Preseason

Figure 11.4 Periodized training program for a strength/power athlete playing a team sport (e.g., football).

unloading phase that significantly reduces the intensity and volume of training to prepare the body for the next phase of training.

During the next mesocycle, the strength phase, the intensity of training is increased and the volume of training is reduced. The primary emphasis during this mesocycle, which may last 6-8 weeks, is increasing maximal strength. The athlete maintains the endurance/conditioning program begun during the preparatory phase. To allow the athlete to adequately recover from this training cycle, another unloading phase may be added. The athlete can then proceed to the next mesocycle or, depending on the team, possibly to spring football.

If there is spring football, the training program goes in a different direction, resembling a maintenance program typically used during the regular competitive season. Training volume is significantly reduced. However, training intensity may remain similar to that performed during the previous strength phase. At the end of spring football, a week of active rest may precede the continuation of the conditioning program.

The next mesocycle (either immediately following the strength phase or after spring football) is the strength/power phase. Olympic exercises (e.g., power cleans, push

presses, high pulls), if not already part of training regimen, may be incorporated into this phase of training. The exercises used have a greater specificity to the movements on the field of play. The intensity of exercise is further elevated and the volume of training (related to the reduced number of repetitions, as the number of sets per exercise might remain constant) is decreased. During this training phase, stretch-shortening exercises may also be included in the program. In addition, sport-specific conditioning, agility, and speed training can also be integrated into the 2- to 3-day per week running program. This mesocycle should also last between 6 to 8 weeks.

The next phase of training, which precedes training camp, is of shorter duration (4-6 weeks) and is designed to bring the athlete to peak strength and condition for the start of the football season. During this peaking phase, training intensity is further elevated and the volume of resistance training is again reduced. This is accomplished by reducing the number of assistance exercises in the resistance training program. By this phase of training, the athlete should be concentrating primarily on getting into the proper physiological condition to play football. The conditioning program emphasizes **anaerobic** training (e.g., intervals, both long and short sprints, and agility exercises). The stretch-shortening exercises incorporated into the previous mesocycle should still be included in the training program.

The preseason period, which lasts until the start of the regular season, begins when the athlete reports to training camp. During this period and for the remainder of the competitive season, the resistance training program may be reduced to a 2-day per week maintenance program. The maintenance phase generally comprises the core exercises plus several assistance exercises that work the antagonist muscle groups. Training intensity is similar to what was used during the strength phase, and training volume may be similar to what was used during the peaking phase. Similarly, sport-specific conditioning also needs to be maintained at a reduced volume and intensity. Some form of anaerobic training (e.g., sprints or intervals) should be continued 2-3 days per week to maintain the athlete's peak condition. This can be easily incorporated into the practice schedule. Practices (in which conditioning drills are included as part of the practice routine) and games have been shown to effectively maintain the athlete's conditioning level dur-

ing the competitive season of an anaerobic sport (Hoffman et al. 1991a).

Periodized Training Program for a Strength/Power Athlete Preparing for a Specific Event

An example of an annual periodized training program for a strength/power athlete preparing to peak for a single event can be seen in figure 11.5. This training program is also considered a monocycle because the athlete is preparing to reach peak condition only once during the competitive year. In the previous example, dividing the year into precise mesocycles was easily accomplished for an athlete participating in a sport that has a well-defined off-season, preseason, and season. However, to prepare an athlete to peak for a specific event that often occurs at the end of the competitive season, precise control of the training variables is required. Unlike the team sport in which there is some room for maneuverability, a mistake in the training prescription for an athlete preparing for a single competition could result in an undesirable outcome. The athlete may not reach peak condition by the time of the contest, or the athlete may peak too early

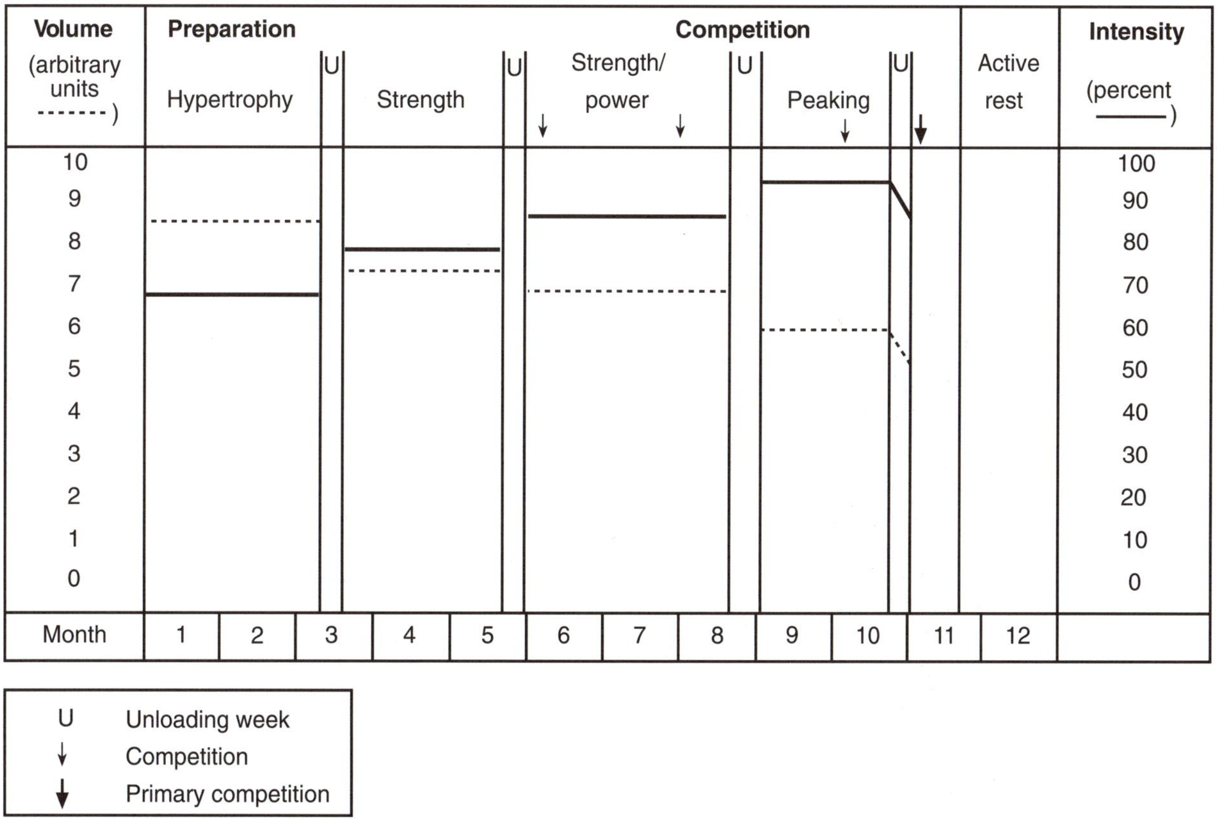

Figure 11.5 Periodized training program for a strength/power athlete preparing to peak for a single event.

and possibly overtrain in an attempt to maintain that high performance level for an extended period of time.

During the initial phase of training, the program is similar to what is typically seen for the strength/power athlete participating in a team sport. However, the competitive phase may be long, with many of the earlier competitions considered of lesser importance. In this instance the athlete trains through those early competitions, preferring instead to peak for the more important competition at the end of the year. During this competition period, the major difference between this athlete and the athlete participating in the team sport is the absence of a maintenance phase. The athlete preparing for a single competition may have several mesocycles during the early to midcompetitive year and then enter a peaking phase to maximize performance before the competition.

Many sports (e.g., track and field) have both an indoor and an outdoor season. In this situation there are two competitive seasons, each having a competition that the athlete primarily focuses on. The training program for two distinct competitive phases is called a **bi-cycle** (Bompa 1999). However, when athletes compete in three or more competitions in a single year (e.g., gymnastics or boxing), these training plans are referred to as **tri-cycle** or **multi-cycle.** Figure 11.6 shows an example of a bi-cycle periodized training program for a strength/power athlete.

A bi-cycle consists of two monocycles that are linked through a short unloading period. The approach to each monocycle is similar to what was previously described for an athlete preparing for a single competition. However, if greater importance is placed on the competition during the second monocycle, then the volume of training would be higher in the preparatory phase during the first monocycle. During this scenario, the condition of the athlete is slightly lower during the first monocycle in comparison with where it should be during the second monocycle. In tri-cycle programs, the last competition is generally the most important. As such, the highest volume of training is seen during the preparatory phase of the first monocycle. In addition, the preparatory phases of the second and third monocycles are relatively short in duration compared with the preparatory phase of the first monocycle. Changes in intensity of training are similar during each monocycle. The challenge of multi-cycle training programs is the reduction in the preparatory phase

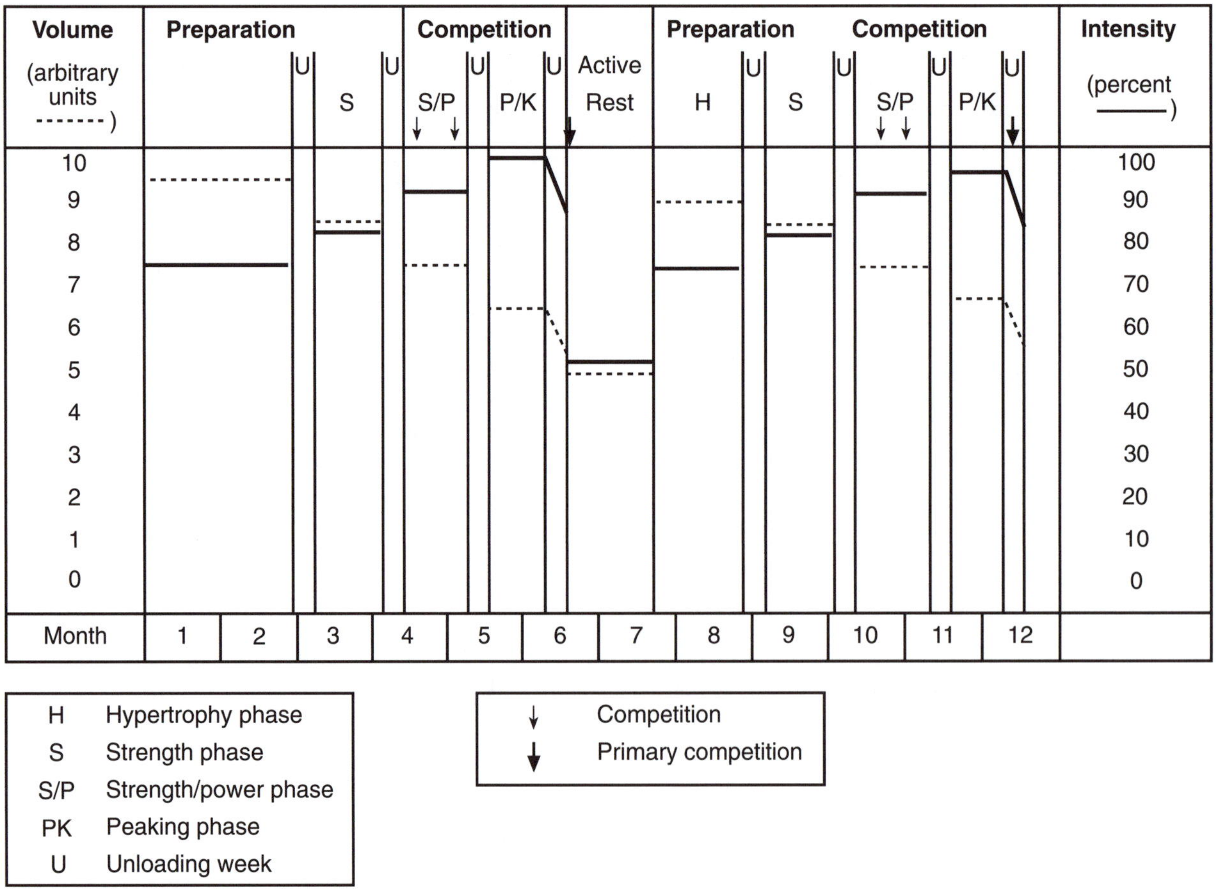

Figure 11.6 Bi-cycle periodized training program for a strength/power athlete.

of training. The higher intensities of training performed more frequently during the year place the athlete at a greater risk of overtraining. It has therefore been recommended that tri-cycle or multi-cycle periodized training programs be limited to more advanced athletes (Bompa 1999). It is thought that the experience and ability of these athletes would give them a better opportunity to adapt to this highly stressful training program.

Periodized Training Program for an Endurance Athlete

The primary difference in developing a training program for an endurance athlete is the high volume of training that is maintained throughout the preparatory and competitive phases (see figure 11.7). If volume of training is not sufficient, it may affect athletic performance. The volume of training is always higher than training intensity in endurance sports even when there are several monocycles during the year. Bompa (1999) has suggested that periodization for an endurance sport is accomplished in three main phases: **aerobic endurance,** aerobic endurance plus **specific endurance,** and finally specific endurance. Aerobic endurance development, which is the

enhancement of the athlete's cardiorespiratory system, occurs during the transition and early preparatory phases. Training is performed at a moderate intensity, and the volume of training steadily increases as the athlete progresses. During the preparatory phase, the endurance athlete continues to emphasize aerobic endurance but also begins to perform some higher-intensity training (e.g., long and medium interval training). The volume of training continues to increase as the athlete completes the preparatory phase. As the athlete enters the competition phase, a greater emphasis on specific endurance training is seen. Intensity of training is high, and the athlete often exercises at intensities that exceed racing intensity. In addition, training volume is also high. Finally, as the athlete enters the unloading (tapering) phase, training intensity is reduced to a much greater extent than that seen of training volume. The greater reduction in training intensity during the unloading period is considered to be of primary importance because of the potential impact that prolonged high-intensity training has on the development of the overtraining syndrome (Bompa 1999). Training volume may also be reduced but not to the extent that is typically seen in the strength/power athlete.

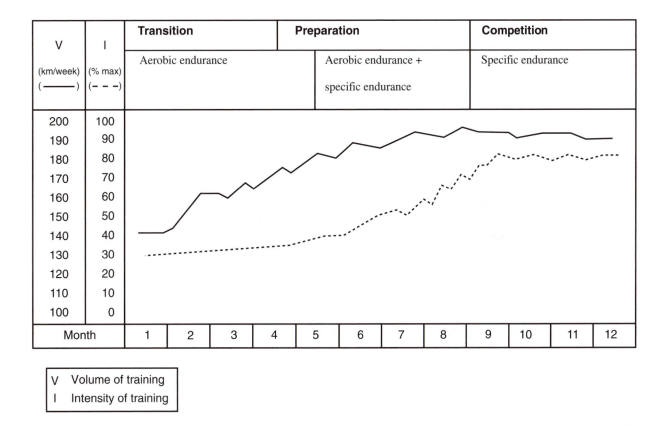

Figure 11.7   Periodized training program for an endurance athlete.
Adapted from Bompa 1999.

©Human Kinetics

SUMMARY

The goal of periodization is to maximize the potential of the athlete to reach peak condition by manipulating both training volume and training intensity. Through the proper manipulation of these training variables, not only will the athlete peak at the appropriate time, but the risk of overtraining is reduced, because the periodized training program is also designed to ease the strain on the athlete. The limited number of studies examining the efficacy of periodized training has demonstrated it to be more effective than nonperiodized training programs for maximizing athletic performance. However, this may be more of a factor for the experienced athlete.

CHAPTER 12

PLYOMETRICS

©Human Kinetics

Plyometrics is a term for exercises that stretch and then shorten the muscle to accelerate the body or limb. As such, plyometrics is often referred to as **stretch-shortening exercises,** a description that may be more appropriate. Plyometrics, or stretch-shortening exercises, has received much attention on the part of coaches since the early 1970s. Initially, track and field coaches were first to incorporate such training regimens into their training programs. This was primarily in response to the superior performances by European Eastern bloc athletes in international track and field, gymnastics, and weightlifting. It was thought that, because these exercises were a staple of the training programs of these athletes, they must contribute to their success and superiority (Chu 1992).

American track and field coaches quickly began to incorporate many of these exercises into the training programs of their athletes, primarily those that jumped, lifted, or threw. Shortly thereafter, coaches in other sports such as volleyball, football, and basketball began to see the potential applicability of this form of training to their own training programs. This chapter discusses the efficacy of plyometric training. In addition, discussion also focuses on how to incorporate plyometric exercises into the **strength** and **power** training programs of athletes. Although a brief description of a number of plyometric exercises is provided, this chapter is by no means an exhaustive catalog of all plyometric exercises available to the athlete and coach.

SCIENTIFIC BASIS FOR PLYOMETRIC TRAINING

During plyometric, or stretch-shortening, exercises the muscle is rapidly stretched **(eccentric contraction),** as in a countermovement jump, and then shortened to accelerate the body upward (see figure 12.1). This stretch-shortening cycle has been demonstrated to enhance power performance to a greater extent than **concentric** training only (Bobbert et al. 1996; Bosco and Komi 1979; Ettema, Van Soest, and Huijing 1990). Differences of 18-20% have been shown in a comparison of the countermovement jump (using a **prestretch**) with the squat jump (concentric contraction only) (Bosco et al. 1982). The improved performance seen in the countermovement jump has been attributed to a greater amount of stored **elastic energy** acquired during the eccentric phase that is able to be recruited during the upward movement of the jump (Bosco and Komi 1979). In addition, the prestretch during the countermovement results in a greater neural stimulation (Schmidtbleicher, Gollhofer, and Frick 1988) as well as an increase in the joint moment (a turning effect of an eccentric force, also referred to as torque at the start of the upward movement (Kraemer and Newton 2000). The greater joint moment results in a greater

force exerted against the ground with a subsequent increase in **impulse** (greater force applied over time) and **acceleration** of the body upward. Bobbert et al. (1996) have suggested that this latter mechanism may be the primary reason for the greater jump height observed during a countermovement jump, whereas the other mechanisms may play more of a secondary role.

The performance of a stretch-shortening cycle requires a finely coordinated action of **agonist, antagonist,** and **synergistic muscle groups.** During the rapid action of the stretch-shortening cycle, the agonist and synergistic muscle groups must apply a great deal of force in a relatively short period of time. To maximize this action, the antagonist muscle groups must be relaxed during the time the agonists and synergists are active. However, a novice to stretch-shortening movements needs some training to coordinate these movements. Through training, contraction of the antagonist muscle groups is reduced, which allows for a greater coordination of these muscle groups and produces a more powerful and effective vertical jump (Schmidtbleicher, Gollhofer, and Frick 1988). In addition, during the initial workouts, the EMG (electrical activity of the exercising muscles, indicating extent of activation) of the agonist muscle groups appears to be reduced (Schmidtbleicher, Gollhofer, and Frick 1988). This is likely the result of activation of the **golgi tendon organ** (a muscle proprioceptor) to protect the muscle from excessive stretch. As the training program continues, these inhibitory effects exhibited by the muscle proprioceptors may be reduced, allowing for improved stretch-shortening performance (Schmidtbleicher, Gollhofer, and Frick 1988).

It has been clearly demonstrated that the use of stretch-shortening exercises provides a distinct advantage for producing greater power than concentric-only movements. The primary question when discussing plyometric training is whether plyometric drills should be considered a supplement to the normal training regimen or an alternative way of training. Specifically, will plyometric training itself be as effective in improving power and strength performance as **resistance training?** Will including these drills as part of the overall training program provide any additional benefit for the athlete?

Three methods of training are generally used to improve the power and athletic performance of athletes who participate primarily in dynamic, explosive sports such as football, basketball, baseball, volleyball, and various track and field events. The primary training method has used traditional resistance training programs with a relatively high **intensity** of training (4-6 RM) performed at a relatively slow **velocity** of movement. The benefits of such training for improving power output and other components of sports performance have been detailed in chapter 7. Plyometric training is another training method that is used to enhance power performance. Most plyometric

Figure 12.1 Countermovement jump.

exercises, although not all drills, require the athletes to rapidly accelerate and decelerate their body weight during a dynamic movement. The body weight of the athletes is most often used as the overload, but the use of external objects such as medicine balls also provides a good training stimulus for certain plyometric exercises. The final method of training to enhance muscular power and explosive sports performance is a combination of traditional resistance training and plyometric training. This form of resistance training, shown to be beneficial for improving power production, is performed at a much lower intensity of training (approximately 30% of 1 RM) using a much higher velocity of movement (Wilson et al. 1993). Referred to as **ballistic training,** its effect on the components of sports performance (e.g., speed and vertical jump height) were discussed in chapter 7.

A number of different studies have demonstrated the effectiveness of plyometric training for improving power, generally expressed as increases in vertical jump height (Adams et al. 1992; Bosco et al. 1982; Brown, Mayhew, and Boleach 1986; Ford et al. 1983; Hakkinen and Komi 1985b; Wilson et al. 1993). Traditional resistance training has been shown to improve vertical jump performance as well (Adams et al. 1992; Wilson et al. 1993; Young and Bilby 1993). However, these improvements may be

limited in experienced strength-trained individuals (Hakkinen and Komi 1985a) or in athletes who have a high pretraining vertical jump ability (Hoffman et al. 1991b). When plyometric drills are combined with a traditional resistance training program, vertical jump performance appears to be enhanced to a significantly greater extent than if performing either resistance training or plyometric training alone (Adams et al. 1992). Neither of these methods may be as effective in improving vertical jump performance as ballistic training.

Wilson and colleagues (1993) compared all three methods of training (traditional resistance training, ballistic resistance training, and plyometric training) in recreational athletes with at least 1 year of resistance training experience. After 10 weeks of training, the ballistic resistance group showed improvements in a greater number of variables tested—jump height (both countermovement jump [CMJ] and squat jump [SJ]), isokinetic leg extensions, 30-m sprint time, and peak power on a 6-s cycle test—than either the traditional resistance training (improvements in CMJ, SJ, and 6-s cycle test) or plyometric training groups (improvements in CMJ only). A closer examination of the effect of each of these training methods on jump performance can be seen in figure 12.2.

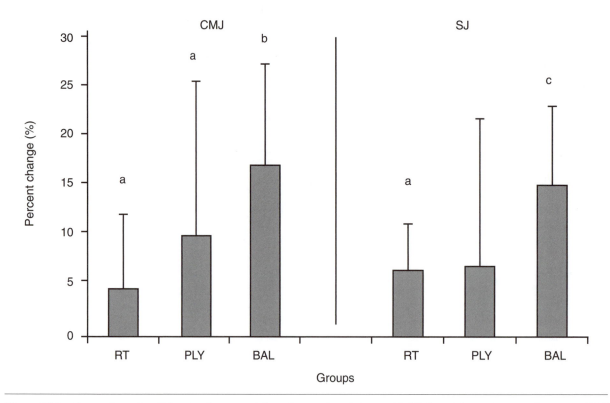

Figure 12.2 Comparison of traditional resistance training, plyometric training, and ballistic training on jump performance. CMJ = countermovement jump; SJ = squat jump; RT = traditional resistance training group; PLY = plyometric training group; BAL = ballistic training group; a = significant pre to post differences; b = significantly different from RT; c = significantly different from both RT and PLY.

Data from Wilson et al. 1993.

All three training programs resulted in significant improvements in countermovement jump performance. However, the subjects that performed ballistic training improved to a significantly greater extent than the subjects that performed traditional resistance training but not plyometric training. In the squat jump, significant pre- to posttraining improvements were only realized by the subjects in the traditional resistance training and ballistic training groups but not the plyometric group. In addition, the ballistic training group improved significantly more than the subjects in the two other groups. The results of this study suggest that ballistic training might be more effective for improving power performance than either traditional resistance training or plyometric training performed separately. The investigators suggested that the use of only body weight as the training load during plyometric drills might pose a disadvantage for optimizing power and sports performance improvements. The use of ballistic training (see figure 12.3), which is a combination of both traditional resistance training and plyometric training, may be the most effective training method to optimize power and sports performance improvements.

INTEGRATION OF PLYOMETRIC TRAINING IN THE ATHLETE'S TRAINING PROGRAM

The use of traditional resistance training programs that require lifting a heavy load at a slow velocity of movement has generally been considered the primary method of increasing power production. This has been based on the notion that because power is equal to force multi-plied by velocity, increasing maximal strength enhances the ability to improve power production. However, to maximize power production, it is imperative to train both the force and velocity components (Kraemer and Newton 2000).

In novice resistance-trained athletes, large increases in strength are common during the beginning stages of training. Improvements in various power components of athletic performance, such as vertical jump height and sprint speed, may also be evident. This is primarily the result of the athlete being able to generate a greater amount of force. As the athlete becomes stronger and more experienced, the rate of strength development decreases and eventually reaches a plateau. At this stage of an athlete's career, not only are strength improvements harder to achieve, but improving maximal strength does not provide the same stimulus to power performance as it did during the earlier stages of training. In addition, training for maximum force development may have its limitations on improving power performance. An important factor for maximizing power production is exerting as much force as possible in a short period of time. By training for maximal strength through heavy resistance training, the rate of force development does not appear to be enhanced (Kraemer and Newton 2000). This is supported by a number of studies referenced earlier that showed improvements in vertical jump performance in novice or recreationally-trained individuals after heavy resistance training programs but limited improvements in individuals or athletes with substantial resistance training experience. However, if plyometric exercises or a combination of plyometric and resistance training (using a light resistance such as might be used with ballistic training) is added to the training program, the athlete's

Figure 12.3 Ballistic training.

ability to increase the rate of force development may be enhanced. This has also has been demonstrated by a number of studies showing that the incorporation of ballistic training does provide a positive stimulus for improving power production even in the trained athlete, especially resistance-trained athletes (Newton, Kraemer, and Hakkinen 1999; Wilson et al. 1993).

To maximize power development, a number of components of power need to be trained and emphasized at various stages of the athlete's career. Kraemer and Newton (2000) have described each of these components as a **window of adaptation** (see figure 12.4). Each window refers to the magnitude of potential for adaptation. For example, as the athlete's strength level increases, the window of opportunity to improve maximal power production from slow-velocity strength training is reduced. Training must then be aimed at improving performance in the athlete's weakest components, because it is in these components that the athlete has the largest window of opportunity for improvement.

Newton, Kraemer, and Hakkinen (1999) demonstrated this approach in a study on the preseason preparation of a NCAA Division I volleyball team. Sixteen athletes with considerable experience in both traditional resistance training and plyometric training participated in this study. They were randomly divided into two

groups; the control group continued to perform resistance training in the same manner as they previously had, while the treatment group performed 6 sets of 6 repetitions of the jump squat with a countermovement (ballistic training). At the end of the study (8 weeks), the inclusion of the ballistic training program proved to be effective in increasing jumping performance in elite volleyball players, whereas the subjects performing the traditional resistance training program did not realize any changes in jump performance. In addition, no changes in maximal leg strength were seen in either group. The authors suggested that although maximal strength was not improved, the ability of the subjects in the ballistic training group to rapidly contract their muscles while maintaining tension may have contributed to the improved power performance. In addition, these subjects were also able to increase their rate of force development, which is thought to be a major contributor to improvements in vertical jump performance (Hakkinen, Komi, and Alen. 1985). The window of adaptation for using ballistic training appeared to be wide open for these athletes. This study demonstrated the effectiveness of using a mode of training (i.e., window) that had a great potential for adaptation in comparison with using a mode of training with little room for adaptation (slow-velocity resistance training) for causing per-

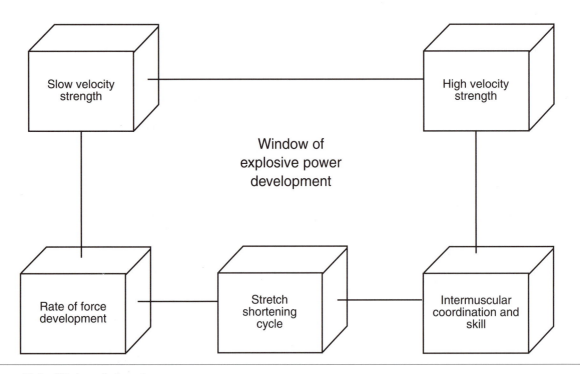

Figure 12.4　Window of adaptation.

Adapted, by permission, from W.J. Kraemer and R.U. Newton, 2000, "Training for musular power," *Physical Medicine and Rehabilitation Clinics of North America* 11(2):361.

formance gains in elite athletes with considerable training experience.

This section has demonstrated that to maximize power performance in an athlete, a number of different methods of training need to be integrated over the course of the athlete's career. As the athlete becomes more highly trained, the relationship between strength and power lessens and a reliance on other methods of power development are needed. It is during this period that the inclusion of plyometric training may have its most profound effect.

PLYOMETRIC TRAINING PROGRAM DESIGN

A primary concern when beginning a plyometric training program is the increased potential for injury, because the drills place high forces on the musculoskeletal system. No epidemiological studies have been conducted that address the issue of whether plyometric training places the athlete at a greater risk for injury in comparison with other forms of training. However, an athlete should be able to meet a fairly high strength base criteria before beginning plyometric training. It has been suggested that athletes be able to perform 1.5 to 2.5 times their body weight in the squat exercise before they can

safely participate in a lower-body plyometric training program (Chu 1992; Wathen 1993). The recommended minimal strength base before beginning upper-body plyometric drills is that athletes weighing more than 200 lb (90 kg) be able to bench press their body weight and athletes weighing less than 200 lb (90 kg) be able to bench press 1.5 times their body weight (Wathen 1993). However, none of these recommendations have been supported by any controlled studies and may appear too conservative, especially for some of the lower-intensity drills. These recommendations may be more valid for performing more complex plyometric drills, such as depth/drop jumps from heights over 18 in. (46 cm). To minimize the risk of injury, some recommendations for participating in plyometric training are listed here:

- Make sure that the athlete has developed a reasonable strength base through a prolonged (> 1 yr) resistance training program.
- Use footwear and landing surfaces with good shock-absorbing qualities.
- Allow for a proper warm-up before beginning the exercise session.
- Use proper progression of drills; master lower-intensity drills before beginning more complex plyometric exercises.

- All boxes used for drills should be stable and have a nonslip top surface.
- Make sure that there is sufficient space for the desired drill. For most bounding and running drills, 33-44 yards (30-40 m) of straightaway are required, whereas for some of the vertical and depth jumps, only 3-4 yards (3-4 m) of space are needed. For jumping drills, ceiling height should be approximately 4 yards (4 m).
- Select exercises that have a high degree of **specificity** within the athlete's sport to enhance performance gains.
- Ensure that all drills are performed with proper technique.
- Allow for sufficient **recovery** between exercise sessions, and do not perform plyometric drills when fatigued.
- Keep track of all **foot contacts** to maintain the proper **volume** and progression of training.

Similar to the development of the resistance training program, the **exercise prescription** for plyometric training involves the control of several acute program variables. These variables (e.g., intensity, volume, **frequency**, and rest) are manipulated based on the ability of the individual athlete, the fitness level of the athlete, and the training phase.

INTENSITY OF TRAINING

In plyometrics, the intensity of training is controlled by the type of exercise performed (Chu 1992). Because most plyometric drills use the athlete's body weight, the complexity of the drill dictates the intensity level. For instance, the intensity continuum for jumping exercises can begin with the two-foot ankle hop as a low-intensity exercise and progress to a more moderate-intensity tuck jump with knees up and progress even further to a high-intensity drill such as the straight spike jump. In certain situations, increasing or adding weights (e.g., weight of medicine ball), may also increase the intensity of training.

Volume of Training

The volume of training is the total work performed in an exercise session. During plyometric training, volume often refers to the number of foot contacts during each session. Foot contacts provide a means of prescribing and monitoring exercise volume, especially for drills involving jumping. For beginners, the volume of training should be between 80 to 100 foot contacts per session. The volume of training increases to 100-150 foot contacts for the intermediate athlete, and the more advanced athlete may perform 150-250 foot contacts during each exercise session. Table 12.1 shows sample exercise volumes for beginning, intermediate, and advanced workouts. In drills that involve exaggerated running exercises such as bounding, the volume of training may be measured by distance. During the early stages of the training program, distances of 30 m per repetition may be used, with progression up to 100 m during the course of the training program (Chu 1992).

The time needed to complete a plyometric exercise session should not exceed 20-30 min for a beginner. This does not include the time needed for warm-up and cooldown. As the experience level of the athlete increases and a greater number of high-intensity drills are incorporated into the training program, the time needed to complete the workout may increase.

Frequency of Training

Frequency is the number of plyometric training sessions performed per week. To date, research has not focused on the optimum number of weekly training sessions. The athlete's need for recovery has been suggested as a basis for the frequency of training for plyometrics (Chu 1992). Low-intensity drills such as skipping or front or diagonal cone hops may be performed daily, whereas more complex drills and higher-intensity exercises such as bounding or depth jumps may require a longer recovery period (48-72 h) before the next exercise session. Most athletic teams (e.g.,

Table 12.1 Number of Foot Contacts by Season for Jump Training

	Beginning	Intermediate	Advanced	Intensity
Off-season	60–100	100–150	120–200	Low–moderate
Pre-season	100–250	150–300	150–450	Moderate–High
In-season	Sport specific	Sport specific	Sport specific	Moderate

Adapted from Chu 1992.

football or track and field) hold two to three plyometric sessions per week (Allerheiligen 1994a). However, this depends on the time of year. During the off-season, the frequency of plyometric exercise sessions may be relatively high because they are integrated into the off-season strength and conditioning program. However, during the season, coaches and players focus primarily on practices and maintaining strength. The number of plyometric sessions, although not necessarily eliminated, is substantially reduced. Nevertheless, common sense should prevail when using plyometric drills during the season. For basketball players who play several games a week, and continually scrimmage during practices, the addition of plyometric exercises may be more likely to cause injury than enhance power performance. In contrast, track and field athletes who compete in a limited schedule often train through several meets to peak at the more important competitions. Maintaining a high frequency and volume of plyometric sessions during the season may be beneficial for these athletes.

Rest and Recovery

Plyometric training is designed to improve power production not to serve as a conditioning tool. As such, adequate recovery between each repetition and set is desired to maximize performance. The quality of each repetition must take precedence over the quantity. Similar to resistance training (see chapter 7), a 2-3 min recovery period should separate each set to permit adequate phosphagen replenishment. Although less information is available for proper rest intervals between repetitions, it has been suggested that the athlete rest 5-10 s between repetitions for the more intense plyometric drills (Allerheiligen 1994a).

Integration of Plyometric and Resistance Training

An athlete is often asked to perform both heavy resistance training and plyometric training together. This combination appears to be more effective in eliciting power performance than performing either modality alone. However, several points should be followed to maximize the effect of these drills and minimize the risk of injury. It is recommended to avoid performing high-intensity resistance training and high-intensity plyometric training on the same day (Chu 1992). However, some athletes (e.g., track and field athletes who are experienced weightlifters) may benefit from such training (performing high-intensity lower-body resistance exercise immediately followed by a plyometric drill), but coaches need to ensure adequate recovery between the plyometrics and other high-intensity lower-body exercises (Chu 1992). In general, the ideal training program consists of the athlete performing the plyometric exercises on the days that upper-body resistance training occurs. If upper-body plyometric exercises are to be included, then they should be performed on days that lower-body resistance training is performed. A sample training program for integrating both strength and plyometric training can be seen in table 12.2.

PLYOMETRIC EXERCISES

Plyometric drills are generally categorized by the movement required and the level of intensity. Plyometric exercises consist of jumps, box drills, bounds, and medicine ball drills. Several examples of each of these categories are provided, with a progression of low to high intensity. As mentioned in the beginning of this chapter, this

Table 12.2   Sample of Integrated Strength and Plyometric Training Program

	Resistance training	Plyometric training
Monday	High intensity upper body	Low intensity lower body exercises
Tuesday	High intensity lower body	
Wednesday		
Thursday	Low intensity upper body	High intensity lower body exercises
Friday	Low intensity lower body	

Plyometric drills do not have to be performed immediately after the resistance exercise session. At times the plyometric drills will be part of an off-season conditioning program that will also include sprint drills and form running, and performed as a team during early morning training sessions (Hoffman et al. 1990). Presently, there is a dearth of information that has determined the ideal order of integrating the resistance training and plyometric training program. Future training studies need to address this issue.

Plyometric Drills

Drill	Intensity	Starting position	Action
Standing long jumps	Low	Stand in semisquat position with feet shoulder-width apart.	With a double arm swing and countermovement with the legs, jump as far forward as possible.
Squat jumps	Low	Stand in squat position with thighs parallel to floor and interlocked fingers behind head.	Jump to maximum height without moving hands. On landing, return to starting position.
Front cone hops	Low	Stand with feet shoulder-width apart at the beginning of a line cones.	Keeping feet shoulder-width apart, jump over each cone, landing on both feet at the same time. Use of a double arm swing and attempt to stay on the ground for as little time as possible.
Tuck jumps with knees up	Moderate	Stand with slight bend in knees and feet shoulder-width apart.	Jump vertically as high as possible, bringing the knees to the chest and grasping them with the hands before returning to floor. Land in a standing vertical position.
Lateral cone hops	Moderate	Stand with slight bend in knees and feet shoulder-width apart beside a row of 3-5 cones stretched 2-3 ft apart.	Jump sideways down the row of cones, landing on both feet. When the row is complete, jump back to starting position.
Double-leg or single-leg zigzag hops	Moderate	Place 6-10 cones about 1.5-2 ft apart in a zigzag pattern. Begin with slight bend in knees and feet shoulder-width apart.	Jump diagonally over the first cone. On landing, change direction and jump diagonally over each of the remaining cones.
Standing triple jumps	Moderate	Stand with feet shoulder-width apart, bending at the knee with a slight forward lean.	Begin with rapid countermovement and jump as far up and forward as possible with both feet, as in the long jump. On landing, make contact with only one foot and immediately jump off. Get maximal distance and land with the opposite foot and take off again. Landing after this jump is with both feet.
Pike jumps	Moderate-high	Stand with slight bend in knees and feet shoulder-width apart.	Jump up and bring the legs together in front of the body. Flexion should occur only at the hips. Attempt to touch the toes at the peak of the jump. Return to starting position.
Split squats with cycle	High	Stand upright with feet split front to back as far as possible. The frontleg is 90° at the hip and 90° at the knee.	Perform a maximal vertical jump while switching leg positions. As the legs switch, attempt to flex the knee so that the heel of the back foot comes close to the buttocks. Land in the split squat position and jump again.
Single-leg hops	High	Stand with one foot slightly ahead of the other, as in initiating a step, with the arms at the sides.	Using a rocker step or jogging into the starting position, drive the knee of the front leg up and out as far as possible while using a double arm action. The nonjumping leg is held in a stationary position with the knee flexed during the exercise. The goal is to hang in the air as long as possible. Land with the same leg and repeat.
Single-leg push-offs with box	Low	Stand on ground in front of a box 6-12 in. high. Place heel of one foot on the box near the closest edge.	Push off the top foot to gain as much height as possible by extending through entire leg and foot. Use double arm action for gaining height and maintaining balance.
Front box jumps	Low-moderate	Stand in front of a box 12-42 in. high (depending on ability) with feet shoulder-width apart and hands behind head.	Jump up and land with both feet on the box and step down. For a more advanced exercise, hop down and immediately hop back on top. Use a variety of box heights.

Drill	Intensity	Starting position	Action
Multiple box-to-box jumps	Moderate	Stand in front of 3-5 boxes 12-42 in. high (depending on ability) with feet shoulder-width apart.	Jump onto the first box then off and jump onto the next box. Continue to the end of the line, using a double arm action for gaining height and maintaining balance.
Multiple box-to-box squat jumps	High	Stand in front of 3-5 boxes 12-42 in. high (depending on ability) in parallel squat position with feet shoulder-width apart and hands behind head or on hips.	Jump onto the first box, maintaining squat position, then jump off and onto the next box. Continue to the end of the line. Keep the hands behind the head or at the hips.
Depth jumps	Low-moderate	Stand on a box 12-42 in. high (the higher the box height, the greater the intensity of the exercise) with toes close to edge and feet shoulder-width apart.	Step from box and drop to ground with both feet. As soon as there is foot contact, jump explosively as high as possible. Try to have as little ground contact as possible.
Depth jumps to prescribed height	Moderate	Stand on a box 12-42 in. high (the higher the box height, the greater the intensity of the exercise) with toes close to edge and feet shoulder-width apart in front of a box of similar height.	Step from box and drop to ground with both feet. As soon as there is foot contact, jump explosively as high as possible onto the second box. Try to have as little ground contact as possible.
Single-leg depth jumps	High	Stand on a box 12-18 in. high with toes close to edge and feet shoulder-width apart.	Step from box and drop to ground, landing with one foot. As soon as there is foot contact, jump explosively as high as possible with that single foot. Try to have as little ground contact as possible.
Skipping	Low	Stand comfortably.	Lift one leg with knee bent to 90° while lifting the opposite arm with elbow also bent to 90°. Alternate between both sides. For added difficulty, push off ground for more upward extension.
Power skipping	Moderate	Stand comfortably.	With a double arm action, move forward in a skipping motion, bringing the lead leg as high as possible in an attempt to touch the hands. Try to get as much height as possible when pushing off on back leg. Each repetition should be performed with alternate leg.
Alternate-leg bounding	Moderate	Begin with one foot slightly in front of the other with arms at the sides.	Using a rocker step or jogging into the starting position, push off the front leg and drive the knee up and out, trying to get maximal horizontal and vertical distance with either an alternate or double arm action. Try to hang in the air for as long as possible. On landing, repeat with opposite leg. Goal is to cover maximal distance with each jump. This is not designed to be a race or sprint.
Single-leg bounding	High	Stand on one foot.	Bound from the one foot as far forward as possible, using other leg and arms to cycle in air for balance and increase forward momentum.
Medicine-ball throws	Low	Stand with a medicine ball overhead.	Step forward and throw ball with both arms to a partner or specified distance. Can also be performed as a chest pass, either sitting or standing.
Power drops with medicine ball	Moderate-high	Lie supine on ground with arms outstretched next to a box 12-42 in. high with a partner standing on top holding a medicine ball.	Partner drops ball. The ball is caught and immediately propelled back to the partner.

is by no means an all-encompassing list of plyometric exercises. However, it should suffice to build a good starting base.

SUMMARY

This chapter demonstrated the benefits of plyometric exercises in the training programs of athletes interested in maximizing power production. However, plyometric training may have a greater impact once the athlete has developed a strength base. Plyometric training can be successfully integrated into the resistance training programs of strength/power athletes. In addition, as the window of adaptation is reduced for slow-velocity strength (traditional resistance training programs) and plyometric training, the combination of resistance training and plyometric training, referred to as ballistic training, may provide a new window for adaptation that further enhances the power production of the athlete.

CHAPTER 13

WARM-UP AND FLEXIBILITY

©Human Kinetics

Before exercise or competition, the athlete generally performs a **warm-up** that includes stretching exercises. The warm-up prepares the athlete for more powerful and dynamic movements that will occur during exercise or competition and possibly minimizes the risk of musculoskeletal injuries. In this chapter, the benefits of the warm-up and how it should be incorporated into the preexercise routine are discussed. In addition, focus is directed on **flexibility** training, including factors that affect flexibility, types of flexibility exercises, and the effect that flexibility training has on performance.

WARM-UP

The goal of the warm-up period is to prepare the athlete both mentally and physically for exercise or competition. It generally consists of 5-10 min of low- to moderate-intensity jogging or cycling. The warm-up may be general in nature and may consist of exercises that are not related to the specific activity that the athlete is preparing for. Besides jogging or cycling, jumping rope, brisk walks, or light calisthenics may also be used. The warm-up can also be more specific to the exercise or sport, and it may include movements that closely simulate actions used during the activity but performed at a reduced intensity.

The benefits of performing a warm-up include increased blood flow to the exercising muscles, which increases muscle temperature and core body temperature. The increase in muscle and core body temperature has a significant positive effect on muscle **strength** and **power** (Bergh and Ekblom 1979) and improves **reaction time** and the rate of **force** development (Asmussen, Bonde-Peterson, and Jorgensen 1976; Sargeant 1987). An increase in muscle temperature during the warm-up session may also increase muscle flexibility by 20% (Wright and Johns 1960). In addition, the warm-up increases heart rate, primes the nervous system by increasing the rate and effectiveness of **contraction** and **relaxation** of both **agonist** and **antagonist** muscle groups **(reciprocal inhibition),** and increases the **elasticity** and mobility of **connective tissue** and joints.

The warm-up is an integral part of the prepractice or pregame routine and does not require a large allocation of time. However, the warm-up should be of sufficient duration to allow the athlete to begin to perspire (an indication of an increase in body temperature). It is important that the warm-up does not fatigue the athlete for the approaching performance. After the warm-up, the athlete generally performs stretching exercises before the training or practice session.

FLEXIBILITY

Flexibility is the ability to move a muscle or a group of muscles through the complete **range of motion.** All stretching exercises should be preceded by a warm-up. The elevated muscle temperature and increased mobility of connective tissue and joints generated by the warm-up allow for a greater range of motion to be reached during each stretching exercise. Flexibility exercises are generally performed before exercise or competition but may also be performed afterward during the cool-down period. Stretching during the postexercise cool-down period should be performed within a short time of the conclusion of practice or competition (5-10 min) to take advantage of the elevated muscle temperatures. Postpractice stretching may also decrease muscle soreness (Prentice 1983); however, there is little experimental evidence to support this contention.

HOW DO MUSCLES STRETCH?

The muscle **proprioceptors, golgi tendon organs** and **muscle spindles** (see figure 13.1), are sensory neural fibers that relay information about the muscle stretch to the upper neural pathways. Their primary responsibility is to protect the muscle from injury. The golgi tendon organs are located in the **tendons** of the muscle fiber, and the muscle spindles are considered **intrafusal fibers** and are situated parallel to the muscle fiber. Golgi tendon organs are sensitive to **tension** development in the **muscle-tendon complex.** They inhibit contraction of agonist muscles and activation of antagonist muscles when tension within the muscle-tendon complex is increased to a level that poses a risk of injury to the muscle. The muscle spindles are responsible for monitoring the stretch and length of the muscle and initiating contraction within the muscle to reduce the stretch if needed.

As the muscle lengthens during a stretching exercise, the muscle spindles become activated, causing a contraction of the muscle that is being stretched. During a rapid stretch that might be seen in a **ballistic** or bouncing type of movement, both the tension in the tendon and the stretch of the muscle are rapidly increased. In response, both the golgi tendon organ and muscle spindles are activated, causing a rapid contraction of the muscle. This **stretch reflex** is easily demonstrated by a light tap to the patellar tendon and the consequent contraction of the quadriceps muscles to ease the tension on the muscle spindles and golgi tendon organs. It is for this reason that **slow static** stretching, which results in a more relaxed and effective stretch, is recommended (Alter 1996).

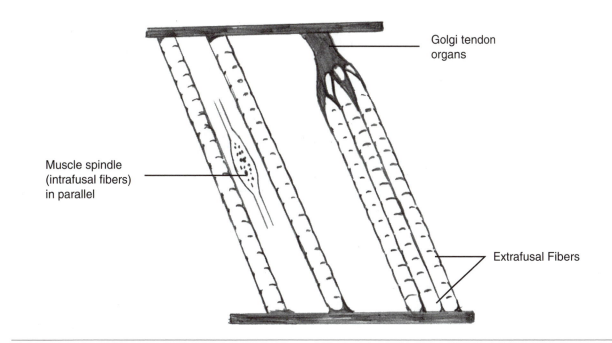

Figure 13.1 Muscle proprioceptors.

TYPES OF STRETCHING TECHNIQUES

There are three types of stretching exercises: slow static, ballistic, and **proprioceptive neuromuscular facilitation (PNF).** All three types of stretching techniques improve range of motion within a muscle (Holt, Travis, and Okita 1970; Worrell, Smith, and Winegardner 1994). However, PNF stretches may result in the greatest improvements (Wallin et al. 1985). The benefits and risks of each of these stretching techniques are outlined in table 13.1.

Slow Static Stretching

Slow static stretching requires the individual to slowly stretch a muscle to its furthest point and hold that position for 10-30 s. Static stretching is the safest of all the stretching techniques and requires very little energy expenditure. In addition, because of its slow movement and long duration, the likelihood of pain and injury is minimized. The static stretch is common to most flexibility programs and is easy to learn.

Ballistic Stretching

Ballistic stretching involves a bouncing, bobbing, or rhythmic movement that does not have an end point.

Table 13.1 Benefits and Risks of Different Stretching Techniques

Factor	Slow static	Ballistic	PNF
Risk of injury	Low	High	Medium
Degree of pain	Low	Medium	High
Resistance to stretch	Low	High	Medium
Ease of performance	Excellent	Good	Poor
Effectiveness	Good	Good	Good

PNF = proprioceptive neuromuscular facilitation

Adapted from Heyward 1997.

This form of stretching produces the greatest amount of pain and risk of injury. Its dynamic movement initiates the stretch reflex, thereby activating muscle proprioceptors and increasing muscle contractility. Because of its short duration, there is insufficient time to cause either tissue or neural adaptation during the stretch (Alter 1996). Ballistic stretching can be performed with the same exercises as static stretching, the difference being the rate of the stretch. Although ballistic stretching may improve range of motion as effectively as static stretching (Holt, Travis, and Okita 1970; Worrell, Smith, and Winegardner 1994), the ballistic nature of the movement forces the muscle to exceed its range of motion, which increases the risk of injury. Nevertheless, ballistic stretching may have its place in the flexibility routines of athletes engaged in activities requiring dynamic movements, such as the martial arts (Alter 1996).

Proprioceptive Neuromuscular Facilitation

Proprioceptive neuromuscular facilitation is a mode of stretching that incorporates two separate techniques and is generally performed with the assistance of a partner. The first technique requires the individual to contract and then relax the stretched muscle, and the second technique requires the muscle to contract, relax, and contract again. Both techniques are effective in enhancing the range of motion in a muscle and are based on the concept of reciprocal inhibition. The contract-relax technique requires the individual to stretch the desired muscle gently. When the muscle is stretched to the point of slight discomfort, the individual isometrically contracts the muscle for 5-15 s against the partner's resistance. This is then followed by a brief period of relaxation before the partner slowly moves the muscle through an extended range of motion. This enhanced range of motion is thought to occur because the isometric contraction causes a reflex facilitation and contraction of the agonist muscles (muscles that are not being stretched) (Heyward 1997). This action suppresses the contraction of the antagonist muscles (muscles being stretched) during the final phase of the stretch, allowing for a greater range of motion. Figure 13.2 provides an example of PNF contract-relax technique using the partner-assisted hamstrings muscle stretch.

The contract-relax-contract PNF technique begins in a similar fashion to the contract-relax technique. However, after the relaxation phase, the individual contracts the agonist muscle group (in the case of the hamstrings being stretched, the quadriceps would be contracted). The partner can also assist with this movement. The theory behind the agonist contraction is that performing a submaximal contraction of the opposite (agonist) muscle group induces additional inhibitory input to the hamstrings through reciprocal inhibition and results in a greater range of motion (Moore and Hutton 1980). However, the exact mechanisms that underlie muscle stretch are still not completely clear.

FACTORS AFFECTING FLEXIBILITY

Flexibility is very individualistic and large variances may be seen within a relatively homogenous population such as a football or basketball team. Although most athletes on the team may be in similar physical condition and of

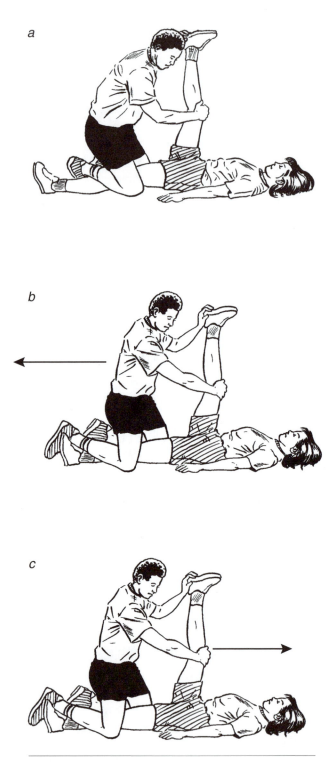

Figure 13.2 Proprioceptive neuromuscular facilitation: Hamstrings stretch. *(a)* First phase: Subject is lying on his or her back with one leg raised 50-60° with knee locked and ankle at 90°. The partner then straddles the subject and with the partner assisting the subject, the hamstrings approach the end of the their range of motion. *(b)* Second phase: Subject isometrically contracts the muscle for 5-15s against the manual resistance of the partner. *(c)* Third phase: Subject briefly relaxes muscle and the partner slowly moves the muscle through an extended range of motion.

similar age, the wide range of flexibility seen among team members suggests that other factors cause differences in muscle flexibility. Flexibility may be related to a host of factors that interact to determine the potential range of motion of a muscle. These factors can be classified as either kinesiological or physiological. Kinesiological factors are primarily associated with **joint structure,** muscle **origin** and **insertion,** muscle **cross-sectional area,** or connective tissue elasticity. Physiological factors may include age, gender, and physical activity level.

Kinesiological Factors

Kinesiology is the study of the mechanics of human motion. It is generally understood that the structure of a joint and the origin and insertion of the muscle have a tremendous influence on range of motion. The range of motion about a joint is highly specific to the type of joint, as well as the muscles, tendons, and **ligaments** that cross that particular joint. Joints that have movement in all three **planes** (frontal, saggital, and transverse), such as a ball and socket joint, generally have the greatest range of motion. The shoulder and hip are ball and socket joints and have a relatively large range of motion in comparison with hinge joints such as the elbow or knee (see table 13.2).

The origin and insertion of the muscle may also affect range of motion. Although anatomical textbooks are precise in describing origin and insertion, the actual location for each individual may be very different. Depending on the extent of the difference and the relationship of the muscle to both bone and joint structure, this may have a large positive or negative effect on the range of motion about the joint.

Connective tissue also plays a large role in determining the range of motion about a joint. Connective tissue may be made up of either **collagen** or **elastic tissue.** When connective tissue is composed primarily of collagen, it is limited in its ability to stretch. In contrast, when connective tissue is composed primarily of elastic tissue, the range of motion has the highest potential. Ligaments account for 47% of the stiffness about a joint, followed by the **fascia** (41%), tendons (10%), and skin (2%) (Johns and Wright 1962). The fascia, which includes the membranes enveloping the individual muscle, muscle fibers, and fasciculi, has the greatest potential for stretch as it is composed primarily of elastic tissue. Tendons and ligaments are less elastic and have a greater amount of collagen. Thus, their ability to adapt is reduced. However, increasing the range of motion of both tendons and ligaments may not be highly desirable. Overstretching these connective tissues may result in joint

laxity and increase the risk of musculoskeletal injuries.

A muscle with a large cross-sectional area may not be as flexible. Although resistance training is not detrimental to muscle flexibility (see chapter 7), individuals with significant bulk (whether through resistance training or obesity) may have a reduced range of motion about a joint. This may affect the ability of that individual to perform certain exercises or movements (Allerheiligen 1994b). For example, an individual with large biceps and deltoids may experience difficulty racking a power clean or performing a front squat exercise.

When considering some of the structural anomalies that naturally occur, flexibility clearly has a large genetic component that may limit the degree of flexibility that a person can achieve. However, although an individual may have poor flexibility, which may or may not be attributed to any structural anomaly, an improvement in flexibility can still be achieved through flexibility training.

Physiological Factors

As we age, the elasticity of the muscle is reduced, resulting in a decrease in range of motion. Reduced elasticity is caused by increased **fibrous cartilage** replacing degenerated muscle fibers, increased **adhesions** and cross-links within the muscle, and increased calcium deposits (Alter 1996). However, flexibility training can still be beneficial in an older population, as demonstrated by the improved range of motion (ROM) in elderly subjects after 10 weeks of stretching exercises performed 3 days per week (Girouard and Hurley 1995).

Gender also appears to affect muscle and joint flexibility. Females tend to be more flexible than males at all ages. This is primarily attributed to gender differences of pelvic structure and hormonal concentrations that may affect the laxity of connective tissue (Alter 1996). However, the advantage that females have over males in flexibility may be joint and motion specific, as males have been reported to have a greater range of motion (ROM) than females in hip extension and spinal flexion/extension in the thoracolumbar region (Norkin and White 1995).

Physical activity is an important determinant of flexibility because active people tend to be more flexible than sedentary individuals (Kirby et al. 1981; McCue 1953). Inactivity causes a tightening or contraction of the inactive muscles. This is easily understood considering the stiffness one feels after sitting for a prolonged period. During long durations of inactivity as a result of deconditioning or immobilization, the connective tissue of the muscles becomes shortened, reducing its range of motion about a joint.

©Human Kinetics

FLEXIBILITY ASSESSMENT

A number of tests can be used to measure the flexibility of a joint or group of muscles. Flexibility can be assessed either directly by measuring the range of joint rotation in degrees or indirectly by measuring static flexibility in linear units (Heyward 1997). The purpose of such testing is to identify joints or muscles that have a poor range of motion and are at a greater risk of injury.

Direct measurements of flexibility include the use of a **goniometer, flexometer,** or **inclinometer,** all of which measure the range of motion about a joint in degrees. The goniometer is a protractor-like device that measures the angle of the joint at both extremes of the range of motion. The center of the goniometer is placed at the axis of rotation (joint) and the arms of the goniometer are aligned with the longitudinal axis of each moving segment. First, the initial joint angle is recorded. The proximal limb is then moved through its complete range of motion to its other extreme, and the angle is again recorded. The difference between both angles indicates the range of motion of that joint. Table 13.2 presents the average range of motion values for healthy adults.

The flexometer and inclinometer also provide direct range of motion measurements of a joint or body segment. Both of these devices can be placed on the subject (either strapped or hand-held), and the range of motion for a particular joint or body segment is easily recorded. Although the validity and reliability of these tests have been established, it appears that reliability may be dependent on the skill of the technician and the joint being measured. Measurements of the upper extremities appear to have a greater reliability than measurements of the lower extremities (Norkin and White 1995).

Indirect measurements of a joint's ROM can be easily assessed. The primary differences are reported as inches or centimeters rather than degrees of range of motion, are also easily assessed. A commonly used indirect measurement of flexibility is the sit-and-reach test (see figure 13.3). Although this test is used to evaluate lower-back and hip flexibility, it appears to have greater validity for assessing hamstring flexibility than lower-back flexibility (Minkler and Patterson 1994). This test can be performed with or without a sit-and-reach box. When a sit-and-reach box is used, the subject sits on the floor with the soles of the feet placed against the edge of the box. With knees extended but not locked, the subject reaches as far forward as possible, with one hand positioned on top of the other, while keeping the back straight. A yardstick or other measuring device is used to mark the zero point. This procedure neutralizes the effect of differences in leg-to-trunk ratio (Hoeger et al. 1990b). Finally, the subject reaches forward as far as possible,

Table 13.2 Average Range of Motion (ROM) Values for Healthy Adults

Joint	ROM (°)	Joint	ROM (°)
SHOULDER		**THORACIC-LUMBAR SPINE**	
Flexion	150–180		
Extension	50–60	Flexion	60–80
Abduction	180	Extension	20–30
Medial rotation	70–90	Lateral flexion	25–35
Lateral rotation	90	Rotation	30–45
ELBOW		**HIP**	
Flexion	140–150	Flexion	100–120
Extension	0	Extension	30
		Abduction	40–45
		Adduction	20–30
		Medial Rotation	40–45
		Lateral Rotation	45–50
RADIO-ULNAR		**KNEE**	
Pronation	80	Flexion	135–150
Supination	80	Extension	0–10
WRIST		**ANKLE**	
Flexion	60–80	Dorsiflexion	20
Extension	60–70	Plantar flexion	40–50
Radial deviation	20		
Ulnar deviation	30		
CERVICAL SPINE		**SUBTALAR**	
Flexion	45–60	Inversion	30–35
Extension	45–75	Eversion	15–20
Lateral flexion	45		
Rotation	60–80		

Data from Heyward 1997.

sliding his or her fingers along the measuring device. The most distant point reached is recorded (either in inches or centimeters). The percentile ranks for the sit-and-reach test for both men and women can be seen in table 13.3. However, this table does not indicate the flexibility requirements for athletes nor predict performance capabilities. It is primarily a tool to evaluate a normal adult population. Scores for specific athletic populations are not presented here.

FLEXIBILITY AND STRENGTH/POWER PRODUCTION

Several recent studies have suggested that stretching before activity may decrease both strength and power performance (Kokkonen and Nelson 1996; Nelson and Heise 1996; Nelson et al. 1998). Although stretching makes the muscle more compliant and potentially reduces the risk of injury, it causes an increase in the muscle-tendon

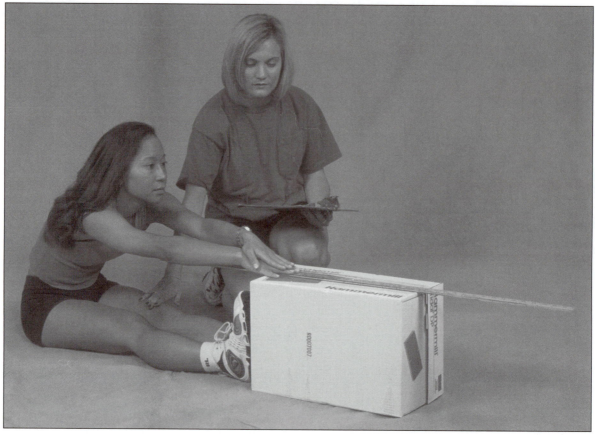

Figure 13.3 Sit-and-Reach test.

Table 13.3 Percentile Ranks for the Sit – and – Reach Test in Inches

Percentile rank	Females				Males			
	18	19–35 yr	36–49 yr	50 yr	18	19–35 yr	36–49 yr	50 yr
99	22.6	21.0	19.8	17.2	20.1	24.7	18.9	16.2
95	19.5	19.3	19.2	15.7	19.6	18.9	18.2	15.8
90	18.7	17.9	17.4	15.0	18.2	17.2	16.1	15.0
80	17.8	16.7	16.2	14.2	17.8	17.0	14.6	13.3
70	16.5	16.2	15.2	13.6	16.0	15.8	13.9	12.3
60	16.0	15.8	14.5	12.3	15.2	15.0	13.4	11.5
50	15.2	14.8	13.5	11.1	14.5	14.4	12.6	10.2
40	14.5	14.5	12.8	10.1	14.0	13.5	11.6	9.7
30	13.7	13.7	12.2	9.2	13.4	13.0	10.8	9.3
20	12.6	12.6	11.0	8.3	11.8	11.6	9.9	8.8
10	11.4	10.1	9.7	7.5	9.5	9.2	8.3	7.8

Adapted from Hoeger 2000.

length (Magnusson et al. 1996). These changes in muscle-tendon length are quickly reversed, but have been shown to decrease force output and rate of force development (Rosenbaum and Henning 1997). Athletes preparing for activities involving maximal strength and power performance should consider refraining from stretching exercises until performance has concluded or should complete flexibility training at least 20 min before exercise or performance. In any situation, it is always vital to maintain the warm-up as a significant part of the prepractice or precompetition routine. The importance of the warm-up, which was described at the beginning of this chapter, has not been diminished by the results of this study. In addition, it should not be misconstrued that flexibility training does not have a role for the strength/power athlete. Only the timing or placement of flexibility training has

been questioned. In consideration of these recent studies, the inclusion of a postexercise cool-down period in which flexibility training is performed appears to have more merit. It will be interesting to observe how the results of future research compare with these recent observations about stretching and strength/power performance.

FLEXIBILITY EXERCISES

The following are examples of stretching exercises for specific muscle groups that can be used as part of a flexibility routine. This is a relatively small sample of the possible exercises available to the athlete and coach. It by no means represents any particular order of importance or effectiveness.

ACHILLES TENDON AND POSTERIOR LOWER LEG

Lying Achilles Stretch

1. Lie on back with legs extended.
2. Flex one leg and keep the foot flat on the ground to relieve strain on the lower back and raise opposite leg toward chest, grasping behind the knee.
3. Slowly flex foot toward chest.
4. Hold stretch and relax.
5. Repeat with opposite leg.

1. Stand with one foot 6-12 in. in front of the toes of the back foot.
2. Lower your upper body by bending at the waist toward front foot and dorsiflex toes of the front foot.
3. Hold stretch and relax.
4. Repeat with opposite leg.

Partner-Assisted Calf Stretch

1. Lie flat on back with one leg raised.
2. Partner straddles lower leg and grasps both the heel and toes of the raised foot.
3. Using PNF technique, partner helps dorsiflex raised foot while leg remains extended.
4. Hold stretch and relax.
5. Repeat with opposite leg.

HAMSTRINGS

Semistraddle (Hamstrings and Gastrocnemius)

1. Sit upright with upper body near vertical and legs straight.
2. Flex one knee and place heel of that foot on inside of opposite knee.
3. Bend forward at the hip, grasp toes of extended leg, and bring toes toward body.
4. Hold stretch and release toes.

5. Grab ankle and push chest toward knee.
6. Hold stretch and relax.
7. Repeat with opposite leg.

Sitting Toe Touch

1. Sit upright with upper body near vertical and legs straight.
2. Partner stands behind with one hand on upper back and the other hand on lower back.
3. Keep both legs straight and bend forward from waist, lowering trunk onto thighs and allowing partner to assist in slowly pushing upper torso onto thighs. Communication with partner to avoid injury is key.
4. Hold stretch and relax.

Straddle (Spread Eagle) (Hamstrings and Adductors)

1. Sit upright with upper body near vertical and legs extended but spread as wide as possible.
2. Keep both legs straight and bend forward from waist, lowering trunk onto floor.
3. Reach as far forward as possible.
4. Hold stretch and relax.

This exercise may isolate hamstrings by bringing the head toward each knee.

Sitting Toe Touch With Partner

1. Sit upright with upper body near vertical and legs straight.
2. Partner stands behind with one hand on upper back and the other hand on lower back.
3. Keep both legs straight and bend forward from waist, lowering trunk onto thighs and allowing partner to assist in slowly pushing upper torso onto thighs. Communication with partner to avoid injury is key.
4. Hold stretch and relax.

Straddle with Partner

1. Sit upright with upper body near vertical and legs extended but spread as wide as possible.
2. Partner stands behind with both hands on lower back.
3. Keep both legs straight and bend forward from waist, lowering trunk onto floor and allowing partner to assist in slowly pushing upper torso toward the floor. Communication with partner to avoid injury is key.
4. Hold stretch and relax.

ADDUCTORS

Butterfly Stretch

1. Sit upright on floor.
2. Flex knees and bring heels and soles of the feet together and pull them toward the buttocks.
3. Place elbows on inside portion of both legs and slowly push legs to the floor.
4. Hold stretch and relax.

QUADRICEPS

Side Quadriceps Stretch (Quadriceps and Iliopsoas)

1. While lying on side, flex one leg and bring heel toward the buttocks.
2. Grasp ankle of flexed leg and pull heel toward the buttocks.
3. Hold stretch and relax.
4. Repeat with opposite leg.

Hurdler's Stretch

1. Sit upright on floor with both legs extended.
2. Bend one leg behind so that the insides of the knee and thigh are resting on the floor with the foot in line with the lower leg.
3. Slowly lean back diagonally onto forearm and elbow opposite of rear leg.
4. Continue until flat on back, being careful not to arch.
5. Hold stretch and relax.
6. Repeat with opposite leg.

Individuals with knee problems may wish to avoid this exercise.

HIP FLEXORS

Forward Lunge

1. Stand upright with legs straddled 2 ft apart.
2. Take a step forward with one leg and flex the knee of that leg until the knee is over the toes.
3. Keep front foot flat on floor and back leg straight; it is not necessary for the heel of the back foot to be on the ground.
4. Keep torso upright and rest hands on hips or forward leg.
5. Slowly lower hips forward and downward.
6. Hold stretch and relax.
7. Repeat with opposite leg.

BUTTOCKS, LOWER BACK, AND HIPS

Single-Leg Lower-Back Stretch

1. Lie flat on back with legs extended and arms out to side.
2. Flex one knee and raise it to chest.
3. Grasp the knee with one hand and pull the knee across the body to floor while keeping elbows, head, and shoulders flat on floor.
4. Hold stretch and relax.
5. Repeat with opposite leg.

Double-Leg Lower-Back Stretch

1. Lie flat on back with legs extended and arms out to side.
2. Flex both knees and raise them to chest.
3. Pull both knees across the body to floor on the same side while keeping elbows, head, and shoulders flat on floor.
4. Hold stretch and relax.

Spinal Twist

1. Sit upright with legs straight and upper body in vertical position.
2. Place right foot outside of left knee and place back of left elbow on outside of right knee.
3. With left elbow, push the right knee while turning the trunk as far to the right as possible.
4. Hold stretch and relax.
5. Repeat with opposite leg.

CHEST

Pectoralis Stretch With Partner

1. Sit upright with both arms flexed and fingers interlocked behind head.
2. Partner stands behind and grasps both elbows.
3. Partner gently pulls elbows backward toward each other.
4. Communication with partner to avoid injury is key.
5. Hold stretch and relax.

SHOULDERS

Lateral Shoulder Stretch

Can be done either sitting or standing with one arm raised to shoulder height.
1. Flex the raised arm across the opposite shoulder.
2. Grasp elbow of raised arm with opposite hand and pull arm across the chest.
3. Hold stretch and relax.
4. Repeat with opposite arm.

Partner-Assisted Internal Rotator Stretch

1. Stand upright with one arm raised to shoulder height and flexed to a right angle.
2. Partner stands in front and grasps the wrist and elbow of raised arm.
3. Partner gently pushes wrist backward and downward. Communication with partner to avoid injury is key.
4. Hold stretch and relax.
5. Repeat with opposite arm.

External Rotator Stretch

1. Sit or stand upright with one arm flexed behind back.
2. Grasp elbow of flexed arm with opposite hand.
3. Pull elbow across midline of back. Grasp wrist if unable to reach elbow.
4. Hold stretch and relax.
5. Repeat with opposite arm.

Shoulder Flexor Stretch

1. Sit or stand upright with arms straight above head.
2. Cross one wrist over the other and interlock hands.
3. Straighten arms and extend them behind head. Elbows should be behind ears.
4. Hold stretch and relax.

NECK

Rotational Stretch of Neck

1. Sit or stand upright.
2. Turn head to the left and hold stretch and relax.
3. Turn head to the right and hold stretch and relax.

Neck Flexion and Extension

1. Sit or stand upright.
2. Flex head anteriorly by bringing chin down toward the chest.
3. Hold stretch and relax.
4. Extend head posteriorly by bringing head as far back as possible.
5. Hold stretch and relax.

TRICEPS

Behind-Neck Stretch

1. Sit or stand upright with one arm flexed and raised overhead next to ear. The hand of the flexed arm should rest on the shoulder blade.
2. Grasp elbow with opposite hand and pull elbow behind your head.
3. Hold stretch and relax.
4. Repeat with opposite arm.

SUMMARY

This chapter discussed the importance of the warm-up and flexibility training. The warm-up is clearly a critical part of the preexercise or competition routine. Further discussion centered on the types of flexibility training and factors that influence flexibility. Flexibility is individualistic and is dependent on factors that are often out of the athlete's control. Finally, new evidence is emerging that performing stretching exercises before strength/power performance may be detrimental. In consideration of this, a greater focus on performing flexibility training during the postexercise cool-down period may become apparent.

CHAPTER 14

ATHLETIC PERFORMANCE TESTING

©Mary Messenger

The growing knowledge base of physiological, biomechanical, and psychological responses to exercise and various training programs has provided coaches with a greater understanding of how to maximize their players' athletic performance. As a result, the development of optimal training programs over the last few years has grown immensely. In addition, exercise scientists have focused their research on **athletic profile** development. The objective for these scientists is determining the type of athlete that would have the greatest potential to achieve success in a respective sport. The development of an athletic profile for each sport not only helps coaches in the selection process for their teams but also provides standards for both athletes and coaches in setting their training goals.

The development of an athletic profile requires a detailed battery of testing to thoroughly analyze all the components of athletic performance (e.g., **strength, anaerobic power, speed, agility, maximal aerobic capacity and endurance, and body composition**). Test results help determine the relevance and importance of each fitness component to a particular sport and permit the appropriate emphasis on that variable in the athlete's training program. In addition, the development of a sport-specific athletic profile helps establish standards for predicting potential success in that sport. Athletes and coaches alike can use these standards as a motivational tool when establishing personal training goals by comparing their results to normative data from similar athletic populations. Performance testing can also be used to provide baseline data for individual exercise program prescription, provide feedback in the evaluation of a training program, and provide information concerning the extent of recovery after injury.

FACTORS AFFECTING PERFORMANCE TESTING

In evaluating performance testing, it is important to recognize that test results may be influenced by several factors, such as body size, **muscle fiber type,** the **training status** of the athlete, the **specificity** and **relevance** of the test to the sport and training program, and the **validity** and **reliability** of the test.

Body Size

The athlete's size may largely influence the results of some performance tests. For instance, in strength testing, absolute strength has a positive correlation to body size, whereas a negative correlation is seen between body size and the strength/mass ratio (Hoffman, Maresh, and Armstrong 1992). Thus, in an evaluation of football play-

ers, it would be expected that linemen would have a greater absolute strength and players in the skill positions (e.g., running backs) would have a greater relative strength.

Fiber-Type Composition

For a given muscle size and architecture, considerable variability is seen in **force production, contraction velocity,** and **fatigue rates.** This variability may be related to the inherent contractile properties of the muscle, which are determined in part by the fiber-type composition (percentage of fibers that are **fast twitch** versus **slow twitch**). Athletes with a higher percentage of fast-twitch fibers appear to have a greater force capability and faster contraction velocity (see chapter 1 for a review). Thus, athletes who have a greater percentage of fast-twitch fibers are likely to be associated with anaerobic sports such as sprinting, basketball, or football. On the other hand, athletes whose fiber-type composition is primarily slow-twitch have a slower rate of fatigue but a lower force capability and a slower contraction velocity. Athletes with predominantly slow-twitch fibers would find success in endurance sports. The ability of athletes to significantly alter their fiber-type composition through training is extremely limited. Therefore, in evaluating the speed or agility of athletes, there might be limitations to the extent of improvement that can be realized in certain athletes based on their physiological limitations. Although it may be possible to make a slow athlete faster, it is highly unlikely that a slow athlete can be made fast.

Training Status of the Athlete

As mentioned in earlier chapters, the more experienced the athlete the smaller the potential to achieve performance gains. During the early stages of a training program, significant improvements in performance can be seen. As the duration of training increases, the rate of performance improvement declines. At some point during the athlete's career, perhaps after several years of training, further performance improvements are harder to achieve and a plateau appears to be reached. In addition, not only is the athlete's training experience a determining factor for performance improvement expectations, but athletes with a high ability level, regardless of training status, may not experience significant performance improvements even when participating in training programs for the first time (Hoffman et al. 1990, 1991b). Thus, to properly evaluate the effectiveness of a training program or to set realistic performance goals, it is imperative that the coach has a clear understanding of the training status of the athlete.

Specificity and Relevance of the Test

To provide any significant information about the athlete's performance, it is imperative that each test used is specific to the athlete's training program. For instance, during strength testing, when training and testing are performed using a similar mode of exercise (e.g., squats), testing results more accurately reflect the magnitude of strength improvements. However, if training and testing are performed on different training modes (e.g., machines versus free weights) or exercises (e.g., squats versus leg presses), the actual strength improvement may not be demonstrated. Fry and colleagues (1991) showed that a 21% increase in squat strength resulted in only an 8% increase in leg-extension strength. In addition, each test chosen for evaluation should have relevance to the specific sport. Selected tests should provide both the athlete and coach with information concerning the athlete's ability to succeed in a specific sport. For example, the Wingate Anaerobic Test is the most widely accepted power test available. However, because it is performed on a cycle ergometer, its relevance to many sports has been questioned. As a result, efforts have been made to develop more specific anaerobic power tests that have a greater relevance to sports that consist primarily of running or jumping movements (Hoffman et al. 2000).

Validity and Reliability of the Test

Two of the most important characteristics of a test are validity and reliability. Validity refers to the degree that each test measures what it is intended to or claims to measure. Reliability refers to the ability of each test to produce consistent and repeatable results. Tests with proven reliability are able to reflect even slight changes in performance when evaluating a conditioning program. If a test is unreliable, differences in testing may reflect only the variation of the test and not the effectiveness of the training program.

The remainder of this chapter discusses tests that are common to each of the performance variables. The most accepted and widely used tests are discussed.

STRENGTH

When selecting an appropriate strength test, it is vital that the test be familiar to the athlete and specific to the athlete's resistance training program. As mentioned earlier, this ensures a clear understanding of the effectiveness of the conditioning program. In addition, using an exercise that is part of the athlete's conditioning program assures proper technique, reduces the potential of injury during testing, and allows for proper selection of the resistance attempted during the strength test. If initial test-

ing occurs before the onset of a conditioning program, which might be the situation for freshmen athletes being tested on the first day of practice, the exercises used to assess strength may be novel to them. As such, it would be appropriate to still use these tests as long as those exercises will be part of their resistance training program and the same tests are used to reassess the athletes at the conclusion of the training program.

If strength tests are to be part of an evaluation to predict potential performance, then the strength test used should incorporate similar movement patterns and involve the same muscle mass that is routinely recruited during actual sport performance (Hoffman, Maresh, and Armstrong 1992). Strength testing generally involves the use of a multijoint, large muscle-mass exercise to evaluate an athlete's adherence to a resistance training program or to make comparisons between athletes. However, there are times when using an exercise that recruits a smaller muscle mass of an isolated joint action may provide additional information. For example, comparing muscle groups of bilateral limbs (e.g., right knee flexors versus left knee flexors) or agonist versus antagonist muscle groups (e.g., knee flexors versus knee extensors) may indicate a potential weakness that can predispose the athlete to injury.

The mode of exercise used to assess strength is dependent on the goals of the testing program. Generally, strength testing is performed with either **dynamic constant resistance** exercises or an **isokinetic testing** device. Isokinetic testing devices are designed to measure joint movements at a constant velocity. The force exerted by a moving body segment is met with an equal and opposite resistance that is constantly altered as the body segment moves through its full range of motion. This force exerted by the body segment to produce rotation about its axis is referred to as **torque** and is expressed in newton-meters. The test-retest reliability of isokinetic devices has been well established (Farrel and Richards 1986). Since isokinetic devices permit only the evaluation of single-joint unilateral movement, its role in strength evaluation is limited primarily to evaluating the athlete's potential for muscle injury as a result of either a bilateral deficit or an agonist/antagonist muscle imbalance (Hoffman, Maresh, and Armstrong 1992).

Isokinetic Testing

Peak torque measures are frequently utilized to establish a ratio between agonist and antagonist muscle groups and bilateral comparisons. They are a reliable indicator of muscle ability in both healthy and injured knee joints. The hamstring to quadriceps ratio (H:Q) is the most prevalent agonist to antagonist relationship reported for the lower body. A 3:5 ratio of H:Q strength has been commonly accepted as the normal strength proportion between these two muscle groups (Hoffman, Maresh, and Armstrong 1992). However, ratios of 2:3 (Fry and Powell 1987) and 3:4 (Knapik et al. 1991) are also acceptable strength ratios. The H:Q ratio can also be reported as a percentage. That is, a 2:3 H:Q ratio could be expressed as the hamstring muscle group being 67% as strong as the quadriceps.

When comparing the H:Q among athletes, a large variation is often seen between different sports and at different testing velocities. Ratios have been reported to range from 52% in college endurance runners (Worrell et al. 1991) to greater than 100% in NCAA Division I basketball players (Hoffman et al. 1991a). This wide variation of H:Q likely reflects the specific demands of each sport as well as the specific speed of contraction.

Differences in H:Q may be a function of participation in a resistance training program. A lower H:Q observed at slow velocities of muscle contraction may reflect the type of strength training regimen employed. For example, performing the squat exercise may disproportionately increase knee extensor strength at a slow velocity of movement (e.g., 60°/s), causing a reduction in the H:Q (Fry and Powell 1987). In addition, the type of athlete being tested also appears to influence the H:Q. Athletes involved in high-intensity anaerobic activities such as basketball, football, or hockey appear to have greater knee flexor strength, resulting in a high H:Q at all speeds of contraction (Hoffman et al. 1991a, 1992; Housh et al. 1988).

Isokinetic testing at different velocities of muscle contraction does have merit. However, testing at fast velocities of contraction in the lower limb may have added importance, considering the greater power development of the knee flexors (hamstring muscle group) at fast speeds of contraction (Read and Bellamy 1990). Knapik and colleagues (1991) reported that female athletes with an H:Q less than 75% at 180°/s were 1.6 times more likely to be injured than athletes with stronger knee flexors. However, in that same study, it was also indicated that the H:Q may not be an independent predictor of injury when bilateral comparisons are also considered.

Initial studies examining bilateral strength differences reported that ratios greater than 10% in the knee flexors were considered an indicator of enhanced likelihood of muscle injury (Burkett 1970). However, subsequent studies have suggested that strength deficits between bilateral muscle groups may be able to reach 15% before a significant increase in the injury rate is seen (Knapik et al. 1991). Athletes with strength imbalances greater than 15% had a 2.6 times greater incidence of muscle injury (Knapik et al. 1991). Still, others have been unable to conclusively determine that bilateral deficits can be used as a significant predictor of lower-leg injury (Worrell et al. 1991). Although bilateral deficits in the lower extrem-

ity exceeding 10-15% may be a cause for concern, normative strength balances are still uncertain.

In some athletes, bilateral deficits are often seen between muscles of the shoulder, elbow, and wrist because of activity patterns that rely predominantly on unilateral arm action (e.g., tennis, baseball pitching) (Cook et al. 1987; Ellenbecker 1991). Strength differences approaching 20% in the upper limb have been seen in tennis players and baseball pitchers. These large bilateral strength differences may be compounded by the non-weight-bearing requirements of the upper-body musculature. Whether this large bilateral strength difference negatively affects performance or increases the risk of injury in these athletes is still not fully understood.

Athletes who rely on unilateral arm action during performance also show significant differences in agonist and antagonist strength balance (shoulder internal/external rotators) (Cook et al. 1987; Ellenbecker 1991; McMaster, Long, and Caiozzo 1991). These imbalances have been attributed to the emphasis placed on shoulder adduction and internal rotation during these activities (e.g., tennis, baseball pitching, or water polo). These movements are characterized by a significant weakness in the external rotators and by significantly stronger internal rotators in the dominant shoulder. Since the external rotators are important for shoulder joint stability, any imbalance may lead to injury in athletes with shoulder capsule laxity (McMaster, Long, and Caiozzo 1991).

Dynamic Constant Resistance Testing

The primary advantage of using dynamic constant resistance testing versus isokinetic testing is the ability to simulate actual sport movement or recruit a larger muscle mass. In addition, most athletes use dynamic constant resistance exercises as part of their resistance training program. The controversy that is frequently encountered in testing maximal strength is whether to test for a 1 repetition maximum (1 RM) or to predict maximal strength from the number of repetitions performed with a submaximal load. Often the motivating factor behind using the latter method is the time difference between achieving a maximal lift versus predicting maximal strength from 1 set. When testing large groups of athletes (as is often the case when strength-testing a football team), the time factor is an important and valid consideration. In addition, the possible risk of injury when performing a maximal lift may also be considered a potential drawback to using the 1 RM as a strength test.

Several studies have shown that the use of submaximal loads to predict maximal strength is highly valid ($r >$ 0.90) (Landers 1985; Mayhew, Ball, and Bowen 1992; Mayhew et al. 1999; Shaver 1970). In fact, many teams in the National Football League routinely use submaximal testing to estimate the maximal strength of their players.

However, several studies have also reported that the number of repetitions performed at selected percentages of 1 RM is variable between different exercises, and the variance within an exercise is also large (Hoeger et al. 1987, 1990). A recent examination of four submaximal equations to predict maximal strength showed that all four of these published equations significantly underestimated or overestimated maximal strength performance in strength-trained athletes (Ware et al. 1995). In addition, a closer examination of the NFL's 225-lb bench press test to assess maximal upper-body strength recently found it to be accurate as long as the number of repetitions performed was below 10 (Mayhew et al. 1999). If greater than 10 repetitions were performed, the equation became less valid and tended to underestimate actual strength ability. This may be a major concern when testing athletes who can bench press more than 300-315 lb (remember from chapter 7 that more than 10 repetitions can generally be performed when exercising at percentages less than 75% of a subject's 1 RM).

The additional concern about using 1 RM tests to assess maximal strength is the possible increased risk of injury. However, no published reports have determined whether a greater risk of injury exists when executing a properly performed 1 RM test versus a submaximal strength test. If proper safety precautions are used (e.g., spotters, proper technique, and qualified supervision), the possibility of injury is minimized. In addition, the resistance selected during a 1 RM is generally based on the training experience of the athlete. That is, based on the training history of the athlete, both the athlete and coach should have a general idea of what the maximum strength level is for a particular exercise. Another factor that should be considered is that when a competitive athlete performs a maximal number of repetitions with an absolute resistance, the last repetition is a RM. Although the RM may be an 8 RM, a 14 RM or even a 25 RM, during each circumstance the athlete is performing the RM with a potentially fatigued muscle. A fatigued muscle may be more susceptible to injury because of changes in neuromuscular recruitment patterns or factors relating to muscle-tendon stress. Thus, when testing an experienced strength-trained athlete, the risk of injury may potentially be lower when performing a 1 RM versus a maximal number of repetitions with a submaximal load. A protocol for assessing a 1 RM is presented in figure 7.2 (page 81). Remember to use proper safety precautions.

The bench press, squat, and power clean are widely used dynamic constant resistance tests for upper-body strength, lower-body strength, and explosive power, respectively. These tests have been demonstrated to have strong test-retest reliability ($r > 0.90$) (Hoffman et al. 1990, 1991a). In table 14.1, strength characteristics of various NCAA football (Divisions I, II, and III) and basketball players are shown.

Table 14.1 Strength Standards for NCAA Collegiate Athletes

	Bench Press (kg)	Squat (kg)
FOOTBALL (Division I data adapted from Black and Roundy 1994; Division II data from Mayhew et al. 1987; Division III data unpublished from the College of New Jersey)		
Division I		
Backs	127.4 ± 19.1	183.6 ± 26.4
Lineman	167.3 ± 26.2	228.6 ± 42.5
Division II		
Backs	115.0 ± 22.3	165.9 ± 29.5
Lineman	126.8 ± 22.3	179.1 ± 43.2
Division III		
Backs	119.7 ± 17.4	158.3 ± 26.5
Lineman	133.2 ± 16.3	173.8 ± 32.2
BASKETBALL (data adapted from Latin et al. 1994)		
Division I Men		
Guards	100.8 ± 17.6	151.1 ± 35.5
Forwards	104.0 ± 21.5	161.9 ± 37.7
Centers	104.4 ± 17.0	138.1 ± 32.1

ANAEROBIC POWER AND ANAEROBIC FITNESS

Anaerobic power can be assessed in both laboratory and field settings. In the laboratory, a variety of tests have been used over the years to assess anaerobic power. These tests have included sprints on a motorized treadmill (Cunningham and Faulkner 1969; Falk et al. 1996), repeated jumps on a force plate or contact mat (Bosco, Mognoni, and Luhtanen 1983), and maximal-effort cycling tests with exercise durations ranging from 7 to 120 s against various breaking forces (Ayalon, Inbar, and Bar-Or 1974; Katch et al. 1977; Sargeant, Hoinville, and Young 1981). Other laboratory tests used to assess anaerobic power have also included isokinetic testing devices (single isolated movements for **peak power** and multiple repetitions to assess anaerobic fatigue) (Thorstensson and Karlsson 1976), and sprints performed with a vertical elevation (Margaria, Aghemo, and Rovelli 1966). Some of these tests assessed peak power (highest power output attained during the test), and others evaluated **mean power** (average power output of entire test). Another variable reported was fatigue rate, which assessed the ability of the athlete to maintain power output through the duration of the test.

The most common laboratory-based anaerobic test used today is the Wingate Anaerobic Test (Bar-Or 1987). It is a 30-s cycling or arm-cranking test performed at a maximal effort against a resistance relative to the subject's body weight. The Wingate Anaerobic Test (WAnT) was first developed at the Wingate Institute in Israel (Ayalon, Inbar, and Bar-Or 1974). Of the available laboratory-based anaerobic power tests, the WAnT has the most extensive research base to date. Test-retest reliability has been consistently shown ($r > 0.90$) (Bar-Or 1987). A testing protocol for performing a WAnT is presented here:

1. Subject is seated comfortably on the cycle ergometer and proceeds to warm up for 4-5 min at a comfortable pace (60-70 rpm) against a resistance equal to 20% of that calculated for the forthcoming test.
2. During the warm-up, the subject performs between 2-4 5-s sprints at the end of each minute of the warm-up.
3. After the warm-up, the subject is given 1-2 minutes to stretch before starting the test. At this time, further instructions can be provided to the subject and last-minute equipment adjustments can be made (e.g., securing toe clips).
4. On hearing the "Go" command, the subject begins to pedal as fast as possible; as the subject reaches full speed, the tester applies the resistance (0.075 kg · kg body mass^{-1} for tests performed on a Monark Cycle Ergometer). The resistance may change depending on the cycle ergometer being used. Some laboratories record maximal pedaling rate (rpm_{max}) during the warm-up period (Hoffman et al. 2000). If this information is available, resistance should be applied when the subject attains 75% of the previously recorded rpm_{max}.
5. Pedaling begins at zero resistance to help the subject overcome the initial inertia. Only when resistance is applied does the 30-s count begin.

The WAnT provides assessments of an individual's peak power, mean power, and **fatigue index.** Peak power is the highest mechanical power output achieved at any stage of the test and represents the explosive capability of the athlete's lower body. Most laboratories report peak power as the highest power output achieved over a 5-s interval; however, some laboratories use a 3-s interval or lower. Mean power is the average power output during the 30-s test. This measure provides an assessment of the athlete's anaerobic endurance, or the ability to maintain high power outputs over a long duration. The fatigue index is often determined by dividing the lowest 5-s power segment (generally the last 5 s of the test) by peak power. Although it is not clear whether the fatigue index is a good indicator of anaerobic fitness, it does appear to correlate highly with the percentage of fast-twitch fibers (Bar-Or et al. 1980). Typically, a higher fatigue index is seen in athletes with a greater percentage of fast-twitch fibers. Athletes who are endurance trained generally have a lower fatigue index. Figure 14.1 depicts a typical performance diagram produced from a 30-s WAnT.

Although the WAnT is often acknowledged as the standard of laboratory-based anaerobic power measurements,

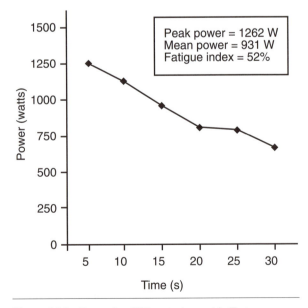

Figure 14.1 Example of Wingate Anaerobic Test.
Data from a football athlete collected at the Human Performance Laboratory, The College of New Jersey.

it has not achieved widespread acceptance among coaches as a performance test for their athletes. This may be related to questions concerning muscle and activity-pattern specificity as well as accessibility to laboratories with such testing capabilities. For example, anaerobic power assessment for a basketball player may be more specific if performed as a vertical jump power test. These tests generally require the athlete to perform repeated countermovement jumps on a force plate or contact mat. The flight time of each jump is recorded (moment subject breaks contact from the mat until contact is made on landing). The time in flight is used to calculate the change in the body's center of gravity (Bosco, Mognoni, and Luhtanen 1983). Using body weight and the calculated jump height, mechanical work is calculated. Finally, using both mechanical work and length of contact time between jumps, anaerobic power can be determined. A vertical jump anaerobic power test does appear to be more sport specific, especially for sports such as basketball and volleyball (Hoffman et al. 2000).

When testing large groups of individuals, administrative concerns (equipment availability and time constraints—only one subject can be tested at a time) may prevent any of the previously mentioned tests from being widely adopted. As a result, many coaches have searched for a field-based test that provides similar assessments to laboratory measures. The line drill is a proposed field test of anaerobic capacity in athletes (Seminick 1994). A description of this test can be seen in figure 14.2. Although recent studies have demonstrated that the line drill is an acceptable measure for anaerobic power performance and may also be indicative of anaerobic fitness (Hoffman et al. 2000; Hoffman and Kaminsky 2000), standards of performance and fatigue have still not been established.

MAXIMAL AEROBIC CAPACITY AND AEROBIC ENDURANCE

Athletes who excel in endurance sports such as cross-country skiing, running, swimming, or cycling generally have a large aerobic capacity (see figure 14.3). Although many factors determine aerobic performance (e.g., capillary density, mitochondrial number, muscle fiber type), the $\dot{V}O_2$max of the athlete does provide important information concerning the capacity of the aerobic energy system (McArdle, Katch, and Katch 1996). Maximal aerobic capacity can be determined either by directly measuring oxygen consumption ($\dot{V}O_2$) while exercising to exhaustion or through submaximal exercise tests. Directly measuring $\dot{V}O_2$ while performing a graded exercise test to exhaustion on a treadmill is considered the standard. The premise behind determining an individual's maximal aerobic capacity was discussed in chapter 9.

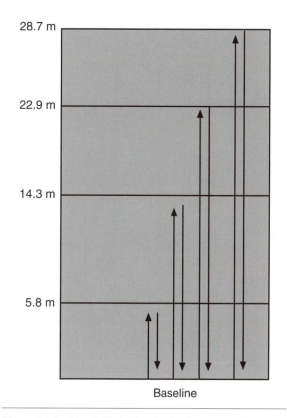

Figure 14.2 Line drill. Athlete begins from a standing position and sprints from baseline to the four cones placed at intervals then back to the baseline. Athlete must touch each cone with their hand.
Adapted from Seminick 1994.

Maximal aerobic capacity can be determined on a treadmill or cycle ergometer or through tethered swimming. The choice of which exercise modality to use depends on the sport that the athlete plays. However, if specificity were not an issue, the treadmill produces the best results. In a study of triathletes, the $\dot{V}O_2$max results from tethered swimming and cycle ergometry were 13-18% and 3-6% lower, respectively, than values obtained from treadmill running (O'Toole, Douglas, and Hiller 1989).

Figures 14.4, 14.5, and 14.6 describe popular treadmill and cycle ergometer testing protocols for assessing maximal aerobic capacity. Many protocols have been developed, and some are population specific. For instance, some exercise protocols are designed primarily for cardiac subjects, whereas others are designed primarily for athletes. The main differences between the two are the initial starting points (elevation and speed of the treadmill) and the increments for each stage of exercise (increases in elevation or speed).

Before the onset of a maximal exercise test, the subject should be allowed to warm up for a minimum of 5 min or until he or she feels ready to proceed. Generally, the warm-up is performed at 0% grade at a speed the subject considers comfortable. After the warm-up, the

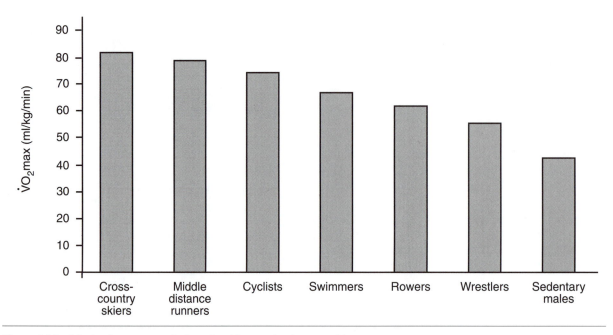

Figure 14.3 Maximal oxygen uptake in male athletes and sedentary subjects.
Adapted from Saltin and Astrand 1967.

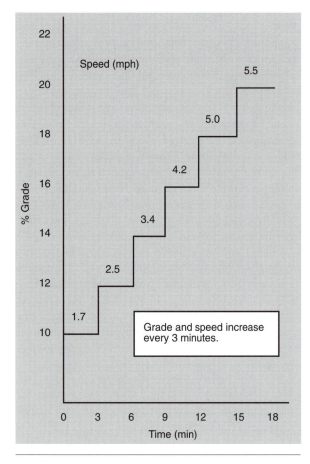

Figure 14.4 Bruce treadmill protocol for assessing maximal oxygen consumption.
Adapted from Bruce et al. 1973.

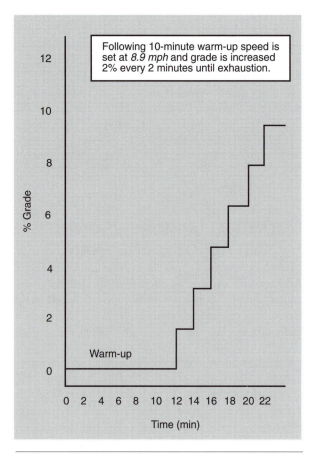

Figure 14.5 Costill and Fox treadmill protocol for assessing maximal oxygen consumption.
Adapted from Costill and Fox 1969.

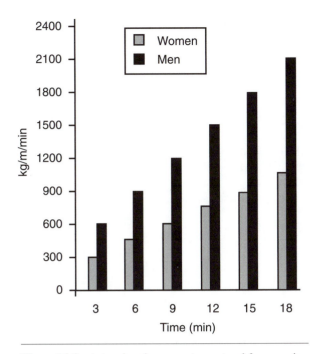

Figure 14.6 Astrand cycle ergometer protocol for assessing maximal oxygen consumption. Men start at 600 kg/m/min and increase 300 kg/m/min every 3 min. Women start at 300 kg/m/min and increase 150 kg/m/min every 3 min. Pedaling speed is set at 50 rpm.
Adapted from Astrand 1965.

subject is attached to the breathing apparatus and the testing protocol begins. The test ends when the subject indicates that he or she has reached exhaustion, or when the subject has met at least two of the first three of the following criteria that determine whether $\dot{V}O_2$max has been reached. The fourth criterion can only be assessed once the test has been completed and is used to confirm that $\dot{V}O_2$max has been attained:

- Increase in oxygen uptake no greater than 150 ml · min^{-1} despite an increase in exercise intensity (plateau criterion)
- Attainment of age-predicted maximal heart rate
- Respiratory exchange ratio ($\dot{V}CO_2/\dot{V}O_2$) greater than 1.10
- Plasma lactate concentration of at least 8 mmol · L^{-1} 4 min after exercise (confirms that $\dot{V}O_2$max has been reached)

Considering the costs associated with the equipment, space, and personnel needed to directly measure oxygen consumption, this methodology of testing is generally reserved for research or clinical settings only. When direct measurement of $\dot{V}O_2$max is not possible, a variety of submaximal tests and maximal exercise tests are available to predict $\dot{V}O_2$max. These tests are generally performed in a controlled environment and are administered

on an individual basis by the exercise technician. The validity of these tests has been well established and is based on several assumptions. There are certain assumptions for using submaximal exercise test to predict $\dot{V}O_2$max (ACSM 2000). If any of these assumptions are not met, the validity of the test may be reduced:

- A steady-state heart rate is obtained for each stage of exercise.
- A linear relationship exists between heart rate and intensity of exercise.
- The maximal heart rate for a given age is consistent.
- The efficiency of exercise (e.g., $\dot{V}O_2$ for the intensity of exercise) is the same for everyone.

Submaximal testing can be performed on either a cycle ergometer or a treadmill. Generally, a submaximal test uses an end point of 85% of age-predicted maximal heart rate. If using a treadmill, the speed and grade of the final stage can be used to estimate $\dot{V}O_2$max (see figure 14.7). The benefit of using a treadmill is related primarily to the familiarity that most individuals have with either walking or running versus riding on a cycle ergometer. However, cycle ergometers may still be a more popular mode of testing because of the ease of performing other measures (e.g., blood pressure and ECG readings) during the test and the non-weight-bearing nature of the test. In addition, the relative inexpense of purchasing a cycle ergometer compared with a treadmill and a greater safety factor (e.g., reduced chance of subject tripping or falling while cycling compared with running on a treadmill) are other reasons that may contribute to a greater use of submaximal cycle ergometer testing. Figure 14.8 describes the popular YMCA submaximal cycle ergometer test.

When large groups of subjects are being tested, it may be more feasible in regard to time to administer a field test. Field tests measuring the time to run 1.0-1.5 mi or the distance that can be run in 12 min have been used to estimate aerobic fitness. The most popular tests are the Cooper 12-min run and the 1.5-mile test for time (ACSM 2000a). The goal of the Cooper test is for the individual to run as great a distance as possible in the allotted time period (12 min). For the 1.5-mile run, $\dot{V}O_2$max can be estimated by the formula:

$$\dot{V}O_2\text{max (in ml · kg}^{-1}\text{· min}^{-1}) = 3.5 + 483/(\text{time in min to run 1.5 miles})$$

In addition to the run tests, $\dot{V}O_2$max can also be estimated for large groups of subjects by using a step test. A variety of step tests have established acceptable reliability and validity standards. The Queens College Step Test was developed by McArdle, Katch, and Katch (1996) to

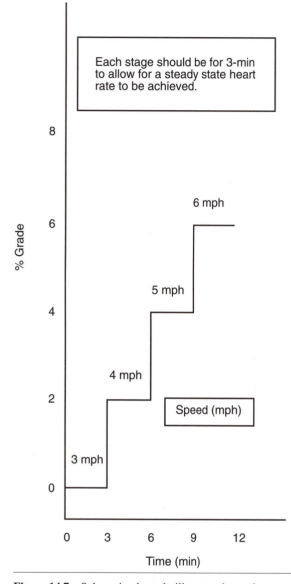

Figure 14.7 Submaximal treadmill protocol to estimate aerobic fitness. To estimate $\dot{V}O_2$max, the following formula may be appropriate to use (from Ebbeling et al. 1991): $\dot{V}O_2$max $(ml \cdot kg^{-1} \cdot min^{-1}) = 15.1 + (21.8 \cdot$ speed in mph) - $(0.327 \cdot$ heart rate) - $(0.263 \cdot$ speed in mph $\cdot$ age) + $(0.00504 \cdot$ heart rate $\cdot$ age) + $(5.98 \cdot$ gender) For gender, 0 = females and 1 = males. This formula is reported to predict $\dot{V}O_2$max within 4.85 ml $\cdot kg^{-1} \cdot min^{-1}$of actual $\dot{V}O_2$max.

predict aerobic capacity in college-aged individuals. Slightly different protocols were established for males and females. Males step to a bench 16.25 in. (41.27 cm) in height (approximate size of a gymnasium bench) at a rate of 24 steps per min. Females step to a bench of similar height but at a rate of 22 steps per min. The duration of the test is 3 min. At the conclusion of the test, the subject remains standing and takes a 15-s heart rate. The

heart rate is then entered into one of the following two equations:

Males: predicted $\dot{V}O_2$max = $111.33 - (0.42 \cdot$ heart rate)

Females: predicted $\dot{V}O_2$max = $65.81 - (0.1847 \cdot$ heart rate)

There is an approximate 16% error associated with this prediction. However, this large error may be acceptable when testing sizable subject populations (McArdle, Katch, and Katch 1996).

SPEED

Speed is the ability to perform a movement in as short a time as possible. It is a relatively easy variable to measure and requires only the use of a stopwatch and track. For programs with large training budgets, electronic timers are available and are becoming more popular. The major issue with using a stopwatch is the potential for measurement error. Even under ideal conditions with an experienced tester, stopwatch times may be 0.24 s more than electronically measured times because of the tester's reaction-time delays in pressing the start and stop buttons as the athlete begins and ends the sprint (Harmon, Garhammer, and Pandorf 2000).

The 40-yd sprint is the most popular distance used to assess the speed of athletes, probably because of the familiarity that most coaches have with sprint times associated with this distance. The 40-yd sprint has achieved tremendous popularity among football coaches and, considering the number of strength and conditioning coaches that have a football background, has become a staple for most athletic testing programs. The justification for the 40-yd distance is not entirely clear. It may have originated as an arbitrary distance that over time has become well accepted.

Coaches of several other sports have used longer or shorter sprint distances depending on the specific requirements of their sport. Some strength and conditioning coaches for basketball have begun to use a 30-yd sprint to assess speed since that is the approximate length of a basketball court. However, the use of that particular distance in basketball has not gained the popularity of the 40-yd sprint (Latin, Berg, and Baechle 1994). Baseball, on the other hand, has used the 60-yd sprint, most likely because it is the distance between three bases (e.g., first to third or second to home).

Table 14.2 provides 40-yd sprint times for NCAA Division I, II, and III football players. In addition, 40- and 30-yd sprint times for basketball players at the NCAA Division I level are included.

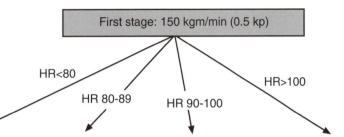

	First stage: 150 kgm/min (0.5 kp)			
	HR<80	HR 80-89	HR 90-100	HR>100
Second stage	750 kgm/min (2.5 kp)	600 kgm/min (2.0 kp)	450 kgm/min (1.5 kp)	300 kgm/min (1.0 kp)
Third stage	900 kgm/min (3.0 kp)	750 kgm/min (2.5 kp)	600 kgm/min (2.0 kp)	450 kgm/min (1.5 kp)
Fourth stage	1050 kgm/min (3.5 kp)	900 kgm/min (3.0 kp)	750 kgm/min (2.5 kp)	600 kgm/min (2.0 kp)

Figure 14.8 YMCA submaximal cycle ergometer test. 1. Set initial workload at 150 kgm/min (0.5 kp) 2. Each stage is 3 min in duration. 3. The second stage is dependent upon the heart rate in the last minute of the first stage. 4. The heart rate measured during the last minute in each stage is plotted against work rate. The line generated from the plotted points is extrapolated to the individuals age predicted maximal heart rate, and a perpendicular line is dropped to the x– axis to determine the work rate that would have been achieved if the person had worked to maximum. 5. $\dot{V}O_2$max can then be calculated with the formula: $\dot{V}O_2$max (ml/min) = (kgm/min · 2 mlkg/m) + 3.5 ml/kgmin · kg)

Adapted, by permission, from *Guidelines for exercise testing and prescription* 6th ed., edited by B. Franklin (Philadelphia, PA: Lippincott, Williams, and Wilkens), 75.

Table 14.2   Sprint Standards for NCAA Collegiate Athletes

	40-yd sprint (sec)	30-yd sprint (sec)
FOOTBALL (Division I data adapted from Black and Roundy 1994; Division II data from Mayhew et al. 1987; Division III data unpublished from the College of New Jersey)		
Division I		
Backs	4.47 ± 0.11	
Lineman	5.04 ± 26.2	
Division II		
Backs	4.91 ± 0.22	
Lineman	5.22 ± 0.26	
Division III		
Backs	4.89 ± 0.15	
Lineman	5.15 ± 0.41	
BASKETBALL (data adapted from Latin et al. 1994)		
Division I Men		
Guards	4.68 ± 0.20	3.68 ± 0.14
Forwards	4.84 ± 0.29	3.83 ± 0.16
Centers	4.97 ± 0.21	3.97 ± 0.21

AGILITY

Agility refers to the ability to change direction rapidly and with accuracy. It is a common testing variable measured during most athletic performance testing. Like speed, it is a relatively easy performance variable to measure. All that is needed are a stopwatch and cones. A variety of different agility tests can be selected; however, agility testing provides more relevant information if the test selected incorporates movements that are similar to the movements the athlete performs during competition and if the test is part of the athlete's training program. Two popular examples of agility tests are the T-test and the Edgren side-step test. The protocols for these tests are outlined in figures 14.9 and 14.10, respectively. Table 14.3 provides normative data on T-test results of NCAA collegiate basketball players.

BODY COMPOSITION

Body composition refers to the proportion of body weight that is fat and the proportion that is lean tissue. The range

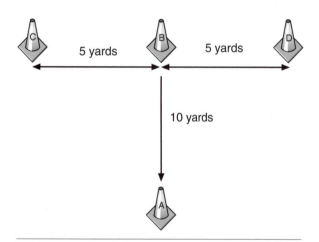

Figure 14.9 T-test. 1. Arrange four cones; A and B are 10 yd apart, C and D are 5 yd from either side of B. 2. Athlete begins by standing at A. 3. Athlete will sprint to B and touch base with hand. 4. Athlete will shuffle to C, facing forward at all times without crossing feet, and touch base with hand. 5. Athlete will then shuffle to D and touch base with hand. 6. Athlete then shuffles back to B, touches base with hand, then runs backwards to A.
Adapted from Seminick 1990.

Table 14.3   Agility Standards (T-test) for NCAA Collegiate Athletes

T-test	Time (sec)
FOOTBALL (unpublished data from the College of New Jersey)	
Division III	
Backs	9.24 ± 0.36
Lineman	9.70 ± 0.65
BASKETBALL (data from Latin et al. 1994)	
Division I Men	
Guards	8.74 ± 0.41
Forwards	8.94 ± 0.38
Centers	9.28 ± 0.81

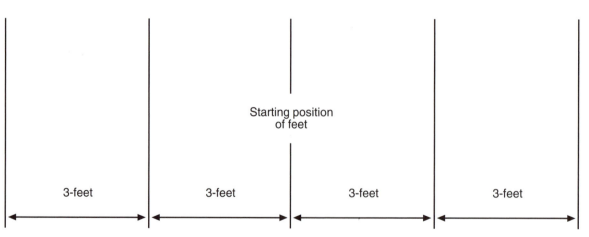

Figure 14.10 Edgren side-step test. 1. Mark five lines three feet apart on floor. 2. Athlete begins astride the centerline then will side step to the right until the right foot has touched or crossed the outside line. 3. The athlete then sidesteps to the left until the left foot has touched or crossed the outside line. 4. The athlete sidesteps back and forth to the outside lines as rapidly as possible for 10 s. 5. The number of lines crossed in 10 s is recorded. Anytime that the athlete crossed his or her feet, a single line or point is subtracted from the total score.
Adapted from Harmon et al. 2000.

in body fat percentage varies among different athletes (see figure 14.11), related primarily to the specific demands of each sport. Endurance athletes or gymnasts are generally on the very lean side, whereas some football players (primarily linemen) may be borderline obese. Standards of body composition for men and women can be seen in table 14.4. Some football players, despite superior athletic skills in other performance variables, may be in the lower-20th percentile of the population in body fat. A recent study reported that the body fat percentage of NFL offensive linemen may exceed 25% (Snow, Millard-Stafford, and Rosskopf 1998). Despite body fat percentages that can be categorized as borderline obese (McArdle, Katch, and Katch 1996), specific needs of the position (e.g., extreme physical contact) require that a higher body fat percentage be deemed acceptable. A concern for the athlete's long-term health should be considered once his playing career is over.

Body composition can be assessed in a number of ways. Methods vary in terms of complexity, cost, and accuracy. The standard of body composition analysis is

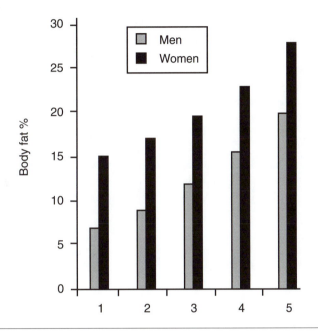

Figure 14.11 Body fat descriptive data for male and female athletes in various sports. 1 = Gymnasts, bodybuilders, wrestlers, endurance runners; 2 = Soccer players, men's basketball players, men and women track and field athletes; 3 = Baseball players, skiers, speed skaters, Olympic-style weightlifters; 4 = Women's basketball players, football skill position players, hockey players, tennis players, volleyball players, softball players; 5 = Football linemen, female shot putters.
Adapted from Harman et al. 2000.

Table 14.4   Body Composition Standards for Men and Women

	Males by age			Females by age		
Percentile	20–29	30–39	40–49	20–29	30–39	40–49
90	7.1	11.3	13.6	14.5	15.5	18.5
80	9.4	13.9	16.3	17.1	18.0	21.3
70	11.8	15.9	18.1	19.0	20.0	23.5
60	14.1	17.5	19.6	20.6	21.6	24.9
50	15.9	19.0	21.1	22.1	23.1	26.4
40	17.4	20.5	22.5	23.7	24.9	28.1
30	19.5	22.3	24.1	25.4	27.0	30.1
20	22.4	24.2	26.1	27.7	29.3	32.1
10	25.9	27.3	28.9	32.1	32.8	35.0

Adapted from American College of Sports Medicine 2000.

hydrostatic weighing. This technique measures body composition based on the volume of water that is displaced when an individual is submerged. As the body is immersed under water, it is buoyed by a counterforce equal to the weight of the water displaced. The loss of mass in water, corrected by the density of water, allows body density to be calculated by the following formula:

body density = body mass in air/{[(body mass in air – body mass in H_2O)/density of H_2O]-residual volume}

To accurately assess body density, calculation of lung **residual volume** is needed. This can be measured directly or predicted through various formulas. Once body density is determined, **percent body fat** can be calculated through various equations. Since body density is variable and is affected by aging, growth and maturation, gender, and ethnicity, an equation that is specific to a population group is often used. A common equation for deriving percent body fat from body density is:

% body fat = (495/body density)-450

Recent technological advances have led to development of a system that measures air displacement rather than water displacement. The use of a plethysmograph (closed chamber that measures body volume by changes in pressure) is promising and may be preferable, especially for individuals who are uncomfortable being fully immersed in the hydrostatic tank.

Other methods often used to compute body composition in athletes include **skinfold measurements** and **bioelectrical impedance analysis.** Both of these methods correlate well with hydrostatic weighing ($r = 0.70$-0.90) and take significantly less time to complete (ACSM 2000a). The principle behind skinfold measurements is that the amount of subcutaneous fat is proportional to the amount of body fat. By measuring skinfold thickness at various sites on the body, percent body fat can be calculated through a regression equation. However, because subcutaneous fat to total amount of body fat varies according to age, gender, and ethnicity (Lohman 1981), the appropriate regression equation must be selected. In addition, regression equations vary in the number of skinfold sites needed. Even when the appropriate regression equation is used, there may be a 3-4% error in the body fat percentage attained from skinfold measurements (Lohman 1981). Thus, care must be taken to select the correct regression equation. Table 14.5 provides several examples

Table 14.5   Examples of Commonly used Regression Equations for Computing Body Fat Percentage from Skinfold Measurements and Description of Skinfold Sites

Reference	Sites	Formula
Durnin and Womersley 1974	Biceps, triceps, subscapular, and suprailiac	D = body density
Males		
17–19		$D = 1.1620 - 0.0630 \times (\log \Sigma)$
20–29		$D = 1.1631 - 0.0632 \times (\log \Sigma)$
30–39		$D = 1.1422 - 0.0544 \times (\log \Sigma)$
Females		
17–19		$D = 1.1549 - 0.0678 \times (\log \Sigma)$
20–29		$D = 1.1599 - 0.0717 \times (\log \Sigma)$
30–39		$D = 1.1423 - 0.0632 \times (\log \Sigma)$
Jackson and Pollock 1985 (7-site) Males	Chest, midaxillary, triceps.	$D = 1.112 - 0.00043499 (\Sigma\ 7$ skinfolds$) + 0.00000055 (\Sigma$ skinfolds$)^2 - 0.00028826$ (age)
	Subscapular abdomen, suprailiac and thigh	$D = 1.097 - 0.00046971 (\Sigma\ 7$ skinfolds$) + 0.00000056 (\Sigma$ skinfolds$^2 - 0.00012828$ (age)

Reference	Sites	Formula
(3-site)		
Males	Chest, abdomen and thigh	$D = 1.10938 - 0.0008267 (\Sigma\ 3\ \text{skinfolds}) + 0.0000016 (\Sigma\ \text{skinfolds})^2 - 0.0002574\ (\text{age})$
Females	Triceps, suprailiac, thigh	$D = 1.1099421 - 0.0009929 (\Sigma\ 3\ \text{skinfolds}) + 0.0000023 (\Sigma\ \text{skinfolds})^2 - 0.0001392\ (\text{age})$

Table 14.6   Population Specific Formula for Conversion of Body Density to % Body Fat

Population	Age	Gender	% Body fat (%BF) formula
White	17–19	Males	$\%\ BF = 4.99/D - 4.55$
		Females	$\%\ BF = 5.05/D - 4.62$
	20–80	Males	$\%\ BF = 4.95/D - 4.50$
		Females	$\%\ BF = 5.01/D - 4.57$
Black	18–32	Males	$\%\ BF = 4.37/D - 3.93$
	24–79	Females	$\%\ BF = 4.85/D - 4.39$

Adapted from Heyward and Stolarczyk 1996.

Table 14.7   Description of Skinfold Sites

Location	Method
Abdominal	Horizontal fold; 2 cm to the right of the umbilcus
Biceps	Vertical fold on the anterior aspect of the arm over the belly of the biceps muscle
Chest/Pectoral	Diagonal fold; 1/2 the distance between the anterior axillary line and the nipple (men), or 1/3 the distance between the anterior axillary line and the nipple (women)
Midaxillary	Horizontal fold on the midaxillary line at the level of the xiphoid process of the sternum
Subscapular	Diagonal fold at a 45° angle, 1–2 cm below the inferior angle of the scapula
Suprailiac	Diagonal fold in line with the natural angle of the iliac crest taken in the anterior axillary line
Thigh	Vertical fold on the anterior midline of the thigh midway between the proximal border of the patella and the inguinal crease
Triceps	Vertical fold on the posterior midline of the upper arm midway between the acromian process of the scapula and the inferior part of the olecranon process of the elbow

of commonly used regression equations and a description of skinfold sites.

Bioelectrical impedance analysis (BIA) is gaining popularity because of its ease of administration and similarity to skinfold measurements in regard to accuracy. BIA is based on the relationship between total body water and lean body mass. Since water is an excellent conductor of electricity, a greater resistance to an electrical current passing through the body would indicate a higher percentage of body fat. Likewise, if resistance was minimal, then a higher percentage of lean tissue would be present. Since BIA is sensitive to changes in body water content, it is recommended that subjects refrain from drinking or eating within 4 h of the measurement, void completely before the measurement, and refrain from consuming any alcohol, caffeine, or diuretic agent before assessment (ACSM 2000a). Failure to comply increases measurement error.

ORDER OF TESTING

When organizing a testing program, several considerations should be made to maximize the athletes' performance. The order of testing is important because of the potential fatiguing effects that certain testing variables may have. For instance, performing an endurance test immediately before a strength test may have a significant detrimental effect on strength performance (see chapter 10). To minimize performance variability, testing order should remain constant through subsequent testing sessions. In addition, because of intertester variability in body composition, sprint, and agility testing, the same tester should perform the same test during repeat testing sessions. This is imperative for testing that is part of a research pro-

tocol; however, in mass-testing situations, common to many athletic teams, this may not always be possible.

SUMMARY

Performance variables are commonly analyzed as part of an athlete's performance testing routine. The use of these performance tests provides an opportunity to assess the athlete's adherence to an off-season training program, evaluate a rehabilitation program, and possibly indicate performance potential by comparing test results to existing performance standards. It is important that the tests selected have both validity and reliability in order for them to have any significant meaning for the coach or athlete. Finally, to minimize measurement error, the testing protocol should follow the same format over repeated test dates. In addition, several performance variables (e.g., speed, agility, and body composition) may be influenced by the test supervisor. To reduce variability between testing sessions, the same tester should perform the same tests. However, it is acknowledged that this may be difficult in an athletic setting. Nevertheless, if testing is part of a research protocol, evaluating the effectiveness of a training program becomes a necessity.

PART III

NUTRITION, FLUID REGULATION, AND ERGOGENIC AIDS

CHAPTER 15

SPORTS NUTRITION

The importance of proper nutrition for athletic performance has been clearly documented over the past 20-25 years. For athletes to maintain a high level of training, they need to have an energy intake that equals their high energy expenditures. For the average person, a **recommended dietary allowance (RDA)** has been established to provide standards for promoting and maintaining good health. However, nutritional requirements are much greater for athletes than the average population. Depending on the needs of the athlete (e.g., size, gender, requirements of the sport), energy intake may be three- to fourfold higher than that recommended for the average individual. The nutritional needs of the athlete have been the focus of much research concerning what to eat, when to eat, and which nutritional **supplements** to take in order to maximize athletic performance. For example, a recommended balance of 55-60% **carbohydrates,** 30% **fat,** and 10-15% **protein** appears to provide a sufficient dietary composition for most people. However, many athletes are concerned that their specific nutritional needs may not be met without altering the recommended balance of protein or carbohydrate intake. This has led to many studies examining the proper diet for athletes and the efficacy of protein and carbohydrate supplementation.

Sports nutrition covers a broad range of topics that may be specific to the individual athlete (e.g., the needs of a marathon runner may be different from the needs of a wrestler). To thoroughly review each topic area would be beyond the scope of this chapter. Thus, this chapter provides a brief review of the nutritional classes, with additional focus on topics considered to be common areas of concern for the general athletic population.

CLASSES OF NUTRIENTS AND THEIR FUNCTIONS

Six classes of nutrients are required for the energy and health needs of the individual: carbohydrates, fats, proteins, **vitamins, minerals, and water.** Carbohydrates, fats, and proteins are the principal compounds that make up our food and provide energy for our bodies. Vitamins and minerals play an important role in energy production and are also involved in bone health and immune function. However, they provide no direct source of energy. Water may be the most important nutrient available. It is needed for nutrient transport, waste removal, body cooling, and most other body reactions. Water will be discussed more thoroughly in chapter 16.

Carbohydrates

There are several types of carbohydrates, all of which are treated differently by the body. **Monosaccharides** are the simplest form of carbohydrate. They consist of a one-unit sugar molecule such as **glucose,** fructose, and galactose. **Disaccharides** are composed of two-unit sugar molecules such as **sucrose** (table sugar), maltose (grain sugar), and lactose (milk sugar). Each of these carbohydrates can be broken down to its simpler form through the process of digestion. Both mono- and disaccharides are considered simple sugars and are a good source of quick energy. Sucrose consists of a glucose and fructose molecule, maltose consists of two glucose molecules, and lactose consists of a glucose and galactose molecule. However, for galactose to be used as energy, it must undergo a secondary conversion to glucose. So, to increase blood glucose levels quickly, a glass of milk may not be ideal. Simple carbohydrates are found in natural foods such as fruits (fresh, dried, and juices) and vegetables, as well as in processed foods such as candies and soft drinks.

Carbohydrates that contain more than two monosaccharides are termed **polysaccharides** and are known as complex carbohydrates. Common polysaccharides include **starch** and **glycogen,** which is made up primarily of chains of glucose molecules. The bonds that bind the monosaccharides of the complex carbohydrate may be either digestible (such as those found in potatoes, pasta, bread, and beans) or indigestible. Indigestible polysaccharides are known as **fiber** and are common in some grains, fruits, and vegetables.

The consumption of simple carbohydrates results in a relatively fast rise in blood glucose. This results in an **insulin** response to move the glucose from the blood into the muscles, where it can be used for immediate energy or stored for later use. Carbohydrates that are not used immediately are stored in the muscles and liver as glycogen. Having full glycogen storage depots is critical for fueling athletic performance. However, if the body's glycogen stores are completely full, the excess carbohydrate is converted to fat and stored in **adipose** sites around the body. The benefit of consuming complex carbohydrates such as starchy foods is that the time required for complete digestion is slower, which results in a more gradual increase of blood glucose. Further discussion of glycogen utilization for athletic performance appears later in the chapter.

Fats

Fat is a highly concentrated fuel that has limited water solubility. Fats, also referred to as **lipids,** exist in the body in several forms. The most common form of lipid is **triglyceride,** which is composed of three **fatty acids** and a **glycerol** molecule. Another common lipid is cholesterol. Although these lipids are frequently linked to heart disease, they also serve the following important functions in the body:

- Provide up to 70% of total energy during the resting state
- Support and cushion vital organs
- Make up essential components of cell membranes and nerve fibers
- Serve as a precursor for steroid hormones
- Store and transport **fat-soluble** vitamins
- Serve as an insulator to preserve body heat

The basic unit of fat is the fatty acid, which is also the part of fat that is used for energy production. Fatty acids occur in one of two forms: saturated or unsaturated. The difference between these two types of fatty acids is the binding between each molecule. Saturated fatty acids have no double bonds, meaning that each carbon atom of the fat molecule is saturated with its full complement of hydrogen atoms. Unsaturated fatty acids have at least one (monounsaturated) or more (polyunsaturated) double bonds and tend to be liquid at room temperature. On the other hand, saturated fatty acids tend to be solid at room temperature and are generally derived from fat of animal origin. Exceptions do exist and examples of common oils high in saturated fats are palm kernel oil and coconut oil, which are liquid at room temperature. The consumption of saturated fats is also associated with a greater risk for cardiovascular disease. Polyunsaturated fatty acids are primarily found in vegetable oils and are the preferred source of fat for lowering the risk of cardiovascular disease.

Protein

Proteins are nitrogen-containing substances that are formed by **amino acids.** Proteins serve as the major structural component of muscle and other tissues in the body. In addition, they are used to produce hormones, enzymes, and **hemoglobin.** Proteins can also be used as energy; however, they are not the primary choice as an energy source. For proteins to be used by the body, they must be broken down into their simplest form, amino acids. There have been 20 amino acids identified that are needed for human growth and metabolism (see table 15.1). Eleven of these amino acids are **nonessential,** meaning that our body can synthesize them, and they do not need to be consumed in the diet. However, the remaining nine amino acids are **essential,** which means that our body cannot produce these amino acids, and they must be consumed in our diets. Absence of any of these essential amino acids from our diet prevents the production of the proteins that are made up of those amino acids. As a result, the ability for tissue to grow, be repaired, or be maintained is compromised.

If the protein portion of the food contains all of the essential amino acids, the protein is called a complete protein. Meats, fish, eggs, and milk are the best sources

Table 15.1 Essential and Nonessential Amino Acids

Essential	Nonessential
Histidine (in children only)	Alanine
Isoleucine	Arginine
Leucine	Asparagine
Lysine	Aspartic acid
Methione	Cysteine
Phenylalanine	Glutamic acid
Threonine	Glutamine
Tryptophan	Glycine
Valine	Histidine (in adults only)
	Proline
	Serine
	Tyrosine

Histidine is not synthesized in children, making it an essential amino acid for that population group. However, in adults it is synthesized, making it a nonessential amino acid for that population.

for complete proteins. Proteins from plant and grain sources do not supply all of the essential amino acids and are described as incomplete proteins. Thus, for vegetarians to receive all of the essential amino acids, they need to combine proteins from several different plant and grain sources.

Vitamins

Vitamins are needed by cells to perform specific functions that promote growth and maintain health, including enabling cells to utilize carbohydrates, fats, and proteins for energy. Vitamins are classified as either **water-soluble** or fat-soluble. There are four fat-soluble vitamins: vitamins A, D, E, and K. Once absorbed, they are bound to lipids and transported throughout the body. Excess fat-soluble vitamins are stored within the fat stores of the body, and excessive intake of fat-soluble vitamins can cause toxic accumulations. The remaining vitamins are water-soluble and, once absorbed, are transported throughout the body in water. In general, any excess of water-soluble vitamins is excreted in the urine. Table 15.2 provides a list of the various vitamins and their major functions, dietary sources, symptoms of deficiency, and RDAs.

Most of these vitamins are important for athletic performance. Their relevance is likely related to their roles in energy metabolism and muscle growth. Many athletes, concerned that their high-intensity workouts require a greater vitamin intake, have invested millions of dollars in purchasing vitamin supplements. However, research

Table 15.2   Vitamin Functions and Requirements

Vitamin	Dietary source	Function	Symptoms of deficiency	RDA*
A (retinol; β-carotene can also be made into Vitamin A)	Foods of animal origin (e.g., liver, egg yolk, butter and milk). β-carotene is found in dark green leafy vegetables and yellow and orange pigmented fruits and vegetables.	Rhodopsin synthesis (necessary for eyesight), health of skin and soft-tissue membranes, bone development, reproduction, and immune system. β-carotene is a powerful antioxidant.	Rhodopsin deficiency, night blindness, frequent infections, poor growth, and skin disorders	800 μg in women 1,000 μg in men
B₁ (thiamine)	Whole grains, legumes, and milk	Carbohydrate and amino acid metabolism, necessary for growth	Beriberi (muscle weakness including cardiac), confusion, and depression	1.0 mg in women 1.2 mg in men
B₂ (riboflavin)	Dairy products, meats, green leafy vegetables, and enriched and whole-grain products	Energy metabolism, vision (especially in bright light), and health of the skin	Bright light sensitivity, skin rashes especially by corners of the mouth	1.1 mg in women 1.3 mg in men
B₃ (niacin)	Dairy products, meats, poultry, fish, and enriched and whole-grain products	Energy metabolism, nerve and digestive system function, health of skin	Pellagra (diarrhea, dermatitis, mental confusion and weakness)	14 mg in women 16 mg in men
B₅ (pantothenic acid)	Found in most foods	Energy metabolism through its involvement as a structural part of coenzyme A (a compound involved in energy metabolism)	Neuromuscular dysfunction and fatigue	5 mg
B₆ (pyrodoxine)	Meats, poultry, fish, green leafy vegetables, and whole-grain products.	Involved in amino acid metabolism	Poor tissue repair and retarded growth, irritability, nausea, dematitis.	1.3 mg
B₁₂ (cobalamin)	Meats, poultry, fish, eggs, and dairy products.	Red blood cell production, amino acid metabolism, and nerve cell maintenance	Anemia, fatigue, and confusion	2.0 μg
Folic acid (folate)	Green leafy vegetables, legumes	Red blood cell production, maintain healthy GI tract	Anemia, GI discomfort (e.g. diarrhea, constipation, etc.)	400 μg
C (ascorbic acid)	Citrus fruits, tomatoes, and green vegetables	Collagen formation, improved resistance to infection, protein metabolism, powerful antioxidant	Scurvy, poor wound healing, muscle soreness, bleeding gums	75 mg in women 90 mg in men
D (cholecalciferol)	All dairy products, eggs, green vegetables, and fish oil	Promotes calcium and phosphorus absorption, and bone mineralization	Rickets, weak bones, joint pain	5 μg
E (tocopherol)	Oils of vegetable origin, green vegetables, nuts, seeds, and whole-grain foods	Powerful antioxidant that protects cells from oxidative damage, also protects vitamin A from oxidative damage	Premature red cell death, possible role in muscular dystrophy	8 mg in women 10 mg in men

Vitamin	Dietary source	Function	Symptoms of deficiency	RDA*
H (biotin)	Available in most foods	Energy metabolism and glycogen synthesis	Mental and muscle dysfunction, fatigue, and nausea	30 μg
K (phylloquinone)	Green vegetables, milk, liver, also made from bacteria in the gut	Involved in the synthesis of clotting factors	Excessive bleeding due to poor blood clotting	70–140 μg

* = RDA based on 1989 RDA and 1998 DRI (dietary reference intakes) from the Food and Nutrition Board, Institute of Medicine - National Academy of Sciences

has been unable to support any need for vitamin supplementation for athletes involved in either power or endurance sports as long as a vitamin deficiency does not preexist (Benardot 2000; Singh, Moses, and Deuster 1992; Telford et al. 1992). It appears that the higher caloric intakes of these athletes more than compensate for any increases in vitamin requirements caused by high-intensity training. However, vitamin supplementation may prove to be beneficial for athletes who do not have a well-balanced diet or are on a calorie-restricted diet.

A topic receiving a lot of attention is the efficacy of both vitamins C and E as **antioxidants,** which prevent the cellular damage that occurs after the release of metabolically generated oxygen **free radicals.** A free radical appears to be produced during periods of oxidative stress when univalent oxygen intermediates leak from the electron transport chain within the mitochondria during oxidative phosphorylation (Kanter 1995). The free radicals are highly reactive and are thought to increase the rate of **fatigue** and contribute to the muscle damage seen after exercise. Although several different foods are known to have antioxidant properties, vitamins C and E have been given the most attention for their antioxidant action.

Vitamin E is generally considered the most important antioxidant in biological systems because of its association with the cell membrane (Bjorneboe, Bjorneboe, and Drevon 1990). Much of the damage from free radicals is characterized by disruption of the cell membrane. Vitamin E is thought to use several mechanisms in its role as an antioxidant. It can act directly on the oxygen-derived free radical by scavenging the potentially damaging singlet oxygen or act indirectly by protecting β-**carotene** and sparing **selenium** usage (Kanter 1995). Selenium is a mineral and is a component of glutathione peroxidase, an antioxidant enzyme. Although these two nutrients cannot be substituted for one another, they do complement each other to provide protection against the free radicals. β-carotene, whose antioxidant ability is well documented, is the most widely distributed carotenoid

compound (Bendich 1989). It also appears to have a direct effect by scavenging singlet oxygen radicals. Vitamin C also serves as an important antioxidant and free radical scavenger (Kanter 1995).

Studies have shown promising results in the ability of vitamin supplementation and antioxidant activity to reduce markers of lipid peroxidation after exercise. Studies by Cannon et al. (1990) and Meydani (1992) have demonstrated that vitamin E supplementation reduced concentrations of malondialdehyde (MDA, a skeletal muscle marker of lipid peroxidation) after exercise, indicating the benefit of vitamin E for reducing exercise-induced lipid peroxidation. Vitamin C supplementation also appears to have antioxidant ability. Several studies have reported reduced muscle soreness and a greater recovery from exercise in subjects taking vitamin C supplements compared with subjects who did not supplement (Jakeman and Maxwell 1993; Kaminski and Boal 1992). Although these studies attributed these differences to the antioxidant activity of vitamin C, they failed to measure any lipid peroxidation markers. Studies that have examined a combination of antioxidant vitamin mixtures have also demonstrated reduced MDA activity (Kanter, Nolte, and Holloszy 1993). However, it is not known whether a combination mixture provides any further protection than supplementing with only one vitamin. In addition, it is still unclear whether the protective mechanism that vitamins E and C appear to provide can be duplicated in all types of physical activity. Kanter (1995) suggested that vitamin supplements are an effective antioxidant during high metabolic activity. However, during exercise with a high mechanical stress (e.g., downhill running or resistance training), the protective mechanisms of these antioxidants may be diminished. Thus, the type of stress imposed (metabolic or mechanical) may determine the magnitude of effect that these vitamins have as antioxidants. Nevertheless, the benefits demonstrated in a number of studies do suggest a positive effect from supplementing with these vitamins during periods of

©Clark Brooks

high-intensity training. In addition, vitamin C supplementation (600 mg per day) has been shown to reduce the incidence of upper respiratory tract infections in marathon runners (Peters et al. 1993).

Minerals

Minerals are inorganic substances needed for normal cell function. Minerals are found everywhere in the body, and the total mineral content of the body is approximately 4% of an individual's body weight. They can act by themselves or function in combination with other minerals or various organic compounds. Minerals that are required by the body at a level greater than 100 mg per day are defined as **macrominerals.** Minerals that are needed in smaller amounts are known as **trace elements,** or **microminerals.** Table 15.3 provides a list of essential macrominerals and trace elements and their major functions, dietary sources, symptoms of deficiency, and RDAs.

Calcium and **phosphorus** are the most abundant minerals in the body. They constitute approximately 40% and 22% of the total mineral content of the body, respectively. They both are essential for bone health, and cal-

cium plays a critical role in muscle function (see chapter 1). However, supplementing any of these minerals in the normal diet does not appear to provide any ergogenic effect on athletic performance. The only mineral that is frequently supplemented is **iron.** Iron is a trace element, found in relatively small quantities in the body, that has an important role in oxygen transportation. Iron is required for the formation of hemoglobin and myoglobin. Hemoglobin is found in red blood cells, and myoglobin is found within the cytoplasm of muscle. Both of these molecules bind oxygen and store it until needed.

Iron deficiency is prevalent in the world today, and it is more common in women than men because of menstruation and pregnancy. The primary problem associated with iron deficiency is **anemia,** a condition in which hemoglobin concentrations are low. As a result, the oxygen-carrying capacity of the blood is reduced, causing fatigue, headaches, and other problems. Poor diet and prolonged endurance exercise may also contribute to iron deficiency anemia. Several studies have shown that prolonged endurance exercise, or high-intensity training over several weeks in previously untrained subjects, causes iron levels to fall (Magazanik et al. 1988; Pattini, Schena, and Guidi 1990). Endurance athletes tend to have a lower

Table 15.3 Mineral Functions and Requirements

Mineral	Dietary source	Function	Symptoms of deficiency	RDA*
Calcium	All dairy products, green leafy vegetables and legumes	Bone and teeth formation, blood clotting, muscle activity, and nerve function	Bone fragility, stunting of growth in children	1,200 mg
Chloride	Table salt	Component of digestive enzymes	Cramps, lethargy	750 mg
Chromium	Whole-grain foods and meat	Involvement in blood glucose control and glucose metabolism	Unknown	50–200 μg
Copper	Meats and most drinking water	Hemoglobin and melanin production, also a component of several enzymes	Anemia and loss of energy	1.5–3.0 mg
Fluoride	Fluoridated water and seafood	Provides extra strength in teeth by creating decay-resistant enamel.	Possible risk of developing dental cavities	3.0 mg in women 4.0 mg in men
Iodine	Iodized salt and seafood	Thyroid hormone production and maintenance of normal metabolic rate	Fatigue, decrease in metabolic rate, a serious deficiency may result in goiter (thyroid gland enlargement)	150 μg
Iron	Meats, poultry, fish, legumes, and dried fruits	Component of hemoglobin, ATP production in electron transport system	Anemia, decreased oxygen transport, fatigue	15 mg in women 10 mg in men
Magnesium	Nuts, legumes, whole grains, dark green leafy vegetables, and seafood	Involved in bone strength, protein synthesis, muscle and nerve function	Nervous system irritability and muscle weakness	320 mg in women 420 mg in men
Manganese	Whole-grain wheats, nuts, seeds, legumes, and fruits.	Hemoglobin synthesis, bone and cartilage growth, and activation of several enzymes.	Associated skeletal problems (osteoarthritis, osteoporosis, increased risk for fracture)	2.5–5.0 mg
Molybdenum	Legumes and whole-grain products	Component of enzymes	Unknown	75–250 μg
Phosphorus	Present in all foods of animal origin and legumes	Bone and teeth formation, energy transfer, and maintenance of body acid/base balance	May see loss of cellular function	700 mg
Potassium	Meats, poultry, dairy products, fruits, vegetables, and legumes	Muscle and nerve function	Muscle weakness and abnormal electrocardiogram	Not established
Selenium	Food content based upon selenium content of soil and water where food was grown	Cellular antioxidant. Component of many enzymes	May be associated with an increased risk for cancer, possible cardiac dysfunction	55 μg
Sodium	Table salt and in most processed foods	Important to body fluid regulation, nerve and muscle function	Nausea, vomiting, exhaustion, and dizziness	Not established; about 2,500 mg
Sulfur	Present in most foods	Component of hormones, vitamins, and proteins	Unknown	Not established
Zinc	Meats, poultry, and fish	Component of several enzymes, necessary for protein metabolism and CO_2 transport	Deficient protein metabolism and CO_2 transport	12 mg in women 15 mg in men

* = RDA based on 1989 RDA and 1998 DRI (dietary reference intakes) from the Food and Nutrition Board, Institute of Medicine - National Academy of Sciences

hemoglobin concentration than the normal population. This is commonly referred to as sports anemia. However, comparisons between athletes and untrained controls were unable to equivocally state that athletes are at a greater risk for iron deficiency (Clarkson 1991). Iron supplementation in athletes who are iron deficient improves performance, particularly endurance performance. Iron supplementation in athletes without any iron deficiency does not have any ergogenic benefit.

Water

Water is second only to oxygen in its importance for maintaining life. It constitutes about 60% of a man's and 50% of a woman's total body weight. Water is critical for dissipating body heat, transporting nutrients in the blood, and removing metabolic waste. Water is also crucial for maintaining athletic performance. Levels of dehydration of only 2% of an athlete's body weight can have significant detrimental effects on athletic performance (Hoffman, Stavsky, and Falk 1995). Chapter 16 provides detailed discussion on the importance of hydration.

NUTRIENT UTILIZATION IN ATHLETIC PERFORMANCE

Although all nutrients work together so an athlete can maintain a high level of performance, several types are of particular value. Carbohydrates, fats, and proteins play critical roles in any athlete's sustenance.

Carbohydrate Utilization in Athletic Performance

Carbohydrate is the critical fuel needed to sustain exercise. The production of ATP during events lasting more than several seconds in duration is dependent on the availability of muscle glycogen or blood glucose. Although protein and fat can also contribute to energy production, they can do so only in the presence of carbohydrate. The oxidation of fat during prolonged exercise depends on Krebs cycle intermediates that are produced from the breakdown of carbohydrate during the process of **glycolysis.** When both muscle and liver glycogen stores are depleted, the ability to provide a sufficient amount of the Krebs cycle intermediates is drastically reduced, thus limiting exercise performance even in the presence of plentiful fat and protein supplies. Marathon runners frequently feel the effects of depleting their glycogen reserves when they feel they "hit the wall" during a race. It is at this point that their ability to sustain exercise at their race pace is drastically reduced.

During the early stages of exercise, muscle glycogen is the primary energy source. As exercise duration becomes prolonged, a greater contribution from blood glucose occurs. As glucose uptake by the active muscle increases, the liver needs to increase the rate at which it breaks down stored glycogen. Unfortunately, the glycogen content of the liver is limited, and the ability of the liver to produce glucose from other substrates (e.g., amino acids) cannot be performed rapidly, thus blood glucose levels decrease. Fatigue quickly sets in as glycogen stores in the muscle become reduced.

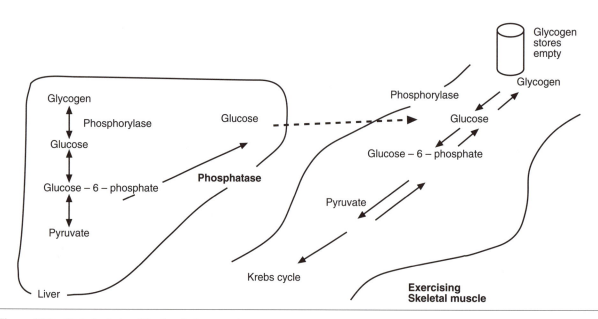

Figure 15.1 Carbohydrate utilization during exercise.

The body does not have the ability to use glycogen reserves from inactive muscles to fuel active muscles. This is related to the lack of the enzyme **phosphatase** within skeletal muscle. This enzyme is present in the liver and to a small extent in the kidneys, thereby permitting glucose to leave those sites and be transported to exercising muscle. This process is depicted in figure 15.1.

Glucose is metabolized through the process of glycolysis. As the glucose molecule enters the cell, it can be used as immediate energy or stored within the muscle as glycogen. The process of glucose molecules being polymerized (linked together) to form glycogen depends on the enzyme **glycogen synthetase.** The process of metabolizing glycogen back to glucose is regulated and limited by the enzyme **phosphorylase.** The activity of this enzyme is influenced to a great extent by the hormone epinephrine (see chapter 2). In the first reaction, glucose is phosphorylated from an ATP molecule to glucose-6-phosphate. In muscle tissue, the phosphorylation of the glucose molecule traps it within the muscle. However, in the liver and to some extent in the kidneys, the enzyme phosphatase can split the phosphate group from glucose, making it available to leave the cell and be transported throughout the body.

At the onset of exercise, muscle glycogen is the primary source of carbohydrate used for energy. The rate of muscle glycogen depletion depends on several factors, including exercise intensity, physical condition, mode of exercise, environmental temperature, and preexercise diet (Costill 1988). As exercise intensity increases, the oxygen uptake may not meet the demands of the exercising muscle. At this point, there is a greater reliance on carbohydrate for energy. This is emphasized in figure 15.2, which shows a greater muscle glycogen use at increasingly higher intensities of exercise. The rate of glycogen

use may be 40-fold greater during sprints than during walking (Costill 1988). When glycogen stores are depressed to very low levels, the intensity of exercise must be reduced to cause a greater reliance on stored fat to fuel muscle activity.

Several chapters of this text discuss the effects of the environment, exercise economy, or mode of exercise on the metabolic demands of activity. Any factor that makes the athlete exercise at a greater percentage of his or her maximal capacity causes an increase in glycogen utilization, resulting in an earlier onset of fatigue. In addition, the time to exhaustion during prolonged endurance exercise is proportional to the preexercise glycogen content of the muscle. Bergstrom and colleagues (1967) showed that subjects exercising at 75% of their $\dot{V}O_2$max under normal conditions (muscle glycogen content 100 mmol · kg^{-1} wet weight) could exercise for 115 min before becoming fatigued. However, if muscle glycogen content was reduced to 35 mmol · kg^{-1} wet weight, exercise duration decreased to only 60 min. When subjects were provided with a carbohydrate-rich diet for 3 days, muscle glycogen content increased to 200 mmol · kg^{-1} wet weight and subsequent exercise performance was prolonged to 170 min. The results of this study are depicted in figure 15.3. It should be noted that glycogen depletion and hypoglycemia cause fatigue in exercise that lasts more than 60 min. Fatigue occurring

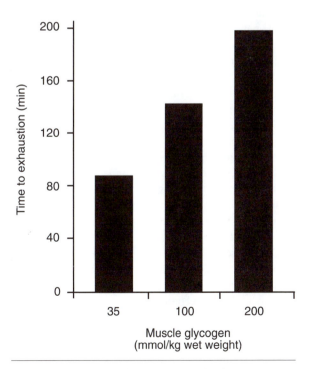

Figure 15.3 Relationship between muscle glycogen content and duration of exercise.
Adapted from data of Bergstrom et al. 1967.

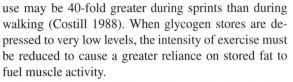

Figure 15.2 Muscle glycogen use at increasing intensity of exercise.
Adapted from Costill 1988.

during athletic events of shorter duration is likely the result of other factors, such as an accumulation of lactic acid and hydrogen ions within the muscle fiber (Costill 1988).

During periods of starvation or inadequate carbohydrate consumption, the body can catabolize muscle or liver protein to form glucose through the process of **gluconeogenesis.** However, it does not appear that a significant amount of carbohydrate can be generated in this manner. Thus, the glycogen storage capability of the muscle and liver depends primarily on the dietary consumption of carbohydrate.

The study by Bergstrom and colleagues (1967) showed that normal muscle glycogen content was 100 mmol · kg^{-1} wet weight for subjects fed a normal carbohydrate diet (55% of total calories). However, when subjects were fed a carbohydrate-rich diet (60-70% of total calories) for 3 days before exercise, muscle glycogen content was doubled. This procedure is known as **glycogen loading** and is a widely used practice of endurance athletes before competition.

A number of studies have shown that a combination of diet and exercise can result in a supercompensation of glycogen replenishment. The idea behind glycogen loading arose during the mid-1960s when scientists were looking for a way to maximize the amount of glycogen within the muscle. Researchers suggested that endurance athletes should deplete existing glycogen stores about 1 week before competition by performing an exhausting training routine and then consume a low-carbohydrate diet for the next 3 days (Wilmore and Costill 1999). The depleted glycogen stores cause an increase in the enzyme glycogen synthetase, which is responsible for glycogen synthesis. The athlete should begin to consume a diet rich in carbohydrate 3 days before competition. Because glycogen synthetase concentrations are increased, the high carbohydrate intake results in a supercompensation and a greater muscle glycogen content. It is important that the exhausting exercise recruits the same musculature that is recruited during the competition. In addition, after the bout of exhausting exercise, the volume and intensity of exercise should be very low for the remainder of the week. Such a regimen may increase muscle glycogen content twofold.

Other studies have shown that similar increases to muscle glycogen content can be attained without the need for exhausting exercise and the subsequent low-carbohydrate, high-fat, and protein diet (Sherman et al. 1981). By reducing training intensity and consuming a normal diet (55% carbohydrate) until 3 days before competition and then performing only a daily 10-15 min warm-up, combined with a high-carbohydrate diet, muscle glycogen content was demonstrated to reach about 200 mmol · kg^{-1} wet weight.

Fat Utilization in Athletic Performance

The two primary fuels that are used to provide energy for muscular activity are carbohydrates and fats. As discussed earlier, the amount of carbohydrate available for energy use is limited. Fat, however, has unlimited availability. During light to moderate exercise, the energy needs of the muscle are met by triglycerides from within the muscle itself and by free fatty acids. Free fatty acids are released from adipose sites around the body and bind to the protein **albumin** in the blood for transport to the active muscle. At the beginning of exercise, energy is used equally from both carbohydrate and fat resources. However, as duration of exercise is prolonged, a greater reliance on fat utilization is seen. When exercise duration exceeds more than 1 h, the carbohydrate reserves become limited and eventually depleted. The utilization of stored fat increases and may supply more than 80% of the total energy required by the end of the exercise session (McArdle et al. 1996). Figure 15.4 shows increases in free fatty acid uptake by the muscle, reflecting a greater utilization of fat as an energy source as duration of exercise increases. A greater reliance on fat is maintained as long as the intensity of exercise remains light to moderate. As exercise intensity increases, the added energy requirements are met by a greater use of blood glucose and muscle glycogen stores.

The increase in **lipolysis** (breakdown of triglycerides into free fatty acids) is a result of sympathetic neural stimulation, which is activated by reduced blood glucose

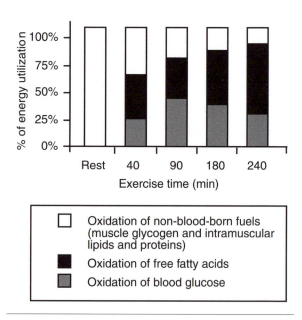

Figure 15.4 Relationship between duration of exercise and substrate utilization.
Adapted from data of Ahlborg et al. 1974.

levels, accompanied by a decrease in insulin and an increase in glucagon concentrations. The sympathetic hormone norepinephrine binds to its receptor site on the adipose tissue, stimulating activation of the enzyme lipase, which regulates the breakdown of triglycerides into free fatty acids. On liberation, the glycerol molecule diffuses out of the adipose tissue and is used for gluconeogenesis in the liver (Bjorntorp 1991). The free fatty acids bind to albumin and are transported to the muscle or other end organs.

Endurance-trained athletes appear to utilize fat for a greater percentage of their energy requirements during exercise (Saltin and Astrand 1993). This is helpful because it allows a greater conservation of both muscle and liver glycogen stores. The mechanisms that bring about the greater utilization of lipids during training are related to the specific physiological adaptations resulting from endurance training. These adaptations lead to a greater concentration of enzymes involved in lipolysis and **β-oxidation** (metabolism of fatty acids for energy) and parallel other adaptations that enhance the delivery system to the muscle (e.g., greater capillary density). Thus, the mechanisms involved in both production (breakdown of triglycerides to free fatty acids) and delivery (transport of free fatty acids from the adipocyte to the exercising muscle) are enhanced through training, providing endurance athletes with a greater concentration of lipids at their disposal for use as energy.

Protein Utilization in Athletic Performance

In general, proteins are not used to any appreciable extent as a fuel during exercise. The primary role of dietary proteins is for use in the various anabolic processes of the body. However, if an insufficient amount of carbohydrates is consumed or if exercise is prolonged, then protein can become a major source of energy. There is no storage depot of protein in the body, unlike carbohydrate (muscle and liver glycogen) and fat (adipose tissue). This is the reason for the atrophy of skeletal muscle mass in people on calorie-reduced diets. When protein is used, it is metabolized for energy primarily from the breakdown of skeletal muscle and liver proteins. During this catabolic process, protein is degraded into its amino acid components. In the liver, the nitrogen is removed from the amino acid molecule through the process of **deamination** and then excreted as urea. In the muscle, the nitrogen is removed from the amino acid molecule through the enzymatic process of **transamination** and then attached to other compounds. In either process, the carbon skeleton can be further metabolized for use as energy. When the body excretes more nitrogen than it consumes, it is said to be in a negative nitrogen balance, indicating

that catabolic processes are occurring within the muscle. On the other hand, if nitrogen intake exceeds nitrogen excretion, the body is in a positive nitrogen balance and anabolic processes are likely to be occurring.

Not all proteins in the body can be used to supply energy. Typically, the proteins from connective tissue and nerves are fixed, meaning that they are unable to be metabolized and used as energy (McArdle et al. 1996). Proteins from skeletal muscle, however, are easier to degrade and can be used as an energy source when carbohydrate stores are near depletion. The **glucose-alanine cycle** is the mechanism that provides amino acids as substrates for energy use (Felig and Wahren 1971). The amino acids that have been metabolized in the muscle are converted to glutamate and then to alanine. Alanine is then transported to the liver, where it is deaminated. The carbon skeleton is converted to glucose through the process of gluconeogenesis and then released back into the circulation, where it is transported to the exercising muscle. Ahlborg and colleagues (1974) have shown that after 4 h of light exercise, the alanine-glucose cycle may account for 45% of the total glucose released from the liver. If exercise intensity is increased, a greater percentage of alanine-derived glucose occurs. Thus, depending on an individual's carbohydrate reserve and both the duration and intensity of training, the reliance on protein as an energy source can become more vital. This magnifies the importance of maintaining a high carbohydrate intake. In addition, for athletes whose training programs are focused on muscle size and strength development, the use of a restricted carbohydrate diet may have severe negative consequences on achievement of their goals.

In general, the protein requirement for an adult is 0.8 $g \cdot kg^{-1}$ body weight. However, many athletes and coaches believe that high-intensity training results in a greater protein requirement. The issue of protein needs for the athlete has been the subject of much debate within the scientific community as well.

Strength/Power Athletes
The belief that strength and power athletes require more protein than the normal sedentary population stems from the notion that additional protein or amino acids available to the exercising muscle promote protein synthesis. This notion does hold some merit considering that a resistance training stimulus plus a greater amino acid pool, which results in a positive nitrogen balance, should theoretically promote greater muscle protein synthesis. The acute benefits from an increased protein intake are not well understood. Tarnopolsky and colleagues (1991), using radioactive-labeled amino acids, showed no changes in whole-body leucine oxidation during a 60-min resistance training session performed at 70% of the subjects' 1 RM (3 sets of 10 repetitions

with nine different exercises). This is not surprising because protein is not considered a significant fuel source during short-duration training. However, in studies of longer duration, an increase in protein consumption has been shown to result in significantly greater gains in protein synthesis, muscle size, and body mass.

Within 4 weeks of protein supplementation (3.3 versus 1.3 g $\cdot$ kg^{-1} $\cdot$ d^{-1}) of subjects in a resistance training program, significantly greater gains were seen in protein synthesis and body mass in the group of subjects with the greater protein intake (Fern, Bielinski, and Schutz 1991). Similarly, Lemon et al. (1992) also reported a greater protein synthesis in novice resistance-trained individuals with protein intakes of 2.62 versus 0.99 g $\cdot$ kg^{-1} $\cdot$ d^{-1}. However, there were no significant differences observed between the two groups in muscle size or strength. This may be related to the novelty of the training stimulus in these subjects. It is generally accepted that strength gains in previously untrained individuals are attributed to neural adaptations and not to any muscle structural changes. It would have been interesting to see the study prolonged to 10 weeks to see if the greater protein synthesis would have resulted in significantly greater structural changes within the muscle.

In studies examining strength-trained individuals, higher protein intakes have generally been shown to have a positive effect on muscle protein synthesis and size gains (Lemon 1995). A comparison of two groups of bodybuilders ingesting a low-calorie diet with differing protein intakes (0.8 versus 1.6 g $\cdot$ kg^{-1} $\cdot$ d^{-1}) showed that the bodybuilders consuming the higher protein intake were in a positive nitrogen balance, and the athletes consuming the RDA for protein were in a constant negative nitrogen balance (Walberg et al. 1988). Tarnopolsky and colleagues (1992) have shown that strength-trained individuals need to ingest a protein intake equivalent to 1.76 g $\cdot$ kg^{-1} $\cdot$ d^{-1} to maintain a positive nitrogen balance, whereas sedentary controls should consume 0.89 g $\cdot$ kg^{-1} $\cdot$ d^{-1}. This is consistent with other studies that have shown that protein intakes between 1.4-2.4 g $\cdot$ kg^{-1} $\cdot$ d^{-1} keep resistance-trained athletes in a positive nitrogen balance (Lemon 1995). However, there does seem to be a limit to the benefit derived from an increased protein intake. When protein intakes exceed 2.0 g $\cdot$ kg^{-1} $\cdot$ d^{-1}, no further increases in protein synthesis appear to occur (Fern, Bielinski, and Schutz 1991; Lemon et al. 1992; Tarnopolsky et al. 1992). Thus, recommendations now suggest that strength and power athletes limit protein intake to 1.4-1.8 g $\cdot$ kg^{-1} $\cdot$ d^{-1}.

Endurance Athletes

The importance of protein intake for the endurance athlete may be similar to that seen for the strength/power athlete. Although protein is not a major fuel source for exercising muscles, an increased reliance on protein as an energy source occurs during prolonged endurance exercise. To prevent a significant loss in lean tissue, endurance athletes have started to increase the protein content of their diet to replace the protein lost during exercise (Lemon 1995). Although the goal for endurance athletes is not necessarily to maximize muscle size and strength, loss of lean tissue can have a significant detrimental effect on endurance performance. Therefore, these athletes need to maintain muscle mass. Several studies have determined that protein intake for endurance athletes should be 1.2-1.4 g $\cdot$ kg^{-1} $\cdot$ d^{-1} to ensure a positive nitrogen balance (Friedman and Lemon 1989; Lemon 1995; Meredith et al. 1989; Tarnopolsky, MacDougall, and Atkinson 1988).

TIMING OF NUTRITIONAL INTAKE

Athletes are concerned that they are properly fueled prior to exercise or competition. Although most realize the importance of eating, many do not understand the timing of the pre-exercise or pre-event meal and what this meal should consist of. However, over the last few years a greater understanding has been attained due to studies examining the effects of food and fluid consumption prior to and during exercise and competition.

Precompetition Meal

It is generally accepted that athletes benefit greatly from having a meal before practice or competition as opposed to performing in a fasted state (ACSM Joint Position Statement 2000). The guidelines for such a meal, as recommended by the American College of Sports Medicine, the American Dietetic Association, and the Dietitians of Canada in their joint position stand, are listed at the end of this paragraph. Ideally, the meal should be eaten approximately 3-4 h before the event and comprise 200-300 g of carbohydrate (Schabort et al. 1999; Sherman et al. 1989). However, there is a concern of a hypoglycemic response and premature fatigue the closer the meal is to either practice or competition (Foster, Costill, and Fink 1979). Recent research, though, has suggested that eating close to competition may not be as detrimental as once thought (Alberici et al. 1993; Devlin, Calles-Escandon, and Horton 1986; Horowitz and Coyle 1993). The meal before competition contributes very little to the glycogen content of the muscle. However, it helps to ensure blood glucose levels and prevent feelings of hunger. If the meal is eaten close to competition, it may be wise for the athlete to consume a liquid meal. Such a meal might be less likely to cause gastric discomfort. These guideline are recommended for the precompetition meal:

- Sufficient in fluid to maintain hydration
- Low in fat and fiber to facilitate gastric emptying and gastrointestinal distress
- High in carbohydrate to maintain blood glucose and maximize filling of glycogen stores
- Moderate in protein intake
- Made up of foods familiar to the athlete
- Eaten 3-4 h before the game or practice

Food Supplements During Competition

The importance of carbohydrate supplementation during competition has begun to attain full acceptance and its efficacy has achieved scientific merit. Several studies have shown that providing carbohydrates during exercise maintains blood glucose levels and improves exercise performance (Ball et al. 1995; Coggan and Coyle 1989; Davis et al. 1997; Jeukendrup et al. 1997). Figure 15.5 shows the benefits of carbohydrate feedings (3 g · kg^{-1} body mass of a liquid comprising 85% glucose polymers and 15% sucrose in a 50% solution) versus placebo on duration of cycling exercise performed at 70% of the subjects' $\dot{V}O_2$max (Coggan and Coyle 1989). In this study, subjects consumed either the carbohydrate drink or placebo at minute 135 of exercise. The exercise time to fatigue averaged 21% longer in subjects drinking the carbohydrate mixture. In addition, the carbohydrate supplement did not cause any elevation in plasma insulin concentrations.

Later research has demonstrated that continually providing carbohydrate at 15-20 min intervals during the first 2 h of activity may be more beneficial than waiting until after 2 h of exercise has elapsed before providing

the supplement (McConnell et al. 1997). This is especially relevant during events in which the athlete may not have carbohydrate-loaded or consumed a precompetition meal. The composition of the carbohydrate should be primarily glucose; however, a combination of glucose plus fructose may be used without gastric problems (Coggan and Coyle 1991). In addition, the carbohydrate consumed can be in liquid, solid, or gel form as long as the athlete drinks adequate fluids (ACSM Joint Position Statement 2000).

The effect of carbohydrate feeding during competitions involving high-intensity, intermittent exercise (e.g., football, basketball, or hockey) has not been investigated to the same extent as endurance sports. However, speculation would suggest that if the athletes enter the competition with an inadequate glycogen supply, they would benefit from carbohydrate supplementation during the event. This may be relevant for athletes participating in tournaments conducted over several days in which glycogen replenishment is incomplete.

Postcompetition Meal

The timing of the postexercise or postcompetition meal is important. It is generally recommended that the postexercise meal be eaten within 2-4 h (Ivy et al. 1988; Volek, Houseknecht, and Kraemer 1997). However, the closer the meal is to the conclusion of the exercise or competition, the greater the opportunity to maximize glycogen loading. Ivy and colleagues (1988) showed that when carbohydrates were given immediately after exercise, muscle glycogen content at 6 h postexercise was significantly higher than when ingestion was delayed by 2 h.

The type of carbohydrate may also be important. It is generally recommended that carbohydrates with a high **glycemic index** be consumed. The glycemic index is a method of classifying food based on its acute glycemic impact (Jenkins et al. 1981). Foods with a high glycemic index are digested quickly and raise blood glucose levels fairly rapidly. Examples of foods with a high glycemic index are baked potatoes, rice cakes, waffles, and instant rice. Foods with a lower glycemic index take longer to be digested. Examples of foods with a low glycemic index include nuts, fruits, dairy products, and pasta. Postexercise carbohydrate meals with a high glycemic index result in a higher muscle glycogen content compared with a similar amount of carbohydrate with a lower glycemic index (Burke, Collier, and Hargreaves 1993).

The primary role of protein consumption postexercise is muscle repair and other anabolic processes within the muscle. It does not appear to have any effect on muscle glycogen replenishment. However, there is evidence that a combined protein and carbohydrate supplement provided

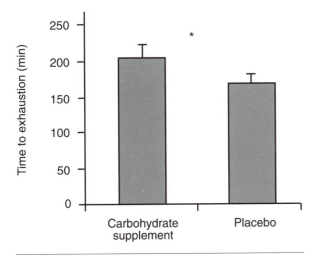

Figure 15.5 Effect of carbohydrate supplementation on time to exhaustion during cycling exercise.
Adapted from data of Coggan and Coyle 1989.

immediately postexercise may enhance the anabolic processes after resistance exercise (Roy et al. 2000). It is thought that the interaction of these two nutrients leads to a more anabolic state because of the combined influence of the hormone insulin (involved in both glucose uptake and protein synthesis) and the exercise stimulus. Roy and colleagues (2000) examined the effect of a combined carbohydrate/protein (CHO/PRO) or carbohydrate only (CHO) postexercise supplement, which was provided immediately after and 1 hr after resistance exercise, on whole-body protein synthesis. Subjects in CHO/PRO consumed a supplement of carbohydrate (~66%), protein (~23%), and fat (~12%), and subjects in CHO consumed a supplement of 56% sucrose and 44% glucose polymer from corn syrup solids. In addition, a third group of subjects was given a placebo (PL). The results depicted in figure 15.6 show that both CHO/PRO and CHO supplementation resulted in significantly greater protein synthesis than PL. No difference in protein synthesis was seen between CHO/PRO and CHO. It appears that the hyperinsulinemia after carbohydrate consumption caused an increase in protein synthesis, likely by enhancing amino acid uptake within the muscle.

SUMMARY

This chapter discussed the importance of the six nutrient classes. It appears that to maintain a high level of performance, athletes need to increase their energy intake to a level that equals their high energy expenditures. With an increase in caloric expenditure, there does not appear to be any need for a change in the vitamin or mineral requirements of the athlete as long as the athlete is not under a restricted diet regimen. The primary difference in the composition of the athlete's diet and that of his or her noncompetitive contemporary is the amount of protein

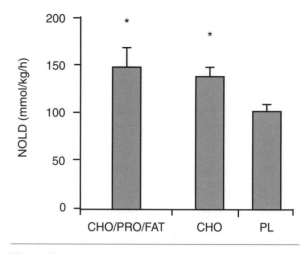

Figure 15.6 Comparison of whole-body protein synthesis after varied postexercise supplements. * = significantly better than PL.
Adapted from Roy et al. 1997.

that appears to be required. The RDA for the normal adult population is $0.8 \text{ g} \cdot \text{kg}^{-1}$ body weight. Competitive athletes appear to need approximately twice that amount to enhance muscle protein synthesis.

A primary strategy for many athletes is to maximize muscle glycogen content. Several loading strategies have been employed to enhance the athlete's ability to delay fatigue, including providing carbohydrate supplementation during and immediately after exercise performance. Such supplementation appears to prolong the exercise duration as well as enhance muscle glycogen replenishment after the exercise session. In addition, a combined carbohydrate and protein meal immediately after exercise may not only enhance glycogen replenishment but also increase protein synthesis, which has tremendous relevance for the strength/power athlete.

CHAPTER 16

HYDRATION

©David Sanders

As mentioned in chapter 15, water is essential for human life. It provides the medium for all biochemical reactions that occur within the body and is essential for preserving blood volume. It therefore plays a critical role in maintaining cardiovascular function and **thermoregulation.** Water constitutes approximately 60% of body weight and about 72% of lean body mass. Approximately two-thirds of the water in our body is found within the cells and is referred to as **intracellular fluid.** The remaining one-third is found in various compartments outside of the cell and is known as **extracellular fluid.** The extracellular fluid includes the fluid surrounding the cells **(interstitial fluid),** blood **plasma,** lymph, and other bodily fluids. Water is second only to oxygen in the necessity for maintaining life. Although the body can withstand a 40% loss in body mass from starvation, a 9-12% loss of body mass from fluid loss can be fatal.

Considering the role that water has in physiological function, it is no surprise that water also has a significant role in maintaining exercise performance. During exercise, the metabolic rate may increase 5-20 times over its resting level. This results in a large increase in body heat that must be dissipated to maintain thermal homeostasis. If exercise is performed in a hot environment, the need for heat dissipation is further magnified. Heat dissipation is primarily accomplished through **evaporative cooling (sweating).** Evaporative cooling may result in a large loss in body water because of sweat loss (this is reviewed more thoroughly in chapter 18). If fluid intake does not match body water loss, the body is said to be **dehydrating,** meaning that a body water deficit is occurring. A state of **hypohydration** (referring to an existing body water deficit) has significant effects on both physiological function and the ability to perform exercise. This chapter focuses on the effects of hypohydration on both physiological function and exercise performance. Fluid replacement during exercise is also discussed.

WATER BALANCE AT REST AND EXERCISE

Under normal conditions, body water is generally in balance (referred to as **euhydration**). Water intake occurs through fluid intake (accounting for approximately 60% of daily water intake), food consumption (approximately 30%), and metabolic processes within the body. Fluid loss occurs through excretion from the kidneys and large intestine and by evaporation of water from the skin surface or from the respiratory tract. During resting conditions, the primary avenue of water loss (approximately 60%) is through excretion from the kidneys (urination). Water loss as a result of evaporation from both the skin and respiratory tract accounts for approximately 35% of

the total water loss. The remaining 5% is the result of water loss in the feces.

Water loss is accelerated during exercise because of the increase in metabolic heat production. Sweat rate can vary greatly and depends on environmental conditions (ambient temperature, humidity, and wind velocity), clothing (insulation and moisture permeability), and the intensity of physical activity (Sawka and Pandolf 1990). Sweat rates generally range between 1.0 to 1.5 L $\cdot$ h^{-1}, which is equivalent to about a 2% decrease in body water per hour in a 155-lb (70-kg) man. The highest sweat rate reported in the literature is 3.7 L $\cdot$ h^{-1}, which was measured for Alberto Salazar during preparation for the 1984 Olympic marathon (Armstrong et al. 1986).

The primary problem during exercise in the heat is countering the large fluid loss from sweating through sufficient **rehydration,** which is related to the reliance on a thirst sensation to begin fluid intake. Studies examining *ad libitum* fluid intakes (the ability to drink at will) have shown that it is common for exercising individuals to voluntarily dehydrate despite the availability of adequate amounts of fluids (Armstrong et al. 1986; Hubbard et al. 1984). Thirst does not appear to be perceived until an individual has incurred a body water deficit of approximately 2% (Rothstein, Adolph, and Wells 1947). In addition, voluntary **dehydration** may impair the ability to rehydrate during later stages of exercise. This impairment may be related to the magnitude of hypohydration. Neufer, Young, and Sawka (1989) reported that **gastric emptying** may be reduced by approximately 20-25% in hypohydrated individuals (5% loss of body weight) performing moderate exercise in the heat (102°F [39°C]). However, at body water deficits of lower magnitudes (below 3% body weight) during exercise at a moderate intensity, gastric emptying or intestinal absorption may not be impaired (Ryan et al. 1998). Nevertheless, it does appear to be prudent practice to force hydration during the early stages of exercise to prevent any compromise in fluid availability to the body at later stages of exercise.

During dehydration, water is lost from both intracellular and extracellular spaces. At body water deficits of ~3% body weight, water is lost primarily from the interstitial spaces. As the body water deficit increases in magnitude, a greater percentage of the lost body fluid comes from intracellular spaces (see figure 16.1). This may be related to the **glycogen** depletion occurring within the cell during prolonged exercise. Approximately 3-4 g of water are bound to each gram of glycogen (Olsson and Saltin 1970). As exercise increases in duration, the large intracellular fluid loss may be partly related to the water released during the breakdown of glycogen (Costill et al. 1981). In addition, when body water deficits become very low, the body redistributes water from both intracellular and extracellular spaces

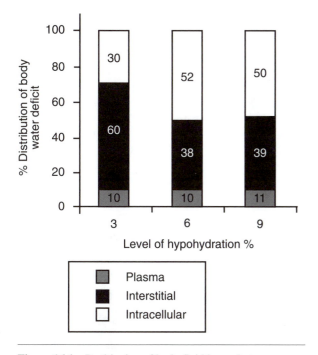

Figure 16.1 Partitioning of body fluid losses between different fluid compartments at rest.

Adapted, by permission, from D.L. Costill, R. Cote, and W.J. Fink, 1995, "Muscle water and electrolytes following varied levels of dehydration in man," *Journal of Applied Physiology* 40: 6-11.

to maintain the water content of the organs necessary to maintain life (e.g., brain and liver) (Nose, Morimoto, and Ogura 1983).

EFFECTS OF HYPOHYDRATION ON PHYSIOLOGICAL FUNCTION

A body water deficit has significant implications for both cardiovascular and thermoregulatory function. Hypohydration results in a reduction in plasma volume. As a consequence, less blood is available to both exercising muscle and the skin. Decreases in plasma volume are also associated with a reduction in **stroke volume** (Nadel, Fortney, and Wenger 1980). This is likely related to a lowering of central venous pressure, which decreases cardiac filling pressure (Kirsch, von Ameln, Wicke 1981). To compensate, heart rate increases to maintain normal blood flow. However, depending on the magnitude of the body water deficit, the increase in heart rate may not be sufficient to fully compensate for the lower stroke volume. For example, if the magnitude of dehydration exceeds 2% of one's body weight and is accompanied by a moderate to severe thermal stress, an increase in heart rate does not appear to fully compensate for the decrease in stroke volume. As a result, **cardiac output** is also reduced (Nadel, Fortney, and Wenger 1980; Sawka,

Knowlton, and Critz 1979). However, it does appear that cardiac output can be maintained at higher degrees of dehydration if the hypohydration occurs in the absence of a thermal strain (Sproles et al. 1976).

A body water deficit also impairs thermoregulation, the severity of which depends on the magnitude of hypohydration. Core temperature, an indicator of thermal strain, is elevated relative to the degree of hypohydration (Sawka, Young, Francesconi, et al. 1985). As the body water deficit increases, the ability of an individual to dissipate heat is reduced. The reduction in blood volume and an increase in blood displacement to the peripheral vasculature, especially during exercise in the heat, causes a reduction in venous return and subsequent cardiac output (Nadel, Fortney, and Wenger 1980; Sawka, Knowlton, and Critz 1979). As a result, the ability to dissipate heat is reduced, reflected by a lower cutaneous blood flow for a given core temperature. Increases in core temperature are also associated with decreases in both sweating rate and blood flow to the skin (Sawka and Pandolf 1990). Similar to changes in core temperature, these changes are also dependent on the magnitude of the body water deficit.

The physiological mechanisms that reduce sweat rate and cutaneous blood flow are not entirely understood but are likely related to increases in plasma **osmolality** and decreases in plasma volume (Sawka and Pandolf 1990). Hyperosmolality has been shown to increase the threshold temperature for sweating and vasodilation, even in the absence of a fall in blood volume (Fortney et al. 1984; Sawka, Young, Francesconi, et al. 1985). The delay in the sweating response because of increases in plasma osmolality is likely related to both central and peripheral mechanisms (Sawka and Pandolf 1990). **Osmoreceptors** within the hypothalamus, sensitive to changes in osmolality, are thought to be the central mechanism responsible for the delay in the sweating response; osmotic pressure changes at the sweat gland have been suggested to be a peripheral mechanism responsible for the reduced sweating response. These changes in the osmotic pressure gradient between the plasma and sweat gland reduce the fluid available to the sweat gland. In addition, the reduction in blood volume causes a decrease in peripheral blood flow, reducing the ability to dissipate heat. A significant relationship ($r = 0.53\text{-}0.75$) between changes in blood volume and changes in sweat rate has been reported during exercise in the heat (Sawka, Young, Francesconi, et al. 1985). It is thought that atrial baroreceptors sensitive to changes in blood volume provide afferent input to the hypothalamus to reduce fluid loss in sweat (Sawka and Pandolf 1990). Although the ability of the body to thermoregulate is reduced, the importance of conserving fluid balance for essential organs needs to take precedence.

The effect of body water deficit on muscle glycogen resynthesis has also been examined (Neufer et al. 1991). The potential for a dysfunction in glycogen resynthesis is related to the association between water and glycogen that has been previously discussed. It was hypothesized that hypohydration would impair muscle glycogen resynthesis because of a permissive role that water might play in this process. However, Neufer and colleagues (1991) were unable to see any difference in glycogen resynthesis between subjects euhydrated or hypohydrated to 5% of their body weight—immediately after 2 h of exhaustive exercise or 15-h postexercise (see figure 16.2)—despite significant differences between the groups in muscle water content. Thus, it does not appear that a body water deficit causes any decline in muscle glycogen resynthesis.

Table 16.1 summarizes the physiological effects and potential warning signs of dehydration. Body water deficits greater than 5% of body weight compromise the health of the individual. If the water deficit exceeds 9% of body weight, the physiological systems of the body may be impaired to the extent that life may become threatened.

ELECTROLYTE BALANCE DURING EXERCISE

In addition to fluid loss as a result of sweating during exercise, a large volume of sweat leads to a loss of **electrolytes** from the body. Sweat is primarily made up of water (approximately 99%) and includes a number of minerals such as potassium (K^-), sodium (Na^+), chloride (Cl^-), magnesium (Mg^{++}), and calcium (Ca^{++}). The concentrations of these minerals in sweat are much lower than that seen in plasma or other body fluids (see table 16.2). Na^+ and Cl^- are considered to be the predominant ions in sweat and account for its salty taste.

Electrolyte loss is variable among individuals and depends on sweat rate, physical condition, and state of

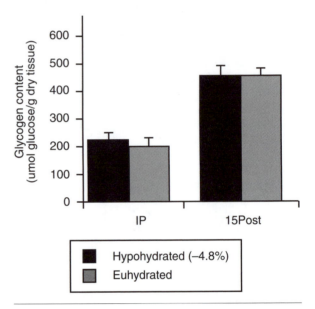

Figure 16.2 Effect of hypohydration on glycogen resynthesis. IP = immediately post exercise; 15Post = 15 min post exercise
Adapted from Neufer et al. 1991.

Table 16.1   Summary of the Physiological Effects and Potential Warning Signs of Dehydration

Body weight loss (%)	Physiological change	Warning signs
0–2	↑ Core temperature	None
2–4	↓ Plasma volume ↓ Muscle water ↓ Stroke volume ↓ Blood flow to the skin and muscle ↑ Heart rate	Thirst, verbal complaints, and some discomfort
4–6	↓ Sweat rate	Flushed skin, apathy, clear loss of muscle endurance, impatience, muscle spasms, muscle cramps, tingling sensation in arms, back, and neck
6–8	↑ Urine acidity ↑ Protein in urine ↓ Blood flow to kidney	Cotton mouth, headache, dizziness, shortness of breath, indistinct speech
8–12		Swollen tongue, spasticity, delirium

Adapted from Armstrong 1988.

Table 16.2   Concentration of Major Electrolytes in Sweat, Plasma, and Muscle

	Electrolyte concentrations (mEq · L^{-1})		
Electrolyte	Plasma	Sweat	Muscle
Sodium (Na$^+$)	137–144	40–80	10
Chloride (Cl$^-$)	100–108	30–70	2
Calcium (Ca^{2+})	4.4–5.2	3–4	0–2
Potassium (K$^-$)	3.5–4.9	4–8	148
Magnesium (Mg^{2+})	1.5–2.1	1–4	30–40

Adapted from Maughan 1991.

heat acclimatization (Maughan and Noakes 1991). During exercise, the large loss of Na$^+$ and Cl$^-$ in sweat results in the release of the hormone aldosterone (for a review of the fluid regulatory hormones, refer back to chapter 2), which acts on the kidneys to increase retention of these electrolytes. As the plasma concentration of these ions increases, plasma osmolality increases as well. This activates osmoreceptors in the hypothalamus, triggering a thirst sensation and stimulating the individual to drink. Electrolyte losses in the sweat cause these ions to be redistributed among body tissues. The resulting change in the electrolyte balance between intracellular and extracellular compartments may affect exercise performance by altering the membrane potential of the motor unit (nerve and muscle fibers that it innervates), possibly leading to potential performance decrements (Sjogaard 1986).

EFFECTS OF HYPOHYDRATION ON PERFORMANCE

The physiological changes that result from a body water deficit may have profound implications for exercise performance. Based on the changes to both cardiovascular and thermoregulatory function during dehydration, it could be assumed that exercise performance, especially endurance performance, would be negatively affected. Investigations examining the effect of moderate to severe levels of hypohydration on aerobic activity have indeed shown significant performance decrements, the extent of which appears to be related to the magnitude of the body water deficit, the environmental conditions where the exercise was performed, and the duration of exercise.

Aerobic Performance

Endurance performance appears to be more adversely affected in hotter environments and at greater levels of dehydration (Sawka 1992). In temperate environments,

maximal aerobic power appears to be maintained if the level of dehydration does not exceed 3% of body weight (Armstrong, Costill, and Fink 1985; Caldwell, Ahonen, and Nousiainen 1984). At body water deficits greater than 3% of body weight, significant reductions in aerobic power are seen (Buskirk, Iampietro, and Bass 1958; Caldwell, Ahonen, and Nousiainen 1984; Webster, Rutt, and Weltman 1990). If exercise is performed in the heat, the combined dehydration and thermal stress can result in performance decrements occurring at a lower level of dehydration. Craig and Cummings (1966) reported that subjects exercising in a hot environment (115°F [46°C]) experienced a significant (10%) decline in aerobic power at only a 2% body weight deficit.

The effect of hypohydration appears to be more pronounced on physical work capacity. Similar to aerobic power, declines in physical work capacity are also related to the magnitude of the body water deficit and the environmental temperature. However, performance decrements appear to occur at a lower level of dehydration. Armstrong, Costill, and Fink (1985) reported a 3.3% ($p > 0.05$) increase in the time recorded for a 1,500-m run at a body water deficit of 1.9% body weight. Although this difference did not result in a statistically significant change, it may have much more practical significance. During a race of relatively short duration (e.g., 1,500 m), the range in performance times among competitors would not be expected to be very large. Thus, a 3% difference in time may be the difference between winning and losing the race. During competitions of greater duration (e.g., 5,000 m or 10,000 m), low levels of dehydration (1.6% and 2.1%, respectively) were shown to result in a significant decline in performance (6.7% and 6.3%, respectively) (Armstrong, Costill, and Fink 1985). These results are depicted in figure 16.3.

Other investigators examining similar levels of dehydration have reported even greater performance decrements. A 31% difference ($p < 0.05$) in the time to **fatigue** was seen in cyclists dehydrated to 1.8% of body weight performing at 90% $\dot{V}O_2$max (Walsh et al. 1994). The difference in the magnitude of performance decrements between these studies may be related to the method of dehydration. The subjects in the first study were dehydrated with a diuretic, whereas the subjects in the cycling study were fluid depleted from 60 min of cycling exercise at 70% $\dot{V}O_2$max immediately before the high-intensity bout of exercise. The accumulated fatigue from the preceding exercise session likely resulted in the large differences in performance between the studies, suggesting that the method of dehydration may also have a significant effect on the extent of performance declines.

As the level of dehydration increases, a greater decline in the physical work capacity of the individual is seen. Caldwell, Ahonen, and Nousiainen (1984) reported more than a threefold drop in cycling performance (7-watt decline versus 23-watt decline) as the level of dehydration went from 2% to 4% in a temperate environment. Dengel and colleagues (1992) reported a 3.7% ($p > 0.05$) decline in cycling duration in subjects with a body water deficit of 3.3%. As the level of hypohydration reached 5.6%, fatigue occurred 6.4% ($p > 0.05$) earlier than under euhydrated conditions. Performance decrements are even further magnified when exercise is performed in the heat. Craig and Cummings (1966) reported that subjects dehydrated to 2% of their body weight reached fatigue 22% faster when walking (3.5 mph [5.6 km · h⁻¹]) to exhaustion in a hot environment (115°F [46°C]). When the level of dehydration was further reduced to 4% of their body weight, the subjects' time to fatigue was 48% faster. It seems fairly clear that the significant impairments to both the cardiovascular and thermoregulatory systems from a body water deficit directly affect the ability of the athlete to perform aerobic exercise.

Anaerobic Performance

The effect of hypohydration on **anaerobic power** performance is less clear. Most studies examining the effect of dehydration on anaerobic exercise performance have been conducted primarily on athletes who reduce weight quickly in order to compete at a given weight class. Most of this research has been conducted on wrestlers or other athletes who participate in high-intensity, short-duration athletic events. These studies have shown that strength, anaerobic power, and anaerobic capacity are not affected by varying levels of hypohydration as long as the duration of these high-intensity activities is less than 40 s (Houston, Marin, et al. 1981; Jacobs 1980; Park, Roemmick, and Horswill 1990; Viitasalo et al. 1987). Jacobs (1980) reported no significant differences in both peak and mean power, even as the magnitude of dehydration increased to 5%. Power outputs tended to increase ($p = 0.06$) when examined relative to body weight (see figure 16.4). As the duration of anaerobic exercise increases (either sustained or intermittent), decreases in anaerobic power and capacity have been reported (Hickner et al. 1991; Horswill et al. 1990; Webster, Rutt, and Weltman 1990). In examining the effect of dehydration on simulated or actual sports performance, Klinzing and Karpowicz (1986) reported significantly slower performance times in wrestlers performing a standardized series of wrestling drills. These studies do suggest that for sports in which strength and power performance are

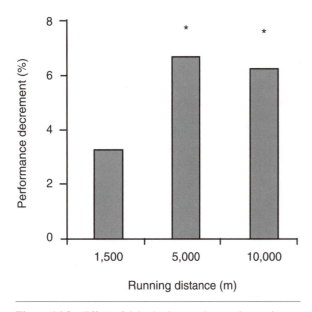

Figure 16.3 Effect of dehydration on changes in running times for 1,500-m, 5,000-m, and 10,000-m races.
Data adapted from Armstrong et al. 1985.

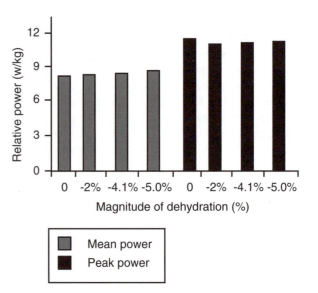

Figure 16.4 Effect of differing magnitudes of dehydration on anaerobic power outputs relative to body weight.
Adapted from Jacobs 1980.

emphasized over a relatively short duration (< 40 s), rapid weight loss may result in an increase of relative strength and power, offering a competitive advantage to the athlete. However, as duration of anaerobic activity increases, the strain of hypohydration does appear to affect performance.

Much less is known about the effects of dehydration in sports that rely on intermittent bouts of high-intensity exercise over a relatively long duration. In such sports (e.g., basketball, football, and hockey), athletes do not voluntarily dehydrate to compete at a lighter weight class. Rather, dehydration may occur because of inadequate fluid intake during actual game performance. When anaerobic power needs to be maintained over a moderate duration of time (e.g., 30-60 min), a body water deficit may hinder such performance. One study examined the effect of fluid restriction on anaero-

bic power and skill performance during a simulated basketball game (Hoffman, Stavsky, and Falk 1995). Fluid restriction during the 40-min game resulted in an average body water deficit of 1.9%. No significant changes in vertical jump height, anaerobic power, or shooting performance were seen. However, an 8% difference in shooting percentage ($p > 0.05$) when subjects were dehydrated was calculated to result in a possible 6-point deficit (average number of shots in a game/ calculated shooting %) during a game (see figure 16.5). Practically speaking, this could clearly affect the outcome of a game. The results of this study suggest that even at low levels of dehydration, in which the athlete may not perceive thirst, performance in activities that require fine motor control may be impaired. Previous research has shown that motor unit recruitment patterns and muscle contraction capabilities may be reduced

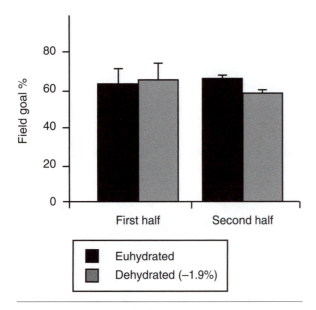

Figure 16.5 Effect of dehydration on basketball shooting performance.

Adapted, by permission, from J.R. Hoffman, H. Stavsky, and B. Falk, 1995, "The effect of water restriction on anaerobic power and vertical jumping height in basketball players," *International Journal of Sports Medicine* 16:214-218.

because of elevated body temperatures and electrolyte imbalances occurring from the combined exercise and hydration stress (Sjogaard 1986).

FLUID REPLACEMENT DURING EXERCISE

The decline in physiological function resulting from a body fluid imbalance and the subsequent decrease in athletic performance emphasize the need to rehydrate during exercise. If the athlete relies on thirst as an indicator to drink, he or she will undoubtedly perform in a dehydrated condition. As discussed previously, the thirst sensation does not occur until body water deficit reaches approximately 2%. Even at this magnitude of hypohydration, physiological changes are apparent and performance decrements can be seen. Although the need for fluid replenishment is obvious, in many athletic events the volume and frequency of fluid consumption may be limited by the rules of the competition (e.g., rest periods or time-outs) or by availability (e.g., distance between drinking stations on a race course). In addition, even when fluid is available, athletes seldom consume more than $0.5 \text{ L} \cdot \text{h}^{-1}$ (Noakes 1993). Thus, athletes need to adopt several strategies to maintain a normal body fluid balance. Several recommendations for maintaining fluid balance are listed here:

- Drink 0.5 L of water 2 h before exercise to promote adequate hydration and allow for excretion of excess fluid.
- During exercise, forced hydration should be practiced (drinking even when the thirst sensation is absent) to prevent physiological and performance impairments.
- Ingested fluids should be between 59°F and 72°F (15°C and 22°C) and flavored to enhance palatability to promote maximal fluid replacement.
- Fluids should be readily available at shorter intervals during a road race, and athletes should be instructed to replace fluids during every time-out or stoppage of play during athletic contests.
- Electrolyte drinks have not been proven to delay fatigue during exercise less than 3-4 h in duration. However, glucose-electrolyte drinks may delay fatigue in exercise of shorter duration.
- Athletes should weigh themselves after each practice and game and replace all lost fluid. The athlete should rehydrate 1 L for each kg of body weight lost.

Fluid Temperature, Fluid Palatability, and Consumption

Water temperature and its palatability are important factors for stimulating fluid intake. Several studies have demonstrated that as water temperature is increased, fluid consumption is significantly reduced (Armstrong, Costill, and Fink 1985; Szlyk et al. 1990). The highest water consumption appears to occur when water temperatures range from 59°F to 72°F (15°C to 22 °C) (Hubbard, Szlyk, and Armstrong 1990). Szlyk and colleagues (1990) demonstrated that subjects walking on a treadmill (3.0 mph [4.8 $\text{km} \cdot \text{h}^{-1}$] at a 5% grade) for 6 h at 104°F (40°C) drank significantly more water when it was chilled to 59°F (15°C) than when its temperature was 104°F (40°C) (see figure 16.6). These results were shown in two types of subjects: avid drinkers (subjects who were able to maintain body weight loss within 2% of initial body weight when given cold water *ad libitum*) and reluctant drinkers (subjects who lost more than 2% of initial body weight even though they were provided water *ad libitum*).

It also appears that water palatability is important for enhancing the drive to drink. If water has a color, odor, or taste that negatively affects human senses, individuals may refuse to drink (Hubbard, Szlyk, and Armstrong 1990). However, water consumption is enhanced if water is flavored, regardless of its temperature (Hubbard et al. 1984; Szlyk et al. 1989). Cherry, raspberry, and citrus appear to be the most popular flavors reported to increase fluid consumption (Hubbard, Szlyk, and Armstrong 1990).

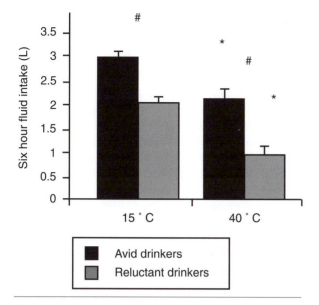

Figure 16.6 Effect of water temperature on drinking. * = significant difference from 15°C. # = significant difference between avid and reluctant drinkers. Data from Szlyuk et al. 1990.

Fluid Consumption and Gastric Emptying

Once the athlete or active individual has established a pattern of replacing lost fluids, the next major concern for ensuring the availability of ingested fluids is gastric emptying. The rate of gastric emptying is determined primarily by the volume and composition of the fluid consumed (Maughan and Noakes 1991). Gastric emptying occurs at a higher rate when the volume of fluid in the stomach is high and falls rapidly as fluid volume decreases (Leiper and Maughan 1988). Although this highlights the importance for athletes to repeatedly drink during exercise, the gastric distension that may occur from a large fluid consumption may also give a sensation of fullness, affecting the athlete's drinking behavior. The primary objective, though, is for the athlete to match fluid intake to fluid loss. Through trial and error, the athlete will find the appropriate fluid volume and the rate of intake that does not impair his or her performance. Gastric emptying also appears to be greater during exercise than at rest (Neufer et al. 1986). This is likely related to the increased mechanical movement of fluid within the stomach during exercise. Therefore, the athlete could possibly consume a greater volume of fluid during exercise without experiencing the feelings of fullness because of the greater emptying rates seen during such activity.

The composition of the ingested fluids has a major effect on the rate of gastric emptying. Both caloric content and osmolality of the drink can affect the ability of fluid to empty from the stomach. Besides replacing lost fluids, the proposed effect of many commercial sports drinks is to spare muscle glycogen and replace lost electrolytes. In regard to the caloric composition of the drink, as the caloric content rises, the rate of gastric emptying declines (Murray 1987). The average gastric emptying rate for fluids containing calories ranges from less than 5 ml · min⁻¹ to 20 ml · min⁻¹. The gastric emptying rate for solutions of plain water or saline is at the higher end of the scale (Brener, Hendrix, and McHugh 1983; Costill and Saltin 1974). Similarly, drinks of high osmolality empty from the stomach at a slower rate than drinks of low osmolality (Murray 1987), likely related to the positive relationship between osmolality and caloric content. In addition, the ingestion of glucose solutions of high concentrations may exacerbate plasma volume loss because of a movement of water into the gut caused by the high osmolality of such concentrated solutions (Maughan, Fenn, and Leiper 1989).

Glucose polymers appear to promote the emptying of glucose-electrolyte solutions from the stomach by reducing the osmolality of the solution while maintaining the total glucose content (Maughan 1991). A number of studies have demonstrated that glucose-polymer solutions had a faster emptying rate than solutions containing free glucose in similar concentrations (Foster, Costill, and Fink 1980; Sole and Noakes 1989). However, results have not been consistent and have often been contrasting. For instance, Foster, Costill, and Fink (1980) showed a significantly greater emptying rate with a 5% glucose-polymer solution compared with a 5% solution of free glucose but no differences between glucose-polymer solutions and free-glucose solutions of 10%, 20%, and 40% concentrations. In contrast, Sole and Noakes (1989) showed that glucose-polymer solutions of higher concentrations (15%) have significantly greater emptying rates than similar free-glucose solutions but did not see any differences when comparing solutions of lower concentrations (5% and 10%). Nevertheless, these studies and others that have examined glucose-polymer solutions have consistently reported a faster emptying rate with the glucose-polymer drink even if the differences were not statistically significant. However, none of these polymer solutions have shown an emptying rate as fast as that of plain water.

Electrolyte Replacement Drinks

The primary objective of electrolyte drinks is to replace electrolytes lost in sweat. Sodium may be lost at a rate of 75 mmol · h⁻¹ during exercise. As discussed earlier, this may contribute to the reduced sweat rate caused by the increase in electrolyte concentrations in the plasma, resulting in a change in the osmotic gradient between the circulation and the sweat gland. By replacing the lost fluid with plain water, the plasma **osmolality** is lowered and sweat rate returns to its normal level (Senay 1979). However, complete restoration of extracellular fluid cannot be accomplished without first restoring the lost sodium (Takamata et al. 1994).

There does not seem to be compelling evidence to support the use of electrolyte drinks during exercise less than 3-4 h in duration. However, in events that are prolonged (> 4 h), the inclusion of electrolyte drinks does appear to prevent **hyponatremia.** Hyponatremia is a condition of low plasma Na^+ (between 117 mmol $\cdot$ L^{-1} and 128 mmol $\cdot$ L^{-1}) that is associated with disorientation, confusion, and grand mal seizures (Noakes et al. 1990). Consuming electrolyte drinks during the event appears to prevent this relatively rare condition.

Most of the studies involving glucose and electrolyte replacements have examined athletes participating in prolonged aerobic activity. The effect of electrolyte supplementation during athletic contests made up of repeated high-intensity, short-duration activities (e.g., football, basketball, and hockey) is unknown. Considering the possible implications of electrolyte loss on motor unit recruitment patterns and muscle contractile capabilities (Sjogaard 1986), it does appear that future research should examine the efficacy of electrolyte drinks in these high-intensity sports.

SUMMARY

This chapter discussed the detrimental effects of hypohydration to both physiological and thermoregulatory systems. These effects are seen beginning at low levels of hypohydration (1% body weight) and increase as the magnitude of the fluid deficit grows. Decreases in athletic performance have been reported primarily in events that are aerobic in nature. However, several studies have indicated that performance decrements may also be seen during high-intensity anaerobic activity that exceeds 40 s in duration or in contests that comprise repeated bouts of high-intensity activity of shorter durations. The significant decline in both physiological function and athletic performance highlights the importance of maintaining a normal fluid balance. Fluid palatability and fluid temperature are factors that enhance fluid consumption. Although significant decreases in electrolytes are seen during hypohydration, the advantages of electrolyte replacement drinks are not realized until exercise durations exceed 3-4 h.

CHAPTER 17

ERGOGENIC AIDS

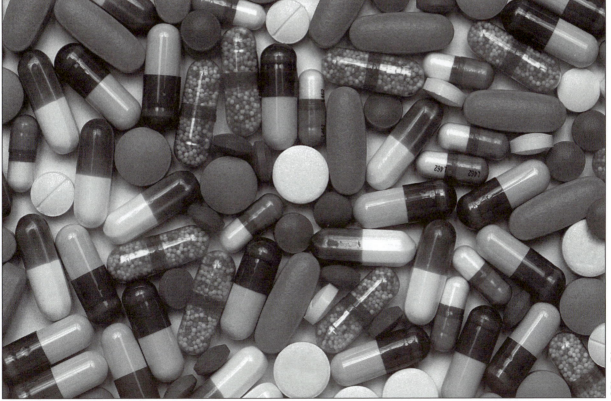

©Photo Disc

An **ergogenic aid** is anything that is used by an athlete to improve athletic performance. There are a number of different categories of ergogenic aids: biomechanical (e.g., lighter running shoes), nutritional (e.g., protein **supplementation,** vitamins), pharmacological (e.g., **anabolic steroids,** diuretics), physiological (e.g., **blood doping**), and psychological (e.g., hypnosis) (Swirzinski et al. 2000; Williams 1992). Ergogenic aids have been used for centuries to improve athletic performance; however, these substances were not closely examined in North America until after the 1954 weightlifting championships in Vienna. Dr. John Ziegler, the American team physician, reported that anabolic steroid use was widespread among the power athletes of the Soviet Union, and much of the success experienced by the Soviet athletes was likely due to their steroid use (Yesalis, Courson, and Wright 1993). This, and other anecdotal reports concerning the prominent use of steroids and their effects on performance enhancement, led to a growth in the use of steroids and other substances by many athletes to get a competitive edge or simply to remain competitive. For the past 50 years, the use of ergogenic aids by athletes has increased tremendously. Unfortunately, the use of such substances has not been without medical tragedy. Through the years a number of deaths have been attributed to various methods (e.g., **amphetamine** use and blood doping) used to enhance performance (Tricker and Connolly 1997).

The medical risks associated with many of these ergogenic aids, as well as ethical considerations, have prompted the major governing bodies in sport (e.g., International Olympic Committee [IOC], National Collegiate Athletic Association [NCAA], National Football League [NFL]) to initiate measures to combat their use. Each of these organizations has defined a list of drugs and methods that are considered illegal; athletes caught using any of these banned substances or procedures are suspended from competition. Despite the known **side effects** and potential risks associated with many of these ergogenic aids, including the risk of being barred from competition, athletes continue to use these substances and search for ways to mask their use to avoid detection.

Why athletes are willing to take a substance that places them at risk is not well understood, but it is likely related to the intense competitive nature of the athlete and the burning desire to be the best. This mindset has been suggested by Bob Goldman in his book *Death in the Locker Room* (1984), in which he reports on the response of 198 world-class athletes to a question concerning the use of performance-enhancing drugs. He reportedly asked these athletes: "If I had a magic drug that was so fantastic that if you took it once you would win every competition you would enter, from the Olympic decathlon to Mr. Universe, for the next five years, but it had one minor draw-back—it would kill you five years after you took it—would you still take the drug?" The response of the athletes was stunning. More than half of the athletes questioned (52%) indicated that they would take the drug. Although this survey was not scientific and its results have not been repeated by other investigators, the mindset that it implies has been acknowledged from a number of other sources (Kerr 1982; Tricker and Connolly 1997; Yesalis 1993).

According to varying sources, the use of ergogenic aids is rampant among all levels of athletes and in all types of sports. Anecdotal reports claim that up to 90% of Olympic athletes use some ergogenic aid to enhance performance, and the incidence of anabolic steroid use in professional football players has been reported to range between 40 to 90% (Yesalis, Courson, and Wright 1993). However, only 1-3% of the athletes tested by accredited laboratories of the International Olympic Committee have tested positively (Laure 1997). This relatively low incidence rate, in contrast to the high anecdotal claims, is likely related to the knowledge that many athletes possess to mask drug use and the use of drugs such as peptide hormones (e.g., **erythropoietin** and growth hormone) that cannot be detected in the urine. In addition, many of the substances used as ergogenic aids are not banned (e.g., **creatine, amino acids,** antioxidants) and would not present as a positive test. In an analysis of published surveys, Laure (1997) reported that when individuals were asked directly about their own use of banned substances,

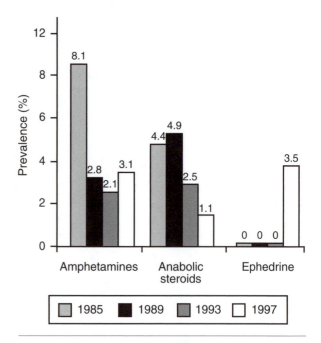

Figure 17.1 Patterns of ergogenic drug use in collegiate athletes.
Data from NCAA 1997.

the prevalence ranged from 5 to 15%. However, when athletes were asked if they knew of other athletes who were using such substances, the range in drug use increased to 15-25%. This difference may suggest that many athletes feel uncomfortable describing their own use of ergogenic substances. In a survey of almost 14,000 NCAA student athletes, the NCAA (1997) reported that the prevalence of anabolic steroid and amphetamine use has declined over the past 12 years (see figure 17.1). However, the prevalence of ergogenic aids may be substantially higher when examining substances that are not banned.

In a study of NCAA Division I athletes, 48% of the males and 4% of the females surveyed admitted to having used creatine (LaBotz and Smith 1999). The sports with the highest percentage of users were baseball (81%) and football (71%) for the males and swimming (19%) and basketball (14%) for the females. Ergogenic aids are also prevalent in high schools. The NCAA (1997) reported that more than 53% of the athletes reporting amphetamine use during college had initially experimented with this ergogenic aid during high school. Twenty-five percent of the anabolic steroid users and 37% of the **ephedrine** users also began taking these ergogenic drugs during their secondary education.

A recent report of high school football players stated that 31% of the athletes surveyed ($n = 170$) were using a performance-enhancing supplement (Swirzinski et al. 2000). Thirteen percent of the athletes surveyed reported taking more than one supplement. The primary reason given for their use of supplements was to increase weight and build muscle. Creatine was by far the most popular ergogenic aid (> 90%) used by the athletes who were taking a nutritional supplement. This study, however, did not provide any data on the use of banned substances. Other studies have reported that steroid use at the secondary level may be as high as 6% (Buckley et al. 1988). Most of the users of anabolic steroids were football players or wrestlers in their senior year of competition. However, 35% of the reported users of anabolic steroids in high school were not participating in any school-sponsored athletic program. Whether the prevalence of anabolic steroid use is similar today to what it was 10 years ago is debatable. Based on the data of the 1997 NCAA study on ergogenic drug use among intercollegiate athletes, there is a trend toward a reduction in anabolic steroid use (see figure 17.1). Athletes are most likely finding more easily accessible alternatives to anabolic steroids (e.g., creatine).

The remainder of this chapter focuses on the more popular ergogenic agents used by today's athletes. This is often a difficult task because of the continual onslaught of ergogenic aids that are flooding the market before scientific research can determine their efficacy. Thus, em-

phasis is directed at creatine, anabolic steroids (including **anabolic precursors** such as **androstenedione**), ephedrine and **caffeine,** blood doping, erythropoietin, and β**-hydroxy-**β**-methylbutyrate (HMB).**

CREATINE

Creatine, in the form of **phosphocreatine (PC),** has an essential role in energy metabolism. PC acts as a substrate for the formation of ATP by rephosphorylating ADP, especially during short-duration, high-intensity exercise. The ability to rapidly rephosphorylate ADP is dependent on the enzyme creatine kinase and the availability of PC within the muscle. As PC stores become depleted, the ability to perform high-intensity exercise declines (Tesch, Thorsson, and Fujitsuka 1989). The following are suggested roles for PC in the skeletal muscle (adapted from ACSM 2000b):

- Triggers the rapid rephosphorylation of ADP through the creatine kinase reaction (primary source for ATP resynthesis during short-duration, high-intensity exercise)
- Enhances transport capacity of high-energy phosphates from the mitochondria (source of ATP synthesis) to sites of ATP utilization (e.g., myofibrils)
- Contributes to buffering of intracellular acidosis during exercise (suggested because net PC hydrolysis consumes hydrogen ions)
- Activates glycogenolysis and other **catabolic** pathways from the products of PC hydrolysis (creatine and inorganic phosphate)

Creatine is a nitrogenous organic compound that is synthesized in the body primarily in the liver. It is also synthesized in smaller amounts in the kidneys and pancreas. The amino acids arginine, glycine, and methionine are the precursors for the synthesis of creatine in those organs (Walker 1979). Creatine can also be obtained though dietary sources. It is found in abundance in both meat and fish. Approximately 98% of creatine is stored within skeletal muscle in either its free form (40%) or its phosphorylated form (60%) (Heymsfield et al. 1983). Small amounts of creatine are also stored in the heart, brain, and testes. Creatine is transported from its site of synthesis to the skeletal muscle by the circulatory system. The normal concentration of creatine in plasma ranges from 50-100 m\gm\mol · L^{-1} (Harris, Soderlund, and Hultman 1992).

Creatine Supplementation

The average PC concentration in muscle is 125 mmol · kg^{-1} dry weight, but levels may range from 90 to 160 mmol · kg⁻¹ dry weight (Juhn and Tarnopolsky 1998). Women may have slightly greater concentrations than men, and vegetarians tend to have lower levels than nonvegetarians (Juhn and Tarnopolsky 1998). Creatine supplementation is reported to increase the creatine content of muscles by approximately 20% (Febbraio et al. 1995; Hultman et al. 1996). However, it does appear that a saturation limit exists for creatine within the muscle. Several studies have noted that once creatine levels in the muscle reach 150-160 mmol · kg⁻¹ dry weight, further supplementation will not increase those levels (Balsom, Soderlund, and Ekblom 1994; Greenhaff 1995). This has important implications for the "more is better" philosophy that governs the thinking of many athletes. It has also helped to develop proper dosing regimens.

A typical creatine supplementation regimen involves a loading dose of 20-25 g daily for 5 days. This is followed by a maintenance dose of 2 g per day. If an individual wishes to dose relative to body weight, a maintenance dose of 0.3 g · kg⁻¹ body weight is generally used (Hultman et al. 1996). Athletes who ingest 3 g daily, without an initial loading dose, reach muscle creatine levels similar to that seen in the athletes who initially used a loading dose. However, it will take longer to reach that same muscle creatine concentration (~30 days versus 5 days). Muscle creatine levels remain elevated as long as the maintenance dose is maintained (2 g · d⁻¹ or 0.03 g · kg⁻¹ · d⁻¹) (Hultman et al. 1996). Once creatine supplementation is stopped, muscle creatine returns to baseline levels after about 4 weeks (Febbraio et al. 1995; Hultman et al. 1996).

Creatine Supplementation and Athletic Performance

Creatine appears to have been the most widely studied ergogenic aid in the past few years. This is likely related to the high prevalence of creatine use by athletes and a desire by both the medical and scientific communities to better understand the risks and benefits of such supplementation. Those athletes who rely primarily on PC as an energy source to fuel exercise (e.g., bodybuilders, power lifters, sprinters, football players, hockey players, and basketball players) would appear to benefit the most from creatine supplementation.

Most studies examining the effect of creatine supplementation on strength performance have consistently shown significant ergogenic benefits (Becque, Lochmann, and Melrose 2000; Brenner, Walberg-Rankin, and Sebolt 2000; Kreider et al. 1998; Pearson et al. 1999; Volek et al. 1999). Figure 17.2 shows the results of a 10-week study by Pearson and colleagues (1999) on a group of Division I intercollegiate football players. The athletes were given a capsule that contained either 5 g of

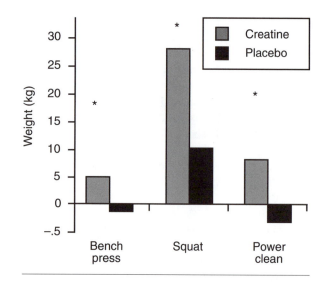

Figure 17.2 Effect of creatine supplementation on muscle strength gains in collegiate football players. * = significant difference between creatine and placebo.
Adapted from Pearson et al. 1999.

creatine or a placebo (no loading phase was used). Strength increases in bench presses, squats, and power cleans were significantly greater in the athletes supplemented with creatine (3.4%, 11.6%, and 6.5%, respectively) when compared with athletes on placebo (–1.1%, 4.5%, and –2.7%, respectively). These results may highlight the benefit of creatine supplementation in experienced resistance-trained athletes whose potential to improve strength may be limited.

Creatine supplementation to improve strength in recreational athletes may not be as effective. Although significant strength improvements have been reported in noncompetitive athletes (Becque, Lochmann, and Melrose 2000), other studies have failed to see any significant differences in strength gains (Syrotuik et al. 2000). The difference between these two studies may be related to the length of time the subjects were taking creatine. The subjects in Becque's study ingested creatine for 6 weeks, whereas the subjects in Syrotuik's study supplemented creatine for only 4.5 weeks. Because subjects in both studies received a similar loading dose for 5 days and then reduced to a typical daily maintenance dose, it is possible that the different results may be related to the additional time for training and supplementation in the first study. In a study by Volek and colleagues (1999), noncompetitive resistance-trained men were given creatine or a placebo for 12 weeks. Although both groups significantly increased their strength, the subjects on the creatine supplementation had an 8% greater improvement in both bench-press and squat strength than subjects on placebo. In addition, it is also likely that recreational athletes have a greater ability to make strength gains without supplementing, compared with experienced resistance-trained athletes who are closer to their genetic strength potential. In the experienced strength athlete, supplementing with creatine may enhance the quality of the workout (less fatigue, enhanced recovery), which may be crucial for providing a greater training stimulus to the muscle.

Most studies that have examined the effect of creatine supplementation on a single bout of explosive exercise (e.g., sprint or jump performance) have not shown any significant performance improvements (Cooke, Grandjean, and Barnes 1995; Dawson et al. 1995; Mujika et al. 1996; Odland et al. 1997; Snow, McKenna, et al. 1998). However, in many of these studies, subjects were supplemented with a loading dose for only 3-5 days (Mujika et al. 1996; Odland et al. 1997; Snow, McKenna, et al. 1998). When subjects are supplemented for an extended duration (28-84 days), significant improvements are seen in jump and power performances (Hafe et al. 2000; Kreider et al. 1998; Volek et al. 1999). It appears that creatine is more effective as a training supplement than a direct performance enhancer. Creatine supplementation may allow the anaerobic athlete to have a better training session, which in turn results in improved performance. It does not appear to be very effective in the short term for improving performance.

Creatine Supplementation and Body Mass Changes

Prolonged creatine supplementation has been generally associated with increases in body weight. Figure 17.3 shows the magnitude of weight gain after varying durations of creatine supplementation. These increases appear to relate primarily to fat-free mass. The mechanism that results in the increase in body mass is believed to be partly related to an increase in total body water. An increase in creatine content within the muscle is thought to increase the intracellular osmotic gradient, causing water to fill the cell (Volek and Kraemer 1996). Increasing the creatine content of the muscle also appears to increase the synthesis of muscle contractile proteins (Balsom et al. 1993; Bessman and Savabi 1990).

Creatine Supplementation and Side Effects

Increases in body mass have at times been referred to as an unwanted side effect (Schilling et al. 2001). However, for many athletes who supplement with creatine, weight gain is often a desired outcome. Side effects are usually potentially debilitating results from a drug or supplement. In regard to creatine supplementation, there have been a host of anecdotal reports of gastrointestinal, cardiovascular, and muscular problems (ACSM 2000b). Muscle cramps have been frequently associated with creatine ingestion. However, no controlled studies have documented any significant side effects from creatine supplementation. Even during prolonged supplementation (10-12 weeks), in either competitive athletes or recreationally trained individuals, no increases in reported side effects were noted in subjects taking creatine (Kreider et al. 1998; Volek et al. 1999).

Another concern for any supplement is the long-term health effects. Schilling and colleagues (2001) performed a retrospective study on health variables in 26 former or current competitive athletes who had taken creatine supplements for up to 4 years. In their study, the only side effect that they could attribute to creatine ingestion was occasional gastrointestinal upset during the loading phase. These disturbances ranged from gas to mild diarrhea. Another major concern of creatine supplementation is kidney strain from the high nitrogen content of creatine and reported increases in creatinine excretion (used to evaluate kidney function, more specifically glomerular filtration rates) during short-term ingestion (Harris, Soderlund, and Hultman 1992). However, no renal dysfunction has been seen from either short-term (5 days) or long-term (up to 5 years) creatine use (Poortmans et al. 1997; Poortmans and Francaux 1999).

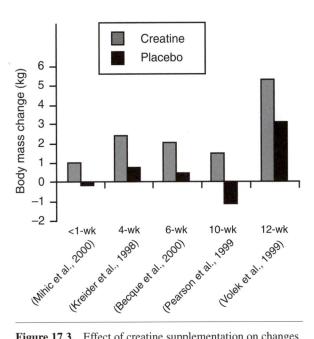

Figure 17.3 Effect of creatine supplementation on changes in muscle mass.

ANABOLIC STEROIDS

Until recently, anabolic steroids were considered to be the most widely used ergogenic aid for enhancing muscle strength and growth. More correctly known as **anabolic-androgenic steroids,** they are synthetic derivatives of the male sex hormone **testosterone.** The physiological actions of testosterone were reviewed in chapter 2. Briefly, its principal anabolic action is to stimulate **protein synthesis,** resulting in large gains in muscle size, body mass, and strength. Its androgenic properties include the full development of the primary sexual characteristics of the male.

Testosterone is secreted primarily in the interstitial cells of Leydig in the testes. Although several other steroid hormones with anabolic-androgenic properties are produced in the testes (e.g., dihydrotestosterone and androstenedione), testosterone is produced in far greater quantities. In addition, testosterone and the other male sex hormones are secreted in significantly smaller amounts from the **adrenal glands** and ovaries. Circulating concentrations of testosterone range from 10.4 to 34.7 nmol $\cdot$ L^{-1} in males and 0.69 to 2.6 nmol $\cdot$ L^{-1} in females (Chattoraj and Watts 1987). The biosynthesis of testosterone, and all other steroid hormones, begins with **cholesterol.** The rate-limiting step for testosterone synthesis is the side-chain cleavage of cholesterol to form **pregnenolone.** The conversion of pregnenolone to testosterone occurs through one of two pathways related to the positioning of a double bond. In the Δ-5 **pathway,** the double bond is between C-5 and C-6, whereas in the Δ-4 **pathway,** the double bond is between C-4 and C-5. In men, the Δ-5 pathway appears to be predominant (Hedge, Colby, and Goodman 1987). The importance of examining the pathway for the biosynthesis of testosterone is related to many of the ergogenic aids on the market today. Many of these ergogenic aids are precursors for testosterone (e.g., androstenedione) and are discussed in further detail later in this chapter.

The physiological changes that testosterone regulates have made it one of the drugs of choice for strength/power athletes or athletes interested in increasing muscle mass.

However, testosterone itself is a very poor ergogenic aid. Rapid degradation is seen when testosterone is given either through oral or parenteral (injectable) administration (Wilson 1988). Thus, it became necessary to chemically modify testosterone to retard the degradation process, thereby maintaining effective blood concentrations for longer periods of time and achieving androgenic and anabolic effects at lower concentrations of the drug (Wilson 1988). Anabolic steroid use through either oral or parenteral administration became possible after these modifications. Examples of commonly used oral and injectable anabolic steroids are listed in table 17.1. Although athletes claim that there are qualitative differences between these drugs, animal studies have not singled out one drug with greater anabolic qualities than the others (Wright and Stone 1993).

Dosing

Exogenous **androgen** administration in physiological dosages (the amount of androgen administered is similar to the concentrations produced by endogenous sources such as the testes and adrenal glands) only serves to shut down the production of androgens in the body through a negative feedback mechanism. Such a dosing schedule simply replaces endogenous production in the body. In addition, in the average man, testosterone receptors within the muscle are generally at or near full saturation from normal androgen production (Wilson 1988). Thus, for exogenously administered androgens to have any ergogenic effect, they need to be given in pharmacological or suprapharmacological dosages. Forbes (1985) has shown that there is a dose response curve of the effect of anabolic steroids on lean body mass (see figure 17.4). The total dose of anabolic steroid ingested has a logarithmic relationship to increases in lean body mass.

Many athletes who take anabolic steroids are reported to use a **stacking regimen** in which several different ste-

Table 17.1   Commonly Used Oral and Parenteral Administered Anabolic Steroids

ORAL COMPOUNDS		PARENTERAL COMPOUNDS	
Trade names	Generic names	Trade names	Generic names
Dianabol	Methandrostenolone	Deca-Durabolin	Nandrolone decanotate
Anavar	Oxandrolone	Delatestryl	Testosterone enanthate
Anadrol	Oxymetholone	Depo-Testosterone	Testosterone cypionate
Winstrol	Stanozolol	Durabolin	Nandrolone phenylproprionate
Maxibolin	Ethylestrenol	Primobolin-Depot	Methenolone enanthate
Halotestin	Fluoxymesterone	Parabolan	Trenbolane acetate

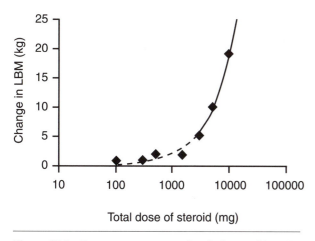

Figure 17.4 Dose response curve of anabolic steroids and changes in lean body mass.
Adapted from Forbes 1985.

roids are taken simultaneously (Alen, Reinila, and Vihko 1985). Most users take these drugs in a cyclic pattern, meaning that they administer the anabolic steroids for several weeks to several months and then alternate with a period of no exogenous ingestion. During the "on" cycles, they may either administer a daily dose that is constant throughout the cycle or follow a pyramid pattern, in which a low dosage is used at the beginning of the cycle and then the dosages are steadily increased over several weeks. Toward the end of the cycle, the athlete may reduce the dosage in a similar step-down routine. The reason for **cycling** anabolic steroids is to reduce the likelihood of negative side effects. The purpose of **pyramiding down** is to slowly reactivate endogenous androgen production, which has been shut down through negative feedback mechanisms. However, these claims have not been confirmed by scientific investigations.

Effect of Anabolic Steroids on Strength and Size Gains

After early reports that anabolic steroids were widely used by Eastern European athletes and were becoming increasingly prevalent in other countries, scientific and medical communities began investigating the effect of anabolic steroids on strength and muscle mass. Initial studies were unable to see any significant differences in strength or body-mass gains in subjects taking anabolic steroids compared with subjects taking a placebo (Fahey and Brown 1973; Fowler, Gardner, and Egstrom 1965; Golding, Freydinger, and Fishel 1974; Loughton and Ruhling 1977; Stromme, Meen, and Aakvaag 1974). As a result, the scientific and medical community at the time suggested that anabolic steroids had little influence on athletic performance. This was contrary to the anecdotal reports emanating from gyms and caused a bit of a cred-

ibility gap between the researchers and athletes. However, examination of these studies revealed several methodological flaws. Several of the investigations used a physiological dosage in contrast to the suprapharmacological dosages typically used by anabolic steroid users. In essence, the subjects were shutting down their endogenous production and replacing it with the exogenous androgens. Another flaw concerned the method of strength assessment. In some of these studies, performance was assessed by a mode of exercise that was different from the training stimulus. This lack of specificity may have masked any possible training effects. In addition, several of these studies used subjects who had minimal resistance-training experience. As discussed earlier, subjects with limited training experience have a large potential for strength gains, negating any need for a performance-enhancing supplement. When anabolic steroids are administered to experienced resistance-trained athletes, in the dosages that are commonly used, results appear to confirm the anecdotal claims of superior performance and size improvements.

The majority of studies examining androgen administration to experienced resistance-trained athletes have reported significant strength and body-mass gains (Alen, Hakkinen, and Komi 1984; Alen, Reinila, and Vihko 1985; Hervey et al. 1981; O'Shea 1971; Stamford and Moffatt 1974; Ward 1973). In such athletes, gains in strength are generally small in comparison with novice lifters, but when they begin an anabolic steroid regimen, the gains in strength may be two- to threefold higher (see figure 17.5) than those of similarly trained athletes who are not supplementing (Hervey et al. 1981; Ward 1973). In these studies, subjects performing a resistance training program for 5 or 6 weeks were administered either the anabolic steroid or a placebo. The strength differences observed were remarkable considering the relatively short training regimen. It should be noted, however, that the athletes in these studies were not elite strength or power athletes. They were experienced resistance-trained, noncompetitive, recreational athletes. Very few investigations have examined anabolic steroid use in competitive strength athletes or bodybuilders. A case study by Alen and Hakkinen (1985) reported on the changes in an elite bodybuilder who was followed for 1 year of self-administration of anabolic steroids. During the phases when the bodybuilder took anabolic steroids, strength levels increased. During cycles in which the athlete trained drug-free, strength levels decreased. Although well-controlled studies using similar subject populations are still lacking, the results reported in that study are consistent with many of the anecdotal reports expressed by these athletes.

When anabolic steroids are given in pharmacological dosages, increases in muscle protein synthesis are seen

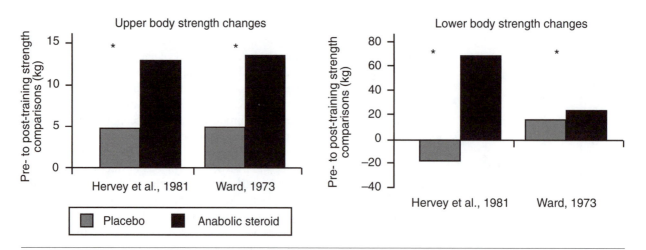

Figure 17.5 Effect of short-term anabolic steroid administration on strength gains. * = significant difference between groups. Adapted from Hervey et al. 1981; Ward 1973.

(Griggs et al. 1989). These increases are likely responsible for the increases observed in body mass and lean body mass in both recreationally trained and competitive athletes given anabolic steroids (Alen, Reinila, and Vihko 1985; Hervey et al. 1981; O'Shea 1971; Stamford and Moffatt 1974; Ward 1973). In a study of competitive power lifters, body mass increased more than 5 kg after 26 weeks of anabolic steroid administration (Alen, Reinila, and Vihko 1985). In comparison, the control group, consisting of both power lifters and bodybuilders, did not see any increase in body mass during the same time period. Although several studies have shown no increase in lean body mass while using anabolic steroids (Crist, Stackpole, and Peake 1983; Fahey and Brown 1973; Fowler, Gardner, and Egstrom 1965), it is generally concluded that resistance training combined with anabolic steroid administration and an adequate diet increases lean body mass (Lombardo 1993).

The principal mechanisms that appear responsible for increasing strength and lean body mass are increases in protein synthesis and inhibition of the catabolic effects of high-intensity training. The anticatabolic activity of androgen administration is reflected by changes in the testosterone to cortisol ratio (**T/C ratio**) (Rozenek et al. 1990). A higher T/C ratio is an indicator of a greater anabolic environment and may give the athlete the ability not only to maintain a higher intensity and volume of training but also to enhance the recovery processes between exercise sessions (Wright and Stone 1993). If the athlete can train harder and for a longer duration, the stimulus that is presented to the muscle will result in a greater physiological adaptation. Thus, the androgens may be considered to have an indirect effect on performance and size gains.

Side Effects Associated With Anabolic Steroid Use

Numerous side effects are reportedly associated with anabolic steroid use. Some of these side effects are listed, in no particular order of importance or prevalence. Many of these risks are based on anecdotal experiences and have been difficult to accurately study. In addition, their severity is diverse, ranging from cosmetic changes (e.g., baldness and acne) to potentially lethal consequences (e.g., liver tumors). Some of the side effects appear to be transient (e.g., alterations to lipid profiles and hypertension), whereas others may become permanent (e.g., deepening of the voice in females). The risk of side effects may also depend on the type of androgen used (Friedl 1993), likely related to their differing metabolites. However, for the most part, the mechanisms causing this variation in side effects from the different androgens are not clear.

Several health risks are associated with anabolic steroid use:

- Decreased HDL (high-density lipoprotein)
- Increased cholesterol
- Increased risk of liver tumors and liver damage
- Gynecomastia
- Elevated blood pressure
- Decreased sperm count
- Acne
- Deepening of voice (females)
- Enlargement of clitoris
- Male pattern baldness
- Facial hair (females)
- Hirsutism

©Mary Messenger

- Testicular atrophy
- Psychotic episodes
- Increased aggression
- Increased risk of AIDS from sharing needles

Among the best-known side effects of anabolic steroid use are the cosmetic changes, which include baldness, acne, and gynecomastia (Friedl 1993). Baldness appears to occur in men that have a genetic predisposition for losing their hair. Acne is often seen as a result of the androgenic effect of increasing sebaceous gland size and secretion rates. Gynecomastia (development of feminine breasts) results when increasing concentrations of testosterone are converted to estrogen.

Psychological and behavioral changes among anabolic steroid users, such as increased aggressiveness, arousal, and irritability, have been reported by various researchers (Pope and Katz 1988; Rejeski et al. 1988; Silvester 1995; Strauss et al. 1983; Yesalis et al. 1990). These psychological effects have been suggested to be both beneficial (e.g., increased arousal, improved self-esteem, general euphoria) and potentially harmful (e.g., aggressiveness, mood swings, increased psychotic episodes). The occurrence of psychological or behavioral changes in anabolic steroid users is not very clear. Silvester (1995) reported that 59% of the athletes interviewed, who were all former anabolic steroid users, subjectively reported behavioral changes (increased aggressiveness and irritability) when taking the androgens. This is similar to the incidence of psychological symptoms reported by Pope and Katz (1988). However, Bahrke (1993) reported that many studies examining the psychological effects of anabolic steroid use have suffered from methodological problems (e.g., small sample size, lack of control group, use of several different steroids, and difficulty recruiting subjects representative of a normal population of steroid users). Wright and Stone (1993) described an unpublished investigation performed by one of their graduate students that showed increased anger and hostility in anabolic steroid users. However, that study also reported that a similar group of strength/power athletes, who weren't taking anabolic steroids, tended to have higher scores for anger and hostility than control subjects. Thus, a cause-effect relationship has yet to be established between anabolic steroid use and psychological changes. In addition, it does appear that individuals who experience some psychological or behavioral changes from anabolic steroid administration recover when steroid use is discontinued (Bahrke 1993).

Anabolic steroids have also been associated with elevated risk of both cardiovascular and hepatic disorders. One of the more frequently reported adverse effects of androgen administration is a decrease in HDL and a possible increase in total cholesterol (Friedl 1993). This is related to androgen stimulation of liver enzymes that regulate serum lipids (Alen and Rahkila 1988). In addition, the liver is a primary target tissue for androgens and the principal site of steroid clearance. Some reports of liver dysfunction have been related to anabolic steroid use (Alen 1985). However, recent evidence indicates that steroid-induced hepatotoxicity may be overstated (Dickerman et al. 1999). The cyclic pattern of anabolic steroid use among athletes has resulted in a transient effect on most of these potential health risks, but the long-term effects associated with these risk factors are not well understood.

TESTOSTERONE PRECURSORS (ANDROSTENEDIONE AND DEHYDROEPIANDROSTERONE)

Testosterone precursors are among the more popular ergogenic aids being used today. The increase in these supplements skyrocketed after reports that baseball player Mark McGwire was supplementing with androstenedione during his record home-run season. Sales for androstenedione, which can be purchased over the counter, increased from about $5 million annually to a projected $100 million annually (Sica and Johns 1999).

The basis for the use of testosterone precursors as an ergogenic aid evolved from a study by Mahesh and Greenblatt (1962) in which healthy women were given 100 mg of either androstenedione or **dehydroepiandrosterone (DHEA).** Both of these precursors resulted in a rapid elevation in testosterone concentrations, with androstenedione causing an approximate threefold greater increase in testosterone than DHEA. Thus, athletes supplementing with these testosterone precursors are hoping to increase testosterone concentrations and achieve performance changes similar to those experienced by individuals taking anabolic steroids. Currently, these testosterone precursors are banned by most professional organizations, including the IOC, NCAA, and NFL. The one notable exception is Major League Baseball. Nevertheless, until recently there was very little research available concerning the efficacy of these androgens as an ergogenic aid. With the increased media attention associated with these supplements, a number of studies have been recently published.

In a recent study, middle-aged men performing a 3-month resistance training program were given either DHEA, androstenedione (100 mg), or a placebo. No significant performance differences (1 RM bench presses and leg presses, aerobic capacity, or body composition) were seen between the groups (Wallace et al. 1999). In addition, no differences between the groups were observed in any of the risk factors normally associated with anabolic steroid use. When DHEA supplementation (150 mg) was examined in a group of younger males (19-29 years) on a 2 weeks on, 1 week off cycle for 8 weeks, no strength or lean tissue gains were seen (Brown et al. 1999). In addition, the investigators were unable to see any changes in serum testosterone, estrone, estradiol, or lipid concentrations in the subjects ingesting the supplement. Even in studies using higher dosages (300 mg) of androstenedione for 8 weeks, in a similar 2 weeks on, 1 week off protocol, no significant effect was observed in strength, muscle size, or testosterone concentrations (King et al. 1999). However, androstenedione supplementation did cause an increase in serum concentrations of estradiol and estrone and a decrease in HDL levels. These results suggest that although performance changes may not occur with athletes taking this supplement, they may be at a higher risk for some of the side effects associated with anabolic steroid use.

These results do not support the claims that testosterone precursors offer similar anabolic and performance effects typically associated with anabolic steroids. In addition, these studies were unable to find any differences in testosterone concentrations following administration of testosterone precursors despite the claims of supplement companies. It should be recalled that these claims are based entirely on a single study that was performed on females (Mahesh and Greenblatt 1962). The effects of testosterone precursors on increasing testosterone concentrations in a male population have not been observed. Nevertheless, it should also be noted that the subjects in these studies were not highly trained athletes. Similar to other anabolic-androgenic supplements provided to noncompetitive individuals, the effects on performance may be varied. In studies examining experienced competitive athletes, the results appear to become more predictable and consistent with the anecdotal reports. Thus, further research is still warranted concerning the efficacy of these testosterone precursors for performance enhancement in a population that is most likely to use this supplement.

CLENBUTEROL

Clenbuterol is a β_2**-agonist** that is generally used to reverse bronchial constriction. However, in recent years it has been used by athletes to increase lean muscle tissue and reduce subcutaneous fat (Prather et al. 1995). This has been based on results from rodent studies that demonstrated increases in muscle protein synthesis (MacRae et al. 1988; Reeds et al. 1986). Though studies in humans are limited, there have been indications of an ergogenic potential of β_2-agonists for strength improvements (Maltin et al. 1993; Martineau et al. 1992). Athletes generally use clenbuterol in twice the recommended daily dosage in a cyclic fashion (3 weeks on alternated with 3 weeks off with a 2 day on, 2 day off cycle during the "on" week) (Prather et al. 1995). It is believed that this cycling regimen will avoid β_2-receptor down-regulation. In contrast to the inhalation route that is often used for relieving bronchial constriction, athletes consume clenbuterol in a capsule form. Although a number of potential side effects from clenbuterol administration have been suggested (e.g., transient tachycardia, hyperthermia, tremors, dizziness, palpitations, and malaise), actual documented occurrences are limited. In addition, the scarcity of data on the ergogenicity of clenbuterol in humans makes it difficult to determine its efficacy.

CAFFEINE/EPHEDRINE

Caffeine is one of the most widely used drugs in the world today. It is found in coffee, tea, soft drinks, chocolate, and various other foods. It is a central nervous system stimulant, and its effects are similar, but weaker, to those associated with amphetamines. Caffeine has been used as a performance enhancer for more than 30 years and is one of the few ergogenic aids that may be equally used by both aerobic and anaerobic athletes. For the aerobic athlete, caffeine is thought to prolong endurance exercise. The mechanisms that have been proposed to cause this effect involve an increase in fat oxidation by mobilizing free fatty acids from adipose tissue or intramuscular fat stores (Spriet 1995b). The greater use of fat as a primary energy source slows glycogen depletion and delays fatigue. During short-duration, high-intensity exercise, the primary ergogenic effect attributed to caffeine supplementation is enhanced power production. This is thought to result from an enhanced excitation-contraction coupling, which affects neuromuscular transmission and mobilization of intracellular calcium ions from the sarcoplasmic reticulum (Tarnopolsky 1994). In addition, caffeine ingestion is thought to enhance the kinetics of glycolytic regulatory enzymes such as phosphorylase (Spriet 1995a).

Initial studies examining the effect of caffeine supplementation on endurance performance reported a 21-min improvement in the time to exhaustion (from 75 min during a placebo trial to 96 min during the caffeine trial) during cycling at 80% of $\dot{V}O_2$max (Costill, Dalsky, and Fink 1978). A number of additional studies demonstrating the ergogenic effect of caffeine during prolonged endurance activity confirmed these results (Essig, Costill, and Van Handel 1980; Graham and Spriet 1995; Ivy et al. 1979; Spriet et al. 1992). These studies have shown that caffeine in doses ranging from 3 to 9 mg · kg^{-1} (equivalent to approximately 1.5-3.5 cups of automatic drip coffee in a 70-kg person) produces a significant ergogenic effect.

Similar performance effects from caffeine have also been demonstrated during short-duration (approximately 5 min), high-intensity exercise (Bruce et al. 2000; Jackman et al. 1996). When the effects of caffeine ingestion on sprint or power performance are examined, the ergogenic benefits become less clear. Several studies have shown that caffeine ingestion does not improve power performance (Collomp et al. 1991; Greer, McLean, and Graham 1998). However, these studies used recreational athletes. When competitive swimmers were examined during two 100-m sprints separated by 20 minutes of rest, caffeine ingestion (250 mg) was shown to improve sprint times 2% and 4% in each sprint, respectively (Collomp et al. 1992). An additional study also demonstrated improved power performance (7% increase, $p < 0.05$) during a series of 6-s sprints (Anselme et al. 1992). The number of studies examining caffeine ingestion on sprint and power performance is limited compared with the volume of work examining its effect on endurance performance. The results of the studies examining the ergogenic benefit of caffeine on power performance are inconclusive. If it does provide any performance benefit, it likely occurs in the trained athlete.

Recently, a group of researchers (Graham, Hibbert, and Sathasivam 1998) examined whether drinking coffee provides the same ergogenic benefits as caffeine tablets. The results of their study suggested that the caffeine in coffee does not provide the same ergogenic benefit as caffeine provided in tablet form. Although no differences were seen in the bioavailability of caffeine from coffee or when ingested alone, it does appear that some compound within coffee prevents the caffeine from enhancing endurance performance.

Many of the side effects of caffeine are well known: anxiety, gastrointestinal disturbances, restlessness, insomnia, tremors, and heart arrhythmias. In addition, caffeine can act as a diuretic and can be dangerous during exercise in the heat. Prolonged caffeine use is also physically addictive and discontinuation can result in some withdrawal symptoms. Caffeine ingestion greater than 9 mg · kg^{-1} appears to result in a greater risk of side effects (Spriet 1995b).

Another β_2-agonist that is reported to be popular, especially among bodybuilders, is ephedrine (Gruber and Pope 1998; Phillips 1997). Ephedrine is thought to have a strong **thermogenic** quality that bodybuilders desire in order to reduce body fat. It is often used as a stacking agent with caffeine to enhance the thermogenic effect (Phillips 1997). Similar to caffeine, ephedrine increases fat oxidation and spares muscle glycogen. Studies examining the ergogenic effect of ephedrine have shown it to be effective only when it is taken in combination with caffeine (Bell, Jacobs, and Zamecnik 1998). These results have consistently shown improved endurance performance (Bell and Jacobs 1999; Bell, Jacobs, et al. 2000). However, side effects (vomiting and nausea) after exercise have been reported in 25% of the subjects ingesting a caffeine/ephedrine mixture of 5 mg · kg⁻¹ caffeine and 1 mg · kg⁻¹ ephedrine (Bell and Jacobs 1999). A subsequent study performed by the same investigators showed that at a lower dosage (4 mg · kg⁻¹ caffeine and 0.8 mg · kg⁻¹ ephedrine), similar ergogenic benefits are realized with no side effects (Bell, Jacobs, et al. 2000). The combined caffeine/ephedrine mixture appears to have a greater benefit than either supplement taken alone.

β-HYDROXY-β-METHYLBUTYRATE

β-Hydroxy-β-Methylbutyrate (HMB) is a derivative of the amino acid leucine and its metabolite α-ketoisocaproate. It is a relatively new ergogenic aid that is achieving some popularity, especially among bodybuilders, who believe that it has both anabolic and lipolytic effects (Phillips 1997). However, there is limited research available to support their claims. Studies that have examined HMB administration demonstrated possible anticatabolic properties of the supplement in both animal and human subjects (Sapir et al. 1974; Tischler, Desautels, and Goldberg 1982). This is reflected by decreased muscle damage and enhanced recovery periods after HMB administration during periods of high muscular stress (Nissen and Abumrad 1997). Studies examining HMB administration and performance in human subjects have been limited. The first study to examine HMB supplementation in humans reported significant strength and lean body-mass increases in previously untrained subjects after 4 weeks of resistance training (Nissen et al. 1996). However, a subsequent study, also studying untrained individuals, was unable to duplicate these results after 8 weeks of resistance training (Gallagher et al. 2000a). Although the results of this latter study did not reach statistical significance, overall improvement of 1 RM strength in 10 exercises increased 32.5% in the placebo group and between 43.5 to 45.5% in two groups of subjects taking different doses of HMB ($p > 0.05$). There was clearly a trend toward performance enhancement in the group taking the HMB supplement. In addition, no adverse effects on hepatic enzyme function, lipid profile, renal function, or the immune system were noted

after 8 weeks of HMB supplementation (Gallagher et al. 2000b). Considering the growing popularity of this supplement, further research is warranted to examine its alleged ergogenic effects and to continue to monitor any adverse results.

SODIUM BICARBONATE

During high-intensity exercise, accumulation of hydrogen ions (H^+), primarily through increases in lactic acid within the muscle, lowers muscle pH. The increase in intracellular acidosis results in a decrease in force production, possibly through the impairment of excitation-contraction coupling and cross-bridge cycling and the inhibition of glycolytic enzymes (e.g., phosphofructokinase) needed for the resynthesis of ATP (Heigenhauser and Jones 1991). In an effort to counter this high level of intracellular acidosis, athletes have attempted to augment the bicarbonate buffering system of the body by ingesting **sodium bicarbonate.**

There has been much variability in studies examining the efficacy of **bicarbonate loading** on improving sprint and power performances. While some investigations have shown a reduced fatigue rate during maximal-effort runs (Costill et al. 1984; Goldfinch, McNaughton, and Davies 1988; Wilkes, Gledhill, and Smyth 1983), an impressive number of studies have also failed to show any efficacy of bicarbonate loading in activity of similar duration and intensity (Gaitanos et al. 1991; Horswill et al. 1988; Katz et al. 1984; Kozak-Collins, Burke, and Schoene 1994; Tiryaki and Atterbom 1995). Several researchers have also examined the ability of sodium bicarbonate to delay fatigue during a resistance exercise session (Portington et al. 1998; Webster et al. 1993). These studies were unable to demonstrate any effect from sodium bicarbonate ingestion on the total number of repetitions performed during 4-5 sets of the leg press exercise (subjects were required to perform approximately 12 repetitions at 70-85% of 1 RM). It should be noted, however, that none of those subjects were competitive weightlifters or bodybuilders. Whether this supplement may be beneficial for competitive weightlifters is not known.

The variability of these studies may be related to differences in the metabolic demands of the training stresses or to differences in the dosing regimen. It appears that a dose of 0.3 g · kg⁻¹ of sodium bicarbonate is necessary to have any beneficial effect during high-intensity performance between 1.0 to 7.5 min (Heigenhauser and Jones 1991). In addition, many of these studies examined noncompetitive athletes. As has often been demonstrated, there is a large degree of variability in the efficacy of ergogenic agents among the noncompetitive subject population.

Several minor side effects are associated with sodium bicarbonate use, primarily gastrointestinal disturbances such as nausea, diarrhea, and bloating. Taking sodium bicarbonate with large volumes of fluid reduces the incidence of these gastrointestinal disturbances (Heigenhauser and Jones 1991). Although there are no long-term side effects associated with its use, one of the physiological effects of sodium bicarbonate ingestion is lowered plasma potassium concentrations. This could potentially result in cardiac arrhythmias (Heigenhauser and Jones 1991).

BLOOD DOPING

Over the past 20 years, many anecdotal reports have emanated from the Olympic Games, various world championships, and other athletic events that endurance athletes have been involved in blood doping. Blood doping involves removing a certain volume of blood, freezing the red blood cells, and then reinfusing them in a saline solution at a later date once the red blood cells return to normal concentrations in the body. The goal of blood doping is to increase the number of red blood cells per unit volume of blood, with the intent of delivering more oxygen to exercising muscles and increasing maximal oxygen consumption. As a result, endurance performance is improved.

Studies examining the ergogenic benefit of blood doping have consistently demonstrated a postinfusion increase in $\dot{V}O_2$max and improved endurance performance (Buick et al. 1980; Robertson et al. 1984; Sawka et al. 1987; Thomson et al. 1983). $\dot{V}O_2$max has been reported to be 5-13% greater in subjects reinfused with 2-4 units of blood (each unit is ~ 450 ml) when compared with control subjects. Spriet (1991) analyzed several studies that examined the effect of blood doping on race performance times. His analysis revealed that reinfusion of 2 units of blood may improve run-times in 2-, 6-, and 10-km races by approximately 7, 30, and 68 s, respectively. Figure 17.6 shows improvements in run-times of various race distances after blood doping. The data from these studies are convincing that blood doping enhances both $\dot{V}O_2$max and endurance performance.

A major concern for athletes who participate in blood doping is the medical risks that they may encounter. As with any transfusion, potential risks include hepatitis and acquired immune deficiency syndrome (AIDS). In addition, athletes may decide to infuse blood that is not their own. It has been reported that certain athletes used homologous transfusions (e.g., transfusions from a family member) during the 1984 Olympics (Spriet 1991). This type of practice could place them at a much higher risk of contracting the previously mentioned diseases. In ad-

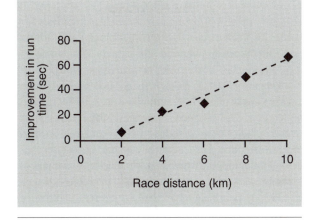

Figure 17.6 Effect of blood doping on run-times.
Data from Goforth et al. 1982; Spriet 1991; Williams et al. 1981.

dition, reinfusion of blood may result in an overload of the cardiovascular system, causing an increase in the viscosity of the blood. A more viscous blood could lead to blood clotting and potentially to heart failure.

ERYTHROPOIETIN

For many endurance athletes, an alternative to blood transfusions is the administration of erythropoietin. Erythropoietin (EPO) is a hormone produced by the kidneys that stimulates red blood cell production. It is this hormone that is responsible for increases in red blood cell volume during exposure to altitude (see chapter 20). Studies of EPO in athletes have been limited. Over a 6-week period, EPO injections increased hemoglobin concentrations 10%, aerobic capacity 6-8%, and time to exhaustion up to 17% (Ekblom and Berglund 1991). Four months before the study, 7 of the 15 subjects investigated had reinfused red blood cells (undergone blood doping). Interestingly, the researchers reported that the increase in aerobic capacity and run-time to exhaustion were almost identical.

The risks of EPO appear to be serious. The deaths of a number of racing cyclists have been related to EPO administration (Gareau et al. 1996). However, this allegation has not been confirmed. The primary risk associated with EPO is its lack of predictability when compared with red blood cell infusion (Wilmore and Costill 1999). Once EPO has been injected into the body, the stimulus for producing red blood cells is no longer under control. The risks are similar to blood doping. The athlete may be at a greater risk for increasing the viscosity of the blood, possibly leading to blood clotting and heart failure.

β-BLOCKERS

β-blockers are a class of drugs that block the β-adrenergic receptors, preventing the catecholamines (e.g., norepinephrine and epinephrine) from binding. β-blockers are generally prescribed by cardiologists to treat a wide variety of cardiovascular diseases, including hypertension. The ergogenic benefit of these drugs may reside in their ability to reduce anxiety and tremors during performance. Thus, athletes who rely on steady, controlled movements during performance (e.g., archers or marksmen) would appear to benefit from these drugs. In addition, β-blockers have been suggested to improve physiological adaptations to endurance training by causing an up-regulation of β-receptors (Williams 1991). This may result in an exaggerated response to sympathetic discharge (e.g., improved contractile response) during intense exercise once the drugs are discontinued.

Several studies have shown that β-blockers improve both slow and fast shooting accuracy (Antal and Good 1980; Kruse et al. 1986; Tesch 1985). In addition, the dose taken appears to have significant effects on the magnitude of improvement. In shooters administered β-blockers in two different doses (80 mg versus 40 mg of oxprenolol), the group taking the higher dosage shot with greater accuracy (Antal and Good 1980). However, in certain sports some degree of anxiety may be important. Tesch (1985) reported that bowlers whose performance was improved during blockade with oxprenolol (a beta-blocker drug) had significantly greater heart rates before, during, and after competition compared with the subjects whose performance did not improve while on β-blockers.

β-blockers may also have an **ergolytic** effect (reduce performance). Studies have shown that β-blockers impair the cardiovascular response to exercise by reducing oxygen and substrate delivery to exercising muscles (Williams 1991). Risks associated with these drugs include bronchospasm in individuals with asthma, lightheadedness (due to decreases in blood pressure), increased fatigue, and hypoglycemia in type 2 diabetics (β-blockers can increase insulin secretion).

SUMMARY

The use of ergogenic aids is prevalent among athletes at all levels and in most sports. Many of these athletes appear to begin their supplementation routines at the high school level. This chapter reviewed a number of the more popular supplements being used today. Several have been well researched, whereas only scant information is available about others. Many of the problems associated with studying ergogenic aids are related to the difficulty of conducting studies with dosages that are commonly used by athletes. Most institutional review boards do not allow research to proceed with substances in dosages that exceed the FDA's recommendations or may potentially place the subject at risk. In addition, many researchers do not have access to the elite-level athletes who are generally using the supplements. Most of the research is performed on recreationally trained subjects, which results in a great deal of variability in the outcome. Consequently, studies have concluded that a particular agent may not have any ergogenic benefit, but this unfortunately may not be applicable to the population that is actually using the supplement.

PART IV

ENVIRONMENTAL FACTORS

CHAPTER 18

EXERCISE IN THE HEAT

©Human Kinetics

During exercise, the **metabolic rate** may increase 5-15 times higher than that seen at rest to provide energy for muscle contraction. This higher metabolic rate creates a large heat production (approximately 80% of all energy that is metabolized is transformed to heat) that needs to be dissipated in order to maintain heat balance within the body. This places a large burden on the **thermoregulatory system** to defend the **core temperature** of the body. When exercise is performed in the heat, the strain experienced by the thermoregulatory system is further magnified. The extent of the physiological stress depends on several factors, including the individual's heat **acclimatization,** level of physical fitness, and **hydration** state. These factors affect the athlete's ability to perform. Under severe environmental conditions, exercising in the heat poses a significant risk to the health and well-being of the athlete.

This chapter reviews the physiological response to exercise in the heat. Further discussion focuses on the factors that influence physiological strain when exercising in the heat and their potential impact on athletic performance. In addition, the heat illnesses are reviewed and mechanisms used to monitor **heat stress** are also examined. Finally, the chapter discusses several necessary precautions the athlete should take to minimize the risk of heat injury.

PHYSIOLOGICAL RESPONSE TO EXERCISE IN THE HEAT

During exercise in the heat, the core temperature of the body increases rapidly. This correlates to the increased metabolic heat that is generated when exercising in hot conditions. Until heat loss equals heat production, core temperature will continue to rise. Once heat loss equals heat production, a steady-state core temperature is achieved (Sawka et al. 1993). The primary goal of the thermoregulatory system during exercise in the heat is to remove the heat from the body. If **hyperthermia** (high body temperature) is not controlled, body temperature could rise to a level that places the individual in grave danger. Thermal signals in the hypothalamus sense increases in the core temperature of the body. The thermal regulatory center in the hypothalamus interprets these signals as an imbalance in temperature **homeostasis** and activates a defense mechanism to return core temperature to normal levels. Sweat glands are stimulated to increase sweat production, and the smooth muscle in the arterioles of the skin are relaxed to allow for **vasodilation** and an increase in blood flow to the periphery. The brain also diverts blood flow from internal organs to the skin to expedite heat **dissipation.**

How Does the Body Dissipate Heat

Body heat can be dissipated to the environment though **evaporative** or **nonevaporative** means. Evaporative heat loss involves the evaporation of water (the result of sweating) from the body. Evaporation accounts for approximately 85-90% of the heat dissipation during exercise in a hot, dry environment (Adams et al. 1975). When the exercise environment is hot and wet (humidity above 50-70%), evaporative heat loss is reduced and a greater amount of heat is stored in the body. During hot, wet conditions, the skin becomes hot and reddish in color, which represents the increased blood that has been diverted to the periphery. The body must rely more on nonevaporative mechanisms to dissipate heat. Nonevaporative heat loss is the result of the combined effects of **conduction, radiation,** and **convection.** Conduction is heat exchange between two solid surfaces that are in direct contact. The rate of conductance depends on the temperature difference between the two surfaces. Conduction may have limited value during exercise in the heat because of the limited surface area that is in direct contact with the ground. It accounts for less than 2% of heat loss in most situations (Armstrong 2000). Conduction is easily illustrated using the example of a person camping out during the winter months. When sleeping on a snow-covered ground, a significant heat exchange occurs between the sleeping camper and the ground. In the morning, the camper wakes up lying in a depression created by body heat melting the packed snow beneath the tent floor.

Radiation is the transfer of energy waves that are emitted by one object and absorbed by another (Armstrong 2000). Convection is heat exchange that occurs between a surface and a fluid medium. Both air and body fluids can dissipate heat through convective means. As mentioned previously, evaporative heat loss is severely reduced during exercise in a hot, wet environment. The body relies more on radiation and convection to dissipate heat under these conditions. Figure 18.1 shows the possible ways that the body regulates heat exchange during exercise.

Unlike evaporative cooling, which always dissipates heat, the ability of radiation and convection to lower body heat depends on the ambient temperature. Skin temperatures generally range from 93 to 98°F (34 to 37°C); if air temperature is 100°F (38°C) or greater, heat would likely be added to the body. During conditions in which both ambient temperature (>100°F [38°C]) and humidity are high, the ability to dissipate heat through either evaporative or nonevaporative mechanisms becomes limited. In such conditions, the athlete will have difficulty dissipating body heat and will be at an increased risk for heat illness.

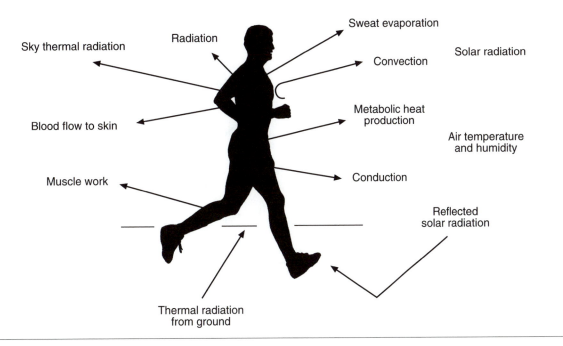

Figure 18.1   Regulation of heat exchange.

Cardiovascular Response to Exercise in the Heat

During exercise in the heat, the volume of blood that is diverted to the skin may reach as high as $7 \text{ L} \cdot \text{min}^{-1}$ (Rowell 1986). This increase in blood flow to the periphery enhances heat dissipation through convection and radiation (a higher blood flow to the periphery increases skin temperature). As blood flow to the skin is increased, the blood vessels of the skin become engorged, creating blood pools in the skin (Sawka et al. 1993). The large volume of blood that remains in the periphery causes a reduced **venous return** and less cardiac filling. The reduced cardiac filling causes a cardiovascular strain, reflected by a reduced stroke volume. As a result, **cardiac output** is lowered. To compensate, the heart rate must increase to maintain cardiac output. In addition, blood flow from splanchnic and renal areas is reduced to compensate for the greater blood flow that is diverted to the exercising muscles and periphery (Rowell 1986). The effect of exercise in the heat on blood distribution is depicted in figure 18.2.

During exercise in the heat, **sweat rate** is increased to enhance evaporative cooling. Sweat rates vary from athlete to athlete; a rate of $1 \text{ L} \cdot \text{h}^{-1}$ is common (Sawka et al. 1993). The highest sweat rate reported in the literature is $3.7 \text{ L} \cdot \text{h}^{-1}$ measured for Alberto Salazar (Armstrong et al. 1986). As sweat loss accumulates, a decrease in blood volume and an increase in plasma tonicity is seen (Sawka and Pandolf 1990). The decrease in blood volume results in a reduction in skin blood flow and a decrease in sweat output. As a result, the ability of the body

to dissipate heat is reduced. Consequently, core temperature rises and, in combination with a reduced blood volume, increases both cardiovascular strain and the risk of heat illness. The importance of maintaining proper hydration levels during exercise in the heat cannot be overestimated. Dehydration exacerbates the physiological strain during exercise in a hot environment. The importance of hydration was reviewed in chapter 16.

HEAT AND PERFORMANCE

Exercising in the heat has significant effects on athletic performance. However, the magnitude and direction of the effect (degree of improvement or decrement) are dependent on the extent of hyperthermia as well as the mode of activity being performed.

Aerobic Exercise in a Hot Environment

When exercise is performed in the heat, the **maximal aerobic capacity ($\dot{V}O_2$max)** of the athlete has been reported to be lowered (Sawka, Young, Cadarette, et al. 1985; Smolander et al. 1986). A reduced $\dot{V}O_2$max appears to occur regardless of the individual's state of heat acclimatization or physical fitness (Sawka, Young, Cadarette, et al. 1985). Not only is **aerobic power** reduced during exercise in the heat, but the time to reach exhaustion is also reduced (MacDougall et al. 1974).

The physiological mechanisms responsible for these performance decrements are likely related to the increased blood flow diverted to the peripheral vasculature. As

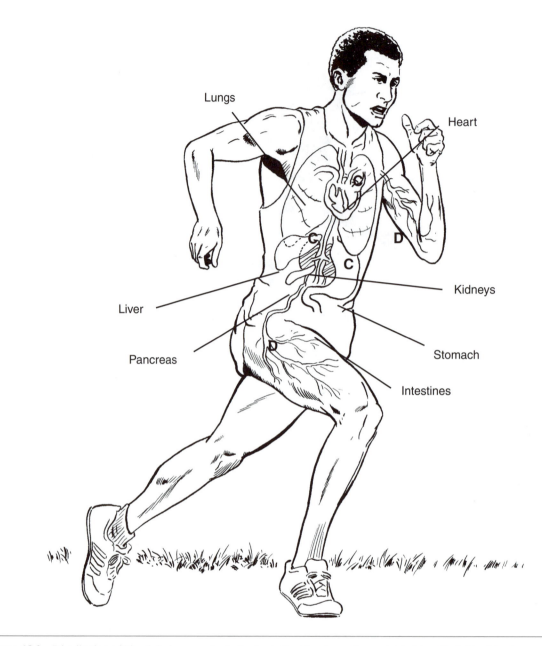

Figure 18.2   Distribution of blood during exercise in the heat. C = blood vessels constricted (smaller); D = blood vessels dialated (larger).

mentioned in the previous section, cardiac output may be lowered because of the reduction in venous return. Cardiac output during exercise in the heat has been reported to be $1.2\,L \cdot min^{-1}$ below that seen during exercise in a temperate environment (Rowell 1986). Dilation of the peripheral vascular beds may also result in a diversion of some blood from the exercising muscles to the skin (Sawka and Young 2000). This would also contribute to the reduced cardiac output available to the muscles, affecting the metabolism of muscle contraction.

During submaximal exercise in the heat, an increase in core temperature causes a shift in metabolism from primarily aerobic to **anaerobic** in the exercising muscles and the liver (Dimri et al. 1980; Young et al. 1985). Using postexercise oxygen-uptake measures, Dimri and colleagues (1980) showed that as ambient temperature increases, metabolic rate also increases, with a larger percentage being derived from anaerobic sources. Young and colleagues (1985) provided further evidence supporting a greater reliance on anaerobic metabolism by demonstrating a greater lactate concentration in both muscle and blood after exercise in the heat. The greater reliance on anaerobic metabolism indicates a greater reliance on carbohydrate stores within the body to fuel exercise. This likely plays a significant role in the quicker rate of **fatigue** seen during exercise in the heat.

Anaerobic Exercise in a Hot Environment

The relatively few studies that have examined anaerobic exercise in the heat have generally shown either no change (Dotan and Bar-Or 1980; Stanley et al. 1994) or an increase in strength or power performance (Falk et al. 1998; Sargeant 1987). Sargeant (1987), using a warm-water (111°F [44°C]) bath immersion, reported significantly greater maximal power performance; however, the increase in maximal power was accompanied by an increase in the rate of fatigue. In contrast, Falk and colleagues (1998), using a study design that simulated a competitive situation (bouts of high-intensity exercise [five 15-s Wingate Anaerobic Tests] interspersed with relatively short **recovery** periods [30 s between each bout]) in the heat (95°F [35°C], 30% relative humidity), reported an increase in power performance but no difference in fatigue rate. The improved power production during exercise in the heat may be related to a warm-up effect. The higher ambient temperatures likely cause a greater increase in muscle temperature. An increase in muscle temperature of 9°F (5°C) has been hypothesized to cause a 10% increase in power development (Binkhorst, Hoofd, and Vissers 1977) in some studies, whereas others have suggested that for each 1.8°F (1°C) rise in muscle temperature, a 4% improvement in power performance may be seen (Sargeant 1987). The higher power performances in the heat may be related to an increase in speed of muscle contraction or a faster metabolic rate (Falk et al. 1998).

In the study by Falk and colleagues, a high ambient temperature did not appear to affect recovery from exercise compared with exercise performed in a thermoneutral condition (72°F [22°C], 40% relative humidity) (see figure 18.3). In that study, subjects performed a series of five bouts of 15 s of the Wingate Anaerobic Test (WAnT) separated by 30 s of active recovery (cycling with no resistance). The 3 min and 15 s exercise/rest period was followed by 60 min of passive recovery in the heat. After the recovery period, another series of five 15-s WAnTs was performed without any observed decrement in performance. The ability to maintain anaerobic power performance in a hot environment may have been related to the ability to keep the subjects euhydrated during the recovery period.

HEAT ACCLIMATIZATION

During initial exposure to the heat, the exercising individual experiences weakness, dizziness, flushed skin, and other signs and symptoms that represent heat stress. However, after several days of heat exposure, a reduction in these symptoms occurs and the heat tolerance of the individual is improved. This is a result of a number of physiological adaptations that improve the thermoregulatory function of the body. When such adaptation occurs in a naturally hot environment, it is termed *acclimatization*. However, if it is accomplished in an artificial environment, such as one that might be created by exercising in a heat chamber in the middle of the winter, it is termed *acclimation*. Both acclimatization and acclimation produce similar physiological adaptations; thus, acclimatization is the inclusive term used to describe both methods (adaptation occurring in either a natural or laboratory setting).

Acclimatization to heat occurs after repeated exposures to a heat stress that is sufficient to raise core body temperature and bring about moderate to profuse sweating (Wenger 1988). This is best accomplished through exercising in the heat at an intensity ranging from 50 to

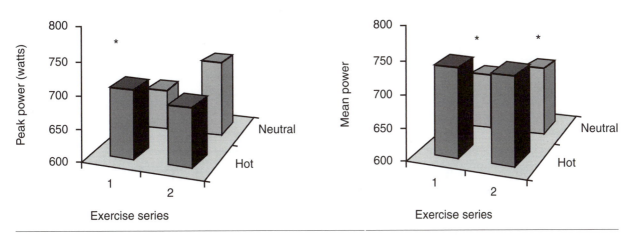

Figure 18.3 Effect of hot, dry conditions on anaerobic power performance. * = significant difference between thermal conditions.

Figure based on data from Falk et al. 1998.

95% of $\dot{V}O_2$max for approximately 1 h a day. Heat acclimatization appears to occur in two stages. After a few days of heat exposure, several adaptations are noted, primarily the result of a reduction in the cardiovascular strain (e.g., reduced heart rate) at a relative level of exercise in the heat (Armstrong and Maresh 1991; Wenger 1988). The reduced cardiovascular strain is also associated with an increase in both plasma volume and exercise tolerance and a decrease in core body temperature and perceived level of exertion. As exposure to the heat is prolonged (up to 14 days), further adaptations may be observed. An increase in sweat rate and sweat sensitivity (i.e., sweat loss expressed for degree rise of core body temperature), as well as a decrease in **electrolyte** losses in both sweat and urine, may also occur. However, these latter adaptations may be dependent on the environmental conditions (Armstrong et al. 1987). Table 18.1 summarizes the adaptations commonly seen during heat acclimatization.

Sweat rate is different when heat acclimatization occurs in a dry versus a humid environment. During dry heat exposure, little to no change in sweat rate is seen; however, during acclimatization in hot, humid conditions, increases in sweat rate do occur (Armstrong and Maresh 1991; Wenger 1988). Under hot, dry conditions, there is a high rate of evaporative cooling. However, in a hot, humid environment, in which ambient water vapor pressure is high, the body needs to increase the wetted skin area to achieve a similar cooling effect as might be seen under hot, dry conditions. The body accomplishes this either by increasing sweating in areas that did not produce much sweat before acclimatization or by increasing the sweating intensity (more profuse sweating)

(Wenger 1988). A greater sweat rate at body surface areas that already produce a large sweat concentration would have a minimal effect on whole-body cooling, because much of the increase in sweat production would be wasted in dripping. In order to maximize efficiency of adaptation, the body selectively increases sweating in areas that previously had poor sweat rates.

Questions about heat acclimatization frequently concern the ability to become acclimatized without exercising (passive heat acclimatization) or whether a high aerobic capacity provides an inherent protection when exercising in the heat. Individuals native to a hot or tropical climate are reported to have a lowered core temperature and a more efficient sweating response to exercise (Wenger 1988). However, for individuals who are not native to such a climate, heat exposure without exercise also induces a heat acclimatization response that is reflected by an improved ability to dissipate heat. Yet to maximize heat acclimatization, the athlete must exercise in the heat. High-intensity exercise was thought to be necessary to increase exercise heat tolerance (Gisolfi and Robinson 1969; Shvartz et al. 1977). However, ample evidence demonstrates comparable heat tolerance when exercise is performed at mild to moderate exercise intensities (>40% of $\dot{V}O_2$max) (Armstrong and Maresh 1991; Houmard et al. 1990). Full heat acclimatization is seen after 14 consecutive days of exercising in the heat, and partial heat acclimatization can be seen after 4 days of exercising in the heat even at a mild exercise intensity (47% of $\dot{V}O_2$max) (Hoffman et al. 1994).

This ability to partially heat acclimatize after only a few days of exercise in the heat has an important practical benefit. As an example, consider a football team based

Table 18.1 Physiological Adaptations to Heat Acclimatization[1]

Adaptation	Days of heat acclimatization
	1–14
Heart rate decrease	(days 4–6)
Plasma volume expansion	(days 4–6)
Rectal temperature decrease	(days 5–7)
Perceived exertion decrease	(days 4–6)
Sweat Na+ and Cl- concentration decrease[2]	(days 7–11)
Sweat rate increase	(days 9–14)
Renal Na+ and Cl- concentration decrease	(days 4–7)

[1]Point at which approximately 95% of the adaptation occurs
[2]While consuming a low NaCl diet

Adapted from Armstrong and Dziados 1986.

in the northern or eastern United States. The team's next game is to be played in a warm climate (e.g., Florida or Arizona) in the middle of November. The players' ability to tolerate the heat will be a strong factor affecting the outcome of the game. This poses a tremendous advantage for the opponent. However, several possible solutions may be tried to partially acclimatize the athletes to enhance their heat tolerance for the game. With only a week to prepare, it is not possible to attain full heat acclimatization. The first potential solution involves flying the team down early in the week to practice in the opposing city. Another potential solution is to practice indoors (if an indoor facility is available) and increase the heat during practice to simulate the climate that is expected at the game. If this is not possible, perhaps the players could be heat acclimated in a laboratory facility that has a heat chamber or some other facility where room temperature could be controlled. Within several days of exercising in the heat, the players will have a greater tolerance. This minimizes the advantage that the warm-weather team would potentially have had.

A high level of physical fitness may provide some benefit for tolerating a heat stress but does not appear to substitute for exercising in the heat to produce heat acclimatization. Most researchers agree that a high level of physical fitness improves the physiological responses to exercise in the heat (Armstrong and Pandolf 1988; Armstrong and Maresh 1991). Although a high fitness level is not as beneficial as full heat acclimatization, it may enhance the rate at which full or partial heat acclimatization can be reached (Pandolf, Burse, and Goldman 1977). Several studies have shown that intense training for 2 weeks in a temperate environment does not produce the same heat-tolerance level as exercising in the heat (Gisolfi and Cohen 1979; Strydom and Williams 1969). However, an interesting study by Armstrong and colleagues (1994) showed that exercise heat tolerance could be maintained in highly trained endurance athletes through the winter months as long as rigorous training was continued. Based on other studies, the heat tolerance of highly trained endurance athletes may be further improved if they undergo a heat acclimation regimen (Gisolfi 1973; Piwonka et al. 1965).

HEAT ILLNESSES

Heat stress places an added burden on the cardiovascular system to maintain thermoregulation. However, when exercise is performed in the heat, the control of hemodynamic stability to maintain exercise performance may take precedence over thermoregulation, even to an extent that places the individual at great risk of heat illness (Hubbard 1990). In other words, during exercise, blood flow to active muscles will not be diverted to peripheral tissues to reduce the risk of heat injury. Four heat illnesses are commonly seen among athletes: **heat cramps, heat exhaustion, exertional heat stroke,** and **heat syncope.** Although some authors have suggested that these heat illnesses may exist in a continuum, it is possible for an individual to exhibit symptoms of any of these heat illnesses without manifesting symptoms of others (Armstrong and Maresh 1993). The incidence of heat illnesses is reduced when the athlete becomes acclimatized to the heat (Armstrong and Maresh 1991).

Heat Cramps

During prolonged or repetitive exercise, the athlete may experience painful contractions within the exercising muscles. These heat cramps are believed to be the result of a large loss of Na^+ and Cl^- in sweat and the replacement of sweat loss with dilute fluid, pure water, or both (Armstrong 2000). Heat cramps should not be confused with exercise-induced muscle cramps. Heat cramps are usually observed in the large muscles of the extremities or in the abdomen. They usually begin as a weak tingling sensation and progress to a localized contraction of several muscle fibers that appear to wander over the muscle as adjacent motor units become activated. The entire muscle mass is generally not affected by heat cramps, unlike exercise-induced muscle cramps in which the entire muscle appears to be affected.

Although the exact mechanisms that induce heat cramps are not clear, it is thought that the loss of NaCl causes a shift in the intracellular/extracellular NaCl and water ratio. This in turn results in an alteration of the electrical properties of the muscle membranes and could alter muscle contraction or relaxation (Armstrong 2000). The importance of electrolytes in both the development and treatment of heat cramps is becoming clearer. A significant reduction of NaCl in the urine of individuals with heat cramps has been reported (Leithead and Gunn 1964) and appears to suggest that the kidneys are reabsorbing sodium because of the whole-body electrolyte imbalance. Further evidence of the involvement of electrolytes has been demonstrated in several recent reports that have shown a rapid and permanent relief of heat cramps with the consumption of fluid-electrolyte beverages (Armstrong and Maresh 1993; Bergeron 1996).

The basic treatment for heat cramps is to restore homeostatic equilibrium by replacing both water and electrolytes. IV solutions are effective and result in rapid relief. Oral NaCl solutions are also a common treatment and involve dissolving two 10-grain salt tablets into 1 L of water. Preventive measures may include consuming salt in the diet or isotonic electrolyte solutions during exercise.

Heat Exhaustion

Heat exhaustion is the most common form of the heat illnesses. It is considered a volume depletion problem and primarily occurs when cardiac output is reduced, resulting in an inability of the cardiovascular system to meet the demands of both the exercising muscle and peripheral tissues. It is defined by an inability to continue exercise in the heat (Armstrong and Maresh 1993; Hubbard and Armstrong 1989). Symptoms may include various combinations of nausea, vomiting, irritability, headache, anxiety, diarrhea, chills, piloerection, hyperventilation, tachycardia, hypotension, and heat sensations in the head and upper torso (Armstrong et al. 1987). Orthostatic intolerance (dizziness) and syncope (fainting) may also occur. Rectal temperature is usually less than 40°C. The signs and symptoms of heat exhaustion may be quite variable and may differ under different exercise-heat scenarios (Armstrong and Maresh 1993).

Since heat exhaustion is the result of volume depletion (e.g., plasma) resulting in inadequate cardiovascular compensation, simply rehydrating will often result in complete recovery within 15-30 minutes (Hubbard and Armstrong 1988). In most cases, fluids can be replenished orally; however in severe cases of heat exhaustion in which the subject is unconscious or is lacking in mental acuity, IV solutions may be the preferred method of fluid replenishment.

During heat exhaustion volume depletion can be attributed to either a salt depletion or water depletion. Often it will be a combination of the two forms with the signs and symptoms being similar for each. However, the treatment for each form of volume depletion will be different. Water depletion occurs quite rapidly and may be the result of a single exercise session. In such a case fluid replenishment, which may include only drinking plain water, would be quite effective in restoring the body to its normal euhydrated state. However, salt depletion is generally the result of a prolonged electrolyte deficit (approximately 3-5 days) brought about by excessive water loss in sweat that was unable to be replenished through dietary salt intake (Armstrong 2000). In a salt depletion situation it is imperative that an appropriate amount of NaCl be included with fluid intake. Armstrong (2000) suggests that salt and water losses can be estimated as 2 g of NaCl and 1.5 L of water per hour of continuous moderate to heavy exercise. Appropriate adjustments to IV solution or oral solution should be made accordingly.

Exertional Heatstroke

Unlike heat exhaustion, in which volume depletion leads to a progressive fatigue and an inability to continue exercise, heatstroke is one of the few illnesses that can threaten the life of a healthy athlete. A rectal temperature greater than 102°F (39°C) and an increase in serum enzymes (e.g., ALT, AST, CPK, and LDH), indicative of cellular damage, are distinguishing characteristics that differentiate heat exhaustion from the more severe and life-threatening exertional heatstroke. In addition, during heat exhaustion, spontaneous body cooling may still be present or perhaps variable in a severe case; however, during heatstroke, spontaneous body cooling is absent (athlete will have hot, dry skin). Another distinguishing characteristic of heatstroke is the bizarre behavior exhibited by the victim. A person who has exertional heatstroke may be disoriented with complete loss of mental acuity or, in a most severe case, may become comatose.

The mortality rate or the extent of multisystem tissue damage is closely related to the patient's core temperature at the point of collapse and the time elapsed between the patient's collapse and the initiation of cooling. If cooling therapy is initiated rapidly, the risk of death or disability is dramatically reduced (Costrini 1990). The method of cooling also affects the recovery rate of the heatstroke patient.

The primary goal of cooling therapy is to lower the core body temperature to a safe and non-life-threatening level (99°F [37°C]). A number of different cooling methods have been used to lower core body temperature, including ice packs placed at the groin, armpits, and neck; ice packs covering the body; air and water sprays at both warm or cool temperatures; and ice-water or cool-water immersion. Some concern has been raised that ice- or cool-water baths may cause a shivering-induced heat production or vasoconstriction of peripheral blood vessels, which would not permit cooling (Yarbrough and Hubbard 1989). In addition, ice- or cool-water baths may hinder specialized treatment (e.g., heart defibrillation) or cause potentially unsanitary conditions (e.g., vomiting or diarrhea in the water) (Armstrong 2000). Nevertheless, Armstrong and colleagues (1996) reported that these issues are minimal or nonexistent in the vast majority of heatstroke cases. The rapid cooling observed with immersion was demonstrated to be the most effective cooling technique and such benefits were seen without any of the previous concerns becoming an issue.

Heat Syncope

Heat syncope (fainting) is rare in a conditioned athlete. In most instances, it occurs when individuals stand for a prolonged period of time in the heat or exercise for a prolonged period in an upright position. Heat syncope is caused by a pooling of blood in the vasculature of the limbs and skin because of excessive ambient temperatures. In response to the hot environment, the cutaneous vesicles of the skin dilate to allow for greater cooling.

The increase in vasodilation reduces the volume of blood that is returned to the heart, which decreases cardiac output and lowers blood pressure. Blood flow to the brain is therefore reduced, resulting in a syncopic episode.

The medical diagnosis of heat syncope is based on a fainting spell with the absence of an elevated rectal temperature (Armstrong 2000). Before the syncopic episode, the patient may experience nausea, weakness, tunnel vision, or vertigo. Treatment for heat syncope is to replace any fluid and electrolyte deficits and have the patient lie in a horizontal position with the feet elevated. The horizontal position allows for a greater venous return to the heart. Subsequently, cardiac output and blood pressure increase, resulting in a return of normal blood volume to the brain.

MONITORING HEAT STRESS

When exercise is performed in the heat, it is imperative to take the necessary precautions to minimize the athlete's risk of heat illness. One of the first steps is to monitor the environment. Weather reports that provide both temperature and humidity for the geographical region of concern help assess the risk of heat illness during training or competition (see figure 18.4). The risk of heat illness is determined by the intersection of both temperature and relative humidity. If the intersection of temperature and relative humidity lies in the "moderate risk" zone, the athlete should be monitored for signs and symptoms of heat exhaustion or exertional heat stroke (see table 18.2). If the intersection lies in the "high risk" to "very high risk" zones, exercise needs to be revised. It may be necessary to postpone exercise until weather conditions become less dangerous or, at the very least, reduce the distance or duration of the workout (Armstrong 2000).

It is possible to monitor the environment at the specific site of activity. The wet bulb globe temperature (WBGT) is a widely used heat stress index that can be measured at the site of activity and used to indicate the level of environmental stress (Armstrong 2000). It consists of a dry bulb temperature (measuring ambient air temperature), a wet bulb temperature (dry bulb thermometer measured under a water-saturated cloth wick), and a black globe temperature (dry bulb thermometer placed inside a black metal sphere). It provides both ambient temperature and relative humidity at the site of activity. The importance of this apparatus over a regular thermometer should not be underestimated. Ambient temperature alone is not sufficient to determine heat stress. A dry bulb

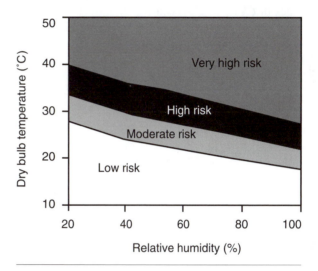

Figure 18.4 Risk of heat exhaustion or heatstroke in hot environments.
Reprinted from Armstrong et al. 1996.

Table 18.2   Warning Signs for Heat Exhaustion and Exertional Heatstroke

Heat exhaustion	Exertional heatstroke
Headache, irritability	Headache
Tingling or numbness on head, neck, back, or limbs	Unconsciousness, coma*
Chills or shivering	Loss of mental clarity*
Great fatigue*	Bizarre behavior*
Rapid, weak pulse	Rapid, strong pulse
Pale, moist, cool skin	Hot, red skin*
Dizziness*	Profuse sweating in most cases
Vomiting, nausea	Fainting
Dehydration*	

* = Important, these occur in most cases
Adapted from Armstrong 2000.

temperature accounts for only 10% of the heat stress index, whereas the wet bulb temperature (indication of relative humidity) accounts for 70% of the heat stress index (Yaglou and Minard 1957). The WBGT index is as follows:

$$WBGT = (0.7 \text{ temp wet bulb}) + (0.2 \text{ temp black globe}) + (0.1 \text{ temp dry bulb})$$

The result of this index corresponds to the four categories seen in figure 18.4. A WBGT index above 82°F (28°C) places the athlete at a very high risk for heat illness. A WBGT index between 73-82°F (23-28°C) places the athlete at a high risk for heat illness. A WBGT index between 64-73°F (18-23°C) places the athlete at a moderate risk, and a WBGT index below 64°F (18°C) is considered a low risk for heat illness. Although a heat stress index of 63°F (17°C) is considered a low risk for heat illness, there is no guarantee that the athlete will experience heat exhaustion or exertional heatstroke. A number of different factors (e.g., sleep deprivation, hydration status, and diet) may interact to increase the individual's risk of heat illness, even at a relatively low heat stress index (Armstrong, De Luca, and Hubbard 1990; Epstein 1990).

SUMMARY

When exercise is performed in the heat, an emphasis is placed on removing heat and maintaining the core temperature of the body. This strains the cardiovascular system, which must distribute blood to the exercising muscles and to the peripheral blood vessels to help cool the body. The adjustments that the body needs to make may have a negative effect on exercise performance (primarily aerobic) and may also place the athlete at a greater risk of heat illness. If the body is unable to control increases in core temperature, the resulting hyperthermia may have lethal consequences for the athlete. Although heat acclimatization causes physiological adaptations that improve the heat tolerance of the athlete, it will not prevent heat illnesses if the heat stress is severe. Regardless of the level of heat acclimatization, an athlete who is not properly prepared for exercising in the heat may still be susceptible to heat illnesses. The following recommendations may help minimize the risk of heat illnesses for individuals preparing to exercise in the heat:

- Before exercise, determine the heat stress index and make appropriate adjustments to the workout.
- Wear lightweight, loose-fitting clothing.
- Exercise at cooler times of the day.
- Avoid long warm-ups on hot, humid days.
- Ensure proper hydration before exercise (body weight should be within 1% of average body weight).
- Ensure proper hydration during exercise by replacing fluids lost through sweating (drink 1 L of water for every kg body weight lost during exercise).
- Ensure proper salt intake at meals.
- Know the signs and symptoms of heat illnesses.

CHAPTER 19

EXERCISE IN THE COLD

©Bob Firth

In contrast to heat illnesses, many of the cold-induced injuries may be eliminated through behavioral regulation during exercise. The ability of humans to insulate themselves from a harsh environment serves as a protective measure against the dangers of the cold. Technological advancements in the design of clothing allow individuals to insulate themselves in a manner that permits ease of movement while causing only minimal increases in energy costs. Nevertheless, athletes may still be subjected to **cold stress** either by self-imposed actions (e.g., not dressing appropriately) or by sudden changes in environmental conditions during an event. Under such circumstances, the physiological response of the individual is altered to combat the specific challenge and maintain **thermal homeostasis.** Consequently, such changes could have implications for both exercise performance and the risk of cold-related injuries. This chapter reviews factors that contribute to heat loss, physiological adaptations to the cold, potential performance effects of exercising in the cold, and the type of injuries caused by cold exposure.

COLD STRESS: FACTORS CONTRIBUTING TO HEAT LOSS

Exposure to the cold, whether it is cold air or cold water, results in a transfer of heat from the body to the environment. The heat transfer progresses from the core of the body to the skin and from the skin to the environment. Most heat loss from the skin occurs through **conductive** or **convective** mechanisms (see chapter 18 for a review of these heat-transfer models). As ambient temperature drops below that of the **core temperature** of the body, a gradient is developed that results in a loss of body heat. The difference between the core temperature and the environment can be further magnified by wind speed. Wind accelerates body-heat loss by removing warm air trapped in insulative clothing, increasing evaporative cooling when insulative material is wet, and increasing evaporative cooling directly from the skin when the skin is wet (Hamlet 1988; Armstrong 2000). The term *windchill* is used to describe the combined effects of cold ambient temperature and air movement. Although a windchill index (see figure 19.1) has been developed to reflect the relative risk of freezing-tissue injuries, several investigators have pointed out some limitations to this index (Danielsson 1996; Kaufman and Bothe 1986; Sawka and Young 2000). The windchill index may overestimate the effect of increasing wind speed and underestimate the effect of lowering skin temperature (Danielsson 1996). In addition, the index estimates only the risk of exposed skin freezing, not well-insulated skin (Kaufman and Bothe 1986). Thus, an individual

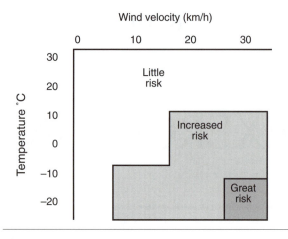

Figure 19.1   Windchill chart: Risk of freezing tissue.
Adapted from Armstrong 2000.

who is well insulated is not at as high a risk in a cold, windy environment as might be predicted by the index. However, the index may still have merit if any part of the body is exposed (e.g., fingers or face) and therefore at risk of freezing.

When the body is in cold water, the individual is at a much greater risk of cold injury than during any other environmental condition. The transfer of body heat through conductive mechanisms is approximately 25 times greater in the water than it is in the air (Toner and McArdle 1988). When immersed in icy water, total body cooling may exceed $6°C \cdot h^{-1}$. At this rate of heat loss, death could occur within 45 min to 3 h (Hayward and Eckerson 1984). Even when exercise is performed in cold water, the **metabolic heat** generated is often not sufficient to compensate for the rapid body cooling. This is witnessed in swimming, when heat loss is augmented by both conductive and convective means (Nadel et al. 1974). The debilitating effects of a cold, wet environment are frequently seen during exercise performed in the rain or during partial water immersion, when the insulative ability of clothing is compromised when wet. This facilitates heat loss through conductive, convective, and **evaporative** means.

PHYSIOLOGICAL RESPONSES TO EXERCISE IN THE COLD

Exercise in the cold, whether it is cold, dry or cold, wet conditions, results in physiological responses that focus on bringing the body back to thermal homeostasis. During exposure to a cold environment, core body temperature falls because of the rapid transfer of heat from the core to the periphery. To defend core temperature, the peripheral blood vessels constrict. **Vasoconstriction** appears to occur when skin temperature falls below 95°F

(35°C) and becomes maximal when skin temperature drops below 88°F (31°C) (Veicteinas, Ferretti, and Rennie 1982). The defense of core temperature is also accomplished by increasing metabolic heat production, primarily through a **shivering** response of the muscles. However, a portion of the increased **thermogenesis** seen during cold exposure may be the result of an increase in metabolism that is not related to muscular contraction (Toner and McArdle 1988).

In certain animals, an increase in metabolic heat production in the cold occurs without the need for muscle activity. This source of heat is thought to be the result of an elevated rate in aerobic metabolism of brown adipose tissue (Toner and McArdle 1988). Although the metabolism of brown fat may also be seen in human infants during a cold stress, its contribution to thermogenesis in adults has not been shown to be of significant value. The primary source of nonshivering thermogenesis in adults appears to be the result of an increase in metabolic heat production stimulated by elevated catecholamines, glucocorticoids, and thyroid hormones (Toner and McArdle 1988). However, the actual contribution of this potential source of increased metabolic heat has not been determined.

Shivering accounts for much of the increased metabolism during a cold stress. After several minutes of cold exposure, the muscles of the torso begin to rhythmically contract, followed by the muscles of the extremities (Horvath 1981). Seventy percent of the total energy liberated during shivering is liberated as heat, with the remainder generating external force (Sawka and Young 2000). As the cold stress becomes more severe, the extent of shivering increases. Shivering has been reported to cause an increase in oxygen uptake to a level that corresponds to 15% of the individual's $\dot{V}O_2$max (Young et al. 1986). Although maximal shivering is difficult to quantify, increases in oxygen consumption approaching 46% of the subject's $\dot{V}O_2$max have been reported during immersion in 54°F (12°C) water (Golden et al. 1979).

Effect of Exercise on Maintaining Thermal Balance

Although shivering significantly contributes to metabolic heat production, the best way to maintain thermal homeostasis is through exercise. The increase in physical activity during a cold stress generates enough metabolic heat so that shivering may not be needed. Claremont and colleagues (1975) showed that core temperature could be maintained within 0.9°F (0.5°C) when exercise is performed at ambient temperatures ranging between 32 and 95°F (0 and 35°C). When properly attired, exercise in temperatures as low as –22°F (–30°C) may be sustained without significant changes in core temperature (Toner

and McArdle 1988). The advantage of exercise for increasing metabolic rate is shown in figure 19.2. Clearly, physical activity is the greatest contributor to metabolic heat production in the cold.

Although exercise has been proven to have a positive effect on maintaining thermal homeostasis, a number of published reports have suggested that prolonged exercise in cold water may increase an individual's risk of **hypothermia** (Centers for Disease Control and Prevention 1983; Danzl, Pozos, and Hamlet 1995; Pugh 1966). The risk of exercise in cold water is well acknowledged. As previously mentioned, the heat loss through conductive, convective, and evaporative mechanisms exceeds the ability of the body to generate metabolic heat through exercise. During exercise in the water, heat loss may be augmented by a number of different factors. The dangers of exercising in cold water were demonstrated by Hayward, Eckerson, and Collis (1975), who reported significantly lower core temperatures in subjects who exercised in cold water versus subjects who remained still. The lower core temperatures of the exercising subjects suggested that they were at a much greater risk for hypothermia than those subjects who did not exercise. Possible mechanisms for the increased risk of hypothermia when exercising in cold water (Toner and McArdle, 1988) are listed here:

- Increased heat transfer from the core to the periphery because of increased blood flow.
- Increased heat production in the extremities versus the trunk when compared with non-exercise condition.
- Reduced insulatory benefit of the boundary layer of the skin/water interface if movement of the upper and lower extremities is increased.
- Increased effective surface area for heat transfer provided by the redistribution of blood from the trunk core to the extremity core.

A combination of these four mechanisms results in a change in the heat gain/heat loss dynamics of the extremities where the surface area to mass ratio is high, causing significant reduction in core temperature.

During exercise in cold weather, some concern has been focused on a **thermoregulatory fatigue** that could blunt the shivering response and reduce vasoconstriction. A study by Young and colleagues (1998) showed that prolonged exercise in the cold accompanied by sleep deprivation and a negative energy balance resulted in a lowered shivering threshold. However, the conditions that these subjects experienced are not commonly encountered by an exercising or competing athlete. An additional concern is the effect of exercise performed in a temperate environment on the thermoregulatory responses during subsequent cold exposure. The thermoregulatory response

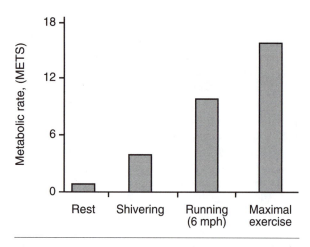

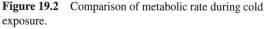

Figure 19.2 Comparison of metabolic rate during cold exposure.

Adapted, by permission, from M.N. Sawka et al., 1987, "Erythrocyte reinfusion and maximal aerobic power: an examination of modifying factor," *Journal of the American Medical Association* 257:1496-1499.

during exercise in a temperate environment is focused on dissipating heat. When exposed to a cold environment, there may be a **thermoregulatory lag** in switching from heat dissipation to heat conservation (Castellani et al. 1999).

Part of this hypothesis has been generated from research demonstrating that the increased profusion of blood to the active muscle during exercise remains elevated for an extended duration (Thoden et al. 1994). Castellani and colleagues (1999) examined this question in a group of subjects who rested for 2 h at 40°F (4.6°C) after either a passive heat exposure or a 1 h exercise period on a cycle ergometer at 55% peak O_2 uptake in 95°F (35°C) water. Their results showed a greater cooling effect for the exercise group. This greater cooling effect was not attributed to a fatigued shivering response but predominantly to a thermoregulatory lag (i.e., an inability of the subjects to change from a situation of heat dissipation to one of heat conservation). The researchers suggested that the postexercise **hyperemia** was still evident in the previously active muscles, which increased convective heat transfer during cold exposure. In addition, a redistribution of body heat from the core to the periphery was thought to be occurring, related to the increased blood flow during and after exercise. For thermoregulatory fatigue to occur, exercise needs to be of much greater duration. Thermoregulatory fatigue appears to set in during exercise exceeding 4 h in severe cold and wet conditions (Thompson and Hayward 1996). These results have important implications for individuals exercising in a temperate environment who are quickly exposed to the cold or for individuals exercising in the cold

who fail to insulate themselves properly after the conclusion of exercise.

Role of Body Composition and Thermal Balance

It is rare to find a situation in which a high percentage of body fat may be beneficial. However, there does appear to be an exception. During cold exposure, individuals with a higher percentage of body fat appear to have a greater ability to maintain their core body temperature than their leaner contemporaries. A number of reports have demonstrated a positive linear relationship between body fat composition and core temperature during cold exposures (Toner et al. 1986; Toner and McArdle 1988). In addition, individuals with a high body fat percentage shiver less than leaner individuals (Toner et al. 1986). The high levels of subcutaneous fat apparently have a greater insulatory ability, which limits the transfer of heat from the core of the body to the periphery by decreasing the rate of heat conduction.

Considering that women have more body fat than men do, it might stand to reason that they could tolerate cold stress better than men can. However, it does not appear that women have a thermoregulatory advantage over men when it comes to cold exposure. Women have a greater body surface area and a smaller total body mass than men do (when comparing men and women of equivalent body fat). The larger surface area seen in women provides for an increased thermal gradient, causing a greater heat loss through convective mechanisms (Sawka and Young 2000). Thus, a greater total heat loss is seen in women than in men when body composition is controlled.

ACCLIMATIZATION TO THE COLD

The extent that an individual can acclimatize to the cold has not been studied as extensively as heat **acclimatization.** It does appear that when individuals are chronically exposed to the cold, some degree of physiological adaptation may occur. Two potential adaptations have been suggested that may contribute to a greater cold tolerance in individuals with prolonged cold exposure. The first adaptation is related to increased metabolic heat production through an exaggerated shivering response or potentially through a nonshivering mechanism (Sawka and Young 2000). Huttunen, Hirvonen, and Kinnula (1981) reported that Finnish workers who are chronically exposed to a cold environment retain brown adipose tissue. This theory is not well accepted, however, because of the alleged inability of humans to retain brown fat as adults. The second possible adaptation to the cold may be an enhanced sympathetic response, which results in a more rapid cutaneous vasoconstriction (Sawka and Young

2000). When compared with heat acclimatization, the physiological adaptations that have been proposed to result from chronic cold exposure appear to develop at a much slower pace, and they may not be as effective in preventing injury.

EXERCISE PERFORMANCE AND THE COLD

To prepare for exercise in a cold environment, the athlete can usually wear insulative clothing that will often negate the harsh conditions. Under these circumstances, the athlete's core body temperature is not only prevented from decreasing, it may even become elevated. However, some athletes may forego proper attire because bulky insulative clothing impedes the fluidity of their movement. Sometimes athletes are caught by a sudden change in the weather in the middle of a training session or competition, leaving them unprepared to combat environmental conditions. As a result, the athlete is exposed to the harshness of the environment and the body is forced to battle the elements to maintain core body temperature. The physiological changes that occur during cold exposure may have important implications for the subsequent performance capability of the athlete.

Aerobic Exercise in the Cold

During prolonged endurance exercise, oxygen consumption appears to be inversely related to the ambient temperature (Beelen and Sargeant 1991; Claremont et al.

1975; Galloway and Maughan 1997). At similar relative exercise intensities, a greater oxygen uptake is seen as the environmental conditions become colder. Galloway and Maughan (1997) reported that when exercise was performed at 39°F (4°C), increases in oxygen uptake were accompanied by increases in **minute ventilation** and carbohydrate utilization when compared with exercise performed at more temperate environments (see figure 19.3). These changes resulted in a decrease in the mechanical efficiency of exercise, as reflected by the significantly shorter time to fatigue (81.4 min) when exercising at 39°F (4°C) compared with exercise per-

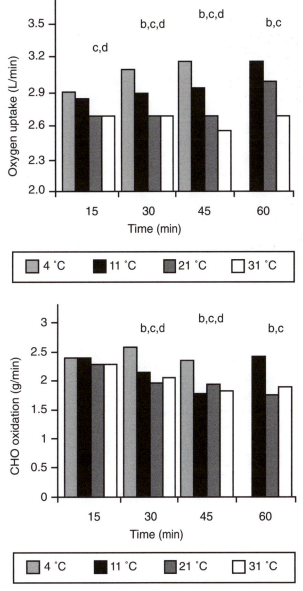

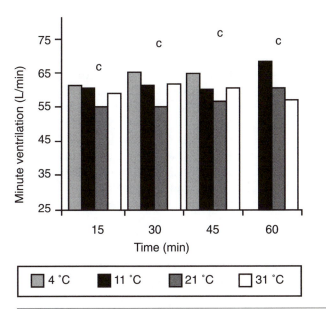

Figure 19.3   Oxygen uptake, minute ventilation, and carbohydrate (CHO) oxidation during exercise. b = significant difference between 4° and 11°; c = significant difference between 4° and 21°; d = significant difference between 4° and 31°
Adapted from Galloway and Maughan 1997.

©Bongarts Photography

formed at 52°F (11°C) (93.5 min to fatigue). However, exercise performed under the cold conditions still appeared to be less exhausting than exercise in more temperate environments (81.2 min and 51.6 min at 70°F [21°C] and 88°F [31°C], respectively). The changes in mechanical efficiency that resulted in these performance decrements appear to be related more to a drop in skin and muscle temperature than to changes in core body temperature. Core temperature of the subjects exercising at 39°F (4°C) was shown to increase to 100°F (38°C) during exercise, indicating that these subjects were not experiencing any cold stress. Thus, changes in the efficiency of exercise may occur when temperature changes are present only at the periphery and do not necessarily reflect a significant threat to the subject's thermal homeostasis.

Not all studies examining prolonged endurance exercise in the cold have been consistent in reporting increases in oxygen consumption. Others have suggested that oxygen uptake during exercise in the cold may be lower or even equal to oxygen consumption during exercise performed in warmer conditions (Young 1990). This is likely related to changes in the core body temperature. If core temperature is significantly lowered, the effect on exercise performance may become more pronounced. Studies examining $\dot{V}O_2$max and the cold have reported that maximal oxygen consumption may not be affected unless core temperature is decreased to a level that reflects a severe cold stress (Horvath 1981; Bergh and Ekblom 1979). Bergh and Ekblom (1979) have suggested that $\dot{V}O_2$max will not become significantly lowered until core temperature is reduced by at least 0.9°F (0.5°C). A reduction in the maximal aerobic capacity of an individual during a severe cold stress is likely related to a change in myocardial contractility and limits on attainment of maximal heart rate (Sawka and Young 2000).

Increases in oxygen consumption during endurance exercise in the cold may also be related to the greater oxygen requirement of the shivering muscles. However, this may be more relevant when exercise is performed at a low intensity of training. At higher intensities of training, the increase in exercise metabolism is sufficient to prevent shivering, and oxygen consumption during such an exercise stimulus may be similar between cold and temperate environments (Sawka and Young 2000).

Anaerobic Exercise in the Cold

The ability to generate maximal **strength** and **power** is related to muscle temperature. When exercise is performed in a cold environment, muscle temperatures may decline if the muscle is not sufficiently warmed. As muscle temperature lowers, significant decrements in performance can be seen. Davies and Young (1983a) reported a 43% decrease in jump power and a 32% decrease in cycling power when muscle temperatures were reduced by more than 14°F (8°C). In addition, cooling of the muscle results in an increase in the time to peak force (Davies, Mecrow, and White 1982; Davies and Young 1983a). The extent of muscle power and force decrements appears to be related to the degree of muscle temperature reduction. Power outputs have been reported to decrease 3-6% for every 1.8°F (1°C) reduction in muscle temperature (Bergh and Ekblom 1979; Sargeant 1987).

The magnitude of force and power decrements during cold exposure may also be related to the velocity of movement. In a study examining the effect of cold immersion on muscle strength (Howard et al. 1994), subjects who were immersed for 45 min in cold water (54°F [12°C]) showed significant decrements in both peak torque and average power during the leg extension exercise when compared with thermoneutral treatments. However, these reductions in peak torque and power occurred only at contraction speeds greater than 180° · s⁻¹. At slower velocities of joint movement (0° · s⁻¹ and 30° · s⁻¹), no significant differences were observed (see figure 19.4). These results were consistent with other studies demonstrating minimal effects on isometric strength after cold

exposure (Binkhorst, Hoofd, and Vissers 1977; Bergh and Ekblom 1979). The performance decrements seen at the higher velocities of contraction may be caused by a variety of factors related to the cold exposure, including decreased rate of cross-bridge formation (Godt and Lindly 1982; Stein, Gordon, and Shriver 1982), reduced nerve conduction velocity (Montgomery and MacDonald 1990), and changed motor unit recruitment patterns (Rome 1990). As muscle temperatures decrease, the rate of muscle enzyme activity may also decrease, resulting in a reduced ability to replenish high-energy phosphates (Ferretti 1992). Most of these mechanisms are speculative and further research is needed to provide a better understanding of muscle performance decrements during exercise in the cold.

MEDICAL CONCERNS

Most injuries related to cold ambient temperatures are the result of prolonged exposure without proper clothing. Cold injuries among athletes engaged in competitive or recreational sports are rare. Individuals performing physical activity in the cold are able to maintain body temperature by wearing layers of clothing as insulation. However, athletes participating in outdoor winter sports (e.g., cross-country skiing, speedskating, and football) may refrain from wearing bulky insulative clothing in order to maintain freedom of motion. This not only alters the physiological response to the physical activity, it also places the athletes at an increased risk of cold injuries.

Cold injuries may occur during all seasons and could result from unexpected cold-water immersions or unexpected storms catching athletes without proper attire. Armstrong (2000) reported on the risk of cold injuries in marathon runners performing in cool weather (39-50°F [4°-10°C]) or in races where a sudden drop in temperature or change in weather conditions occurs midway through the race. During such races, the runner loses a great deal of body heat through convection, radiation, and evaporation. The body cools rapidly, and it is more difficult to maintain core body temperature as fatigue sets in during the latter stages of the race. In addition, a greater risk of cold injury may arise during races that begin under mild conditions with runners seemingly dressed appropriately. Sudden changes in the weather (cold and rain) during the race may result in the runner now being ill-prepared (lack of appropriate clothing) to face the challenges of the new environmental condition, making the athlete more susceptible to cold injury.

Several injuries are common to prolonged cold exposure. Hypothermia and **frostbite** are the most dangerous of the cold injuries. **Immersion foot** and **chilblain** are also common but pose much less threat to survival.

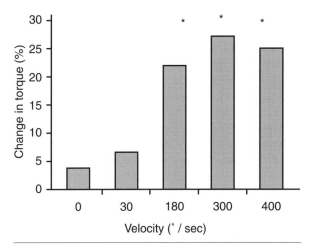

Figure 19.4   Effect of cold immersion on peak torque.
* = significant difference from values performed in a temperate climate.
Adapted from Howard et al. 1994.

Hypothermia

Hypothermia is defined as a lowering of the core body temperature below 95°F (35°C) (Hamlet 1988; Ward, Milledge, and West 1995). It can be classified according to both core temperature and length of exposure. Core temperature between 90 to 95°F (32 to 35°C) is considered to be mild hypothermia, whereas core temperature below 90°F (32°C) is considered to be severe hypothermia. When exposure to the cold is of relatively short duration, but the cold stress experienced by the individual exceeds the ability of the body to maintain core temperature despite maximum heat production, the subsequent hypothermia is classified as **acute hypothermia.** When prolonged activity in the cold is accompanied by exhaustion and depletion of energy reserves, as might occur during mountaineering, the ability to maintain core temperature is diminished and the ensuing hypothermia is classified as **subacute** (Ward, Milledge, and West 1995). The primary difference between acute and subacute hypothermia is that during acute hypothermia the body still has the capability of heat production, but the cold stress far exceeds the ability to maintain warmth. During subacute hypothermia, the body is able to maintain core temperature but is unable to maintain heat production because of prolonged activity and subsequent exhaustion the body. During a mild cold stress over a prolonged period of time (days or weeks), the thermoregulatory response is not overwhelmed but is unable to maintain sufficient body heat. This gradual decrease in body temperature is known as **chronic hypothermia** and is commonly seen in an elderly population (Ward, Milledge, and West 1995).

Clinical features of hypothermia can be seen in table 19.1. During mild hypothermia, the individual begins to shiver uncontrollably and skin color appears grayish. Speech patterns may become slurred and slow, and the person's attitude and demeanor toward the task at hand may become negative and irritable. Muscular coordination becomes impaired and an increase in fatigue becomes evident (Ward, Milledge, and West 1995). As core temperature decreases below 90°F (32°C), mental acuity is diminished. Thinking is slowed, memory deteriorates, and the individual becomes confused. The slurred speech and unresponsive behavior resembles a stroke victim. As core temperature continues to decrease, the individual becomes progressively unresponsive and eventually lapses into a coma.

The principal plan of management for cases of hypothermia is to prevent any further heat loss and restore body temperature to normal. The patient should be removed from the cold environment and placed in a shelter out of the wind, rain, or snow. For mild hypothermia, a slow rewarming of the body is considered acceptable; however, for severe hypothermia, a rapid, active rewarming should be performed (Ward, Milledge, and West 1995).

During a case of mild hypothermia, rewarming should be accomplished through surface rewarming. Replace wet clothing with dry, and cover the body with material that has some insulating ability (e.g., sleeping bag or blankets) or place hot water bottles at the victim's underarms or groin. When no dry clothing is available, the wet clothing may be wrung out and put back on. It is important to cover the clothing with additional insulating material to prevent further heat loss. Rewarming can also be accomplished through conduction from close body-to-body contact. Warm fluids should be given either orally (when victim is alert) or possibly through an IV. In addition, the victim should inhale warm air.

During severe hypothermia, the victim should be evacuated as quickly as possible. Remember that in extreme cases of hypothermia, the patient may appear dead. Core temperature must be elevated to normal levels before this possibility can be considered (Ward, Milledge, and West 1995). Rewarming should begin using the techniques previously discussed for mild hypothermia. Once the patient has been transported to an appropriate facility, warm IV fluids should be provided and airway warming via an endotracheal tube is recommended (Hamlet 1988). In addition, internal organ warming is generally accomplished through gastric and peritoneal lavage (i.e., irrigation with a warm saline solution). Rewarming must be proceeded with caution, and the patient must be treated gently to avoid **ventricular fibrillation.**

Frostbite

As ambient temperature drops toward the freezing point, the skin becomes numb and loses its sense of touch and pain at about 50°F (10°C) (Hamlet 1988). This is accompanied by a transient general vasoconstriction. As ambient temperature continues to drop below freezing, the skin will actually freeze. Several injuries are associated with freezing of the skin. The extent of these injuries depends on the environmental temperature, wind velocity, and duration of exposure. If exposure results in only the superficial layers of the skin becoming frozen, **frostnip** is considered to have occurred. Frostnip is not associated with any subsequent damage or tissue loss and is not considered to be a serious cold injury. During frostnip, the skin may appear red and scaly. Sensation in the affected areas may be lost, but the skin still remains pliable. Once rewarmed, the affected areas may appear similar to first-degree sunburn.

Superficial frostbite is the result of freezing of the skin and subcutaneous tissues. During this injury, the skin becomes white and frozen, but the deep underlying tis-

Table 19.1 Clinical Features of Hypothermia

Temperature		Clinical features
°C	°F	
37.0	98.6	Normal body temperature
36.0	96.8	Increase in metabolic rate to compensate for heat loss
35.0	95.0	Maximal shivering
34.0	93.2	Lowest temperature compatible with continuous exercise
31.0–33.0	87.8–91.4	Severe hypothermia, retrograde amnesia, clouded consciousness
28.0–30.0	82.4–86.0	Progressive loss of consciousness, muscular rigidity, slow respiration and pulse, ventricular fibrillation if heart is irritated
27.0	80.6	Voluntary motion ceases, person appears to be dead, pupils nonreactive to light and deep tendon and superficial reflexes absent
26.0	78.8	Victim seldom conscious
25.0	77.0	Ventricular fibrillation may develop spontaneously
21.0–24.0	69.8–75.2	Pulmonary edema develops
20.0	68.0	Cardiac standstill
17.0	62.6	No measurable brain waves

Adapted from Ward et al. 1995; Armstrong 2000.

sues remain pliable. Rewarming should be rapid in nature. After rewarming, the skin swells and becomes mottled blue or purple. Within several days the affected areas may become gangrenous.

If freezing involves some of the deeper structures (muscle, bone, and tendons), the more serious **deep frostbite** is said to have occurred. The affected area is insensitive and becomes hard and fixed over joints (nonpliable). The color of the skin may be grayish purple or white marble. Actual crystallization of tissue fluids in the skin or subcutaneous tissues occurs. Although the tissue is frozen, movement in the affected body part may still occur because the tendons are not as sensitive to the cold as other tissue (Ward, Milledge, and West 1995), and the primary muscle groups are at a distance from the area of injury. However, the affected areas will still become gangrenous and a permanent loss of tissue is almost inevitable (Ward, Milledge, and West 1995).

The mechanisms of frostbite injuries involve changes of blood flow to the affected area and the freezing of tissue (Foray 1992). During cold exposure, the unprotected body part has a reduced oxygen supply because of an increase in blood viscosity and vasoconstriction from the freezing temperatures. At some point, total cessation of blood flow may occur. Consequently, the tissues become deprived of needed oxygen (**hypoxia**) and metabolic **acidosis** will shortly ensue. In combination with the crystalline formation

occurring within the superficial tissues, the resulting damage to the cells and microscopic blood vessels results in tissue death.

Frostbite injuries may also be associated with hypothermia, which has a priority in treatment. Treatment plans that were discussed earlier for hypothermic victims should be followed. In a situation of frostnip, the affected part should be warmed by covering it (place a glove on an exposed hand or place the hand under the armpit or in the groin area). Sensation and full function should be restored fairly quickly. It should be understood that repeated frostnip predisposes the individual to subsequent frostbite injury (Riddell 1984).

Rapid rewarming is important for frostbite injuries. This is best accomplished by placing the exposed area in warm water. However, under no circumstances should the affected area be beaten, rubbed, or overheated. This includes rubbing the area with snow or ice or heating with excessively hot water (> 111°F [44°C]) because the affected area may suffer additional burn injury from **cold anesthesia** (Flora 1985). In addition, a freeze-thaw-freeze sequence could have disastrous results for future prognosis and should always be avoided. It may be better to keep the injured part frozen than to thaw it and take the chance that it may become frozen again while the victim is transported to a hospital. This is a real concern for injuries occurring in remote mountainous areas.

Immersion Foot

Immersion foot or **trench foot** is a cold injury associated with the feet being cold or wet for a prolonged period of time. It frequently occurs when the feet are in water or snow and may take from several hours to repeated exposures over several days to manifest. This is a nonfreezing injury that may cause lasting damage to the muscles and nerves. Although this type of injury could occur in the hands, it is much easier to check the hands and take appropriate measures to keep the hands warm and dry. Thus, it is more common to see this condition in the feet.

Immersion foot progresses through several stages. Initially, the extremity is cold, discolored, and numb (Hamlet 1988). After several days of this stage, a tingling pain may occur that is associated with swelling, blister formation, desquamation, ulceration, and **gangrene.** This second stage may last between 2 to 6 weeks and may be followed by a final stage, which could last several months or a lifetime, involving a heightened sensitivity to the cold (**Raynaud's disease**) accompanied by severe pain. Amputation is a possibility in severe cases.

The primary preventive measure is wearing heavy socks in well-fitting boots or shoes. Be prepared with a pair of dry socks to replace socks that become wet or damp. Keep the feet out of water or snow, and make sure to keep the feet clean, dry, and warm. Numbness and tingling are signs of trench foot and should serve as a "red flag" to warm the feet immediately.

Chilblain

Chilblain, another nonfreezing injury, appears to affect women more than men. It is an inflammatory condition that develops as a result of cold exposure (Ward, Milledge, and West 1995). This injury appears to be more frequent in cold, humid conditions than cold, dry conditions. Most lesions occur on the back of the extremities, between the joints, after an extended period of cold and dampness. Vasodilation and subcutaneous **edema** and swelling are characteristic of chilblain, and itching is the primary symptom during initial injury (Hamlet 1988). In chronic lesions, vasodilation may no longer be present, but the swelling remains and the itching is replaced by pain. **Pernio** is a more severe form of chilblain and is associated with a greater burning and pain sensation in the affected area. As the duration of the cold exposure lengthens and the severity of the injury increases, the individual is at a greater risk of trench foot. Table 19.2 provides some minimal recommendations for preparing to exercise in a cold environment.

SUMMARY

When exercise is performed in a cold environment, the primary response of the body is to maintain thermal homeostasis by defending core temperature. Athletes can usually prevent unnecessary and potentially dangerous exposure to the cold through behavioral modifications (e.g., wearing appropriate clothing). However, sometimes by choice or by circumstances beyond the athletes' control, they may find themselves in an environmental condition that places them at an increased risk of cold injury. Cold acclimatization does not appear to be easily achieved and knowledge concerning cold acclimatization is still forthcoming. With the technological advances being made in the sport-clothing industry, many athletes can protect themselves from the environment without sacrificing their mobility.

Table 19.2 Recommendations for Exercise in a Cold Environment

Do not wear tight clothing that may restrict circulation.

Cover exposed areas of the body with gloves (mittens preferably), scarves, wool hat, high-top shoes, or boots.

Try to avoid wetting clothing, shoes, and socks and be prepared to change wet clothing with dry socks, shoes, and undergarments.

Avoid exercising in the snow or in wet areas.

Avoid water-based skin lotions, and use oil-based ones instead such as Vaseline or Chap Stick.

Make sure to cover legs and genitals when dressing for outdoor training by wearing sweat pants, long-underwear, lycra-tights, gore-tex pants, or a combination of these garments.

Avoid wearing excessive clothing. A large sweat buildup will decrease the effectiveness of the insulatory ability of the clothing. As exercise concludes make sure to quickly change into dry clothes.

Use proper judgment in deciding to exercise outdoors. Perhaps the workout can occur under safer conditions inside.

Remember that individuals with a previous cold injury may be more susceptible to subsequent injuries.

Dehydration causes added physiological strain and limit your ability to generate metabolic heat.

Be sure to maintain proper water balance.

Adapted from Armstrong 2000.

CHAPTER 20

EXERCISE AT ALTITUDE

©Jurgen Ankenbrand

The acute response to a **hypobaric** environment can cause significant physiological and psychological changes that negatively affect athletic performance. However, prolonged exposure to a hypobaric environment results in **acclimatization.** The physiological adaptations that stimulate the acclimatization process not only help athletes perform at altitude, but may also have an **ergogenic effect** when the athletes return to sea level. This chapter discusses the effect that a hypobaric environment has on physiological processes and the subsequent effect that these changes have on performance. In addition, the potential advantages of acclimatization for subsequent performances at sea level are discussed.

THE HYPOBARIC ENVIRONMENT

As an individual ascends above sea level, **barometric pressure** is reduced relative to the magnitude of the elevation. Because the weight of the upper atmosphere compresses the air of the lower atmosphere, barometric pressure decreases rapidly as one ascends from sea level. Changes in pressure are also influenced by changes in ambient **temperature.** The relationships between pressure, temperature, and **volume** follow some of the basic laws of physics. For a better understanding of the hypobaric environment, it would help to briefly review some of these basic laws.

> **Boyle's law** refers to the relationship between pressure and volume. It states that the pressure of a given gas at a constant temperature is inversely proportional to its volume.
>
> **Charles' law** concerns the relationship between volume and temperature. It states that at constant pres-

sure, the volume of a gas is proportional to its absolute temperature.

> **Dalton's law** states that each gas in a mixture exerts a pressure according to its own concentration, independently of the other gases present.

The pressure of each gas is referred to as its **partial pressure.** Gas molecules, because of their random motion, tend to distribute themselves uniformly in a given space until the partial pressure is the same everywhere. These laws that govern the relationship between pressure and volume were the basis behind the development of the **standard atmosphere model** by the National Oceanic and Atmospheric Administration (1976) (see figure 20.1).

During ascent, the atmospheric pressure continuously decreases; however, the composition of the air remains the same as it was at sea level (20.93% oxygen, 0.03% carbon dioxide, and 79.04% nitrogen). The partial pressure of each gas is reduced in direct proportion to the increase in altitude (see table 20.1). The reduced partial pressure of oxygen (PO_2) results in a reduced **pressure gradient,** which impedes oxygen diffusion from the blood to the tissues.

In addition to reduced oxygen availability at altitude, other environmental issues appear to pose significant hurdles for performance at high elevations. The individual exercising at altitude may be at risk of cold injuries because of the low air temperatures common at high elevations. Ambient temperature drops at a rate of 1.8°F (1°C) for every 490 ft (150 m) of ascent. For example, the average temperature at the summit of Mount Everest is about -40°F (-40°C) in the winter. As a result, most climbs are performed during the summer months when temperature has been recorded at 16°F (-9°C) at the summit (Ward, Milledge, and West 1995). However, considering the effects of windchill, it may feel even colder at these

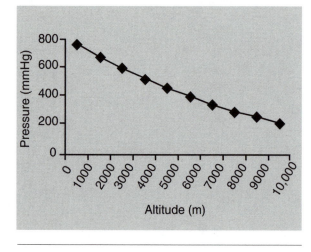

Figure 20.1 Standard atmosphere model.
Data from National Oceanic and Atmospheric Administration 1976.

Table 20.1   Changes in Barometric Pressure (P_B) and Partial Pressure of Oxygen (PO_2) at Varying Altitudes

Altitude (m)	P_B (mmHg)	PO_2 (mmHg)
0	760	159
1000	674	141
2000	596	125
3000	526	110
4000	463	97
5000	405	85
6000	354	74
7000	308	65
8000	267	56
9000	231	48

high altitudes. Wind velocities in excess of 93 mph (150 km · h⁻¹) have been reported on Himalayan peaks (Ward, Milledge, and West 1995), suggesting that windchill may have even more significance on the cold stress associated with altitude.

As ambient temperature is reduced during elevation, the amount of water vapor per unit volume of gas is also reduced. At high altitudes, even if the air were fully saturated with water, the actual amount of water vapor is very small. For example, at 68°F (20°C), the water vapor pressure is 17 mmHg, but at -4°F (-20°C), the corresponding water vapor pressure is only 1 mmHg. Thus, the extremely low humidity seen at altitude results in a large evaporative heat loss caused by **ventilation** of the dry inspired air. The risk of dehydration at altitude is great even at rest, but during exercise when the ventilation rate is further elevated, the risk becomes even more significant. A fluid loss from the lungs equivalent to 200 ml per hour was reported during moderate exercise at elevations of 18,000 ft (5,500 m) (Pugh 1964). Acclimatization seems unable to reduce this chronic volume depletion, even in individuals who have been living at altitude for a prolonged period of time. Blume and colleagues (1984), during an ascent to Mount Everest, reported significantly elevated serum osmolality in subjects residing at 20,700 ft (6,300 m) compared with sea level, even when fluids were readily available and no significant changes in exercise or diet were apparent. This study highlighted the risk of dehydration during exercise at altitude, emphasizing the need for high fluid intake during such environmental extremes, even when the thirst drive is absent.

PHYSIOLOGICAL RESPONSE TO ALTITUDE

The physiological stress associated with altitude is manifested primarily in its effect on oxygen availability to the tissues. When oxygen levels are reduced in arterial blood, inspired gases, or tissues, a situation of **hypoxia** is present. Acute hypoxia, which occurs when an individual is initially exposed to altitude, results in changes in various physiological systems in the body. An overview of these changes is depicted in figure 20.2. Although there is considerable variability, most changes observed at moderate altitudes primarily involve the central nervous system. At 4,900 ft (1,500 m), night vision becomes impaired. At about 6,600 ft (2,000 m), resting heart rate becomes elevated and continues to elevate as ascent continues. During ascents above sea level to approximately 9,800 ft (3,000 m), a subject at rest does not experience any noticeable symptoms. However, performance of novel tasks (e.g., reaction time in complex choice reaction tests) may be affected (Ernsting and Sharp 1978).

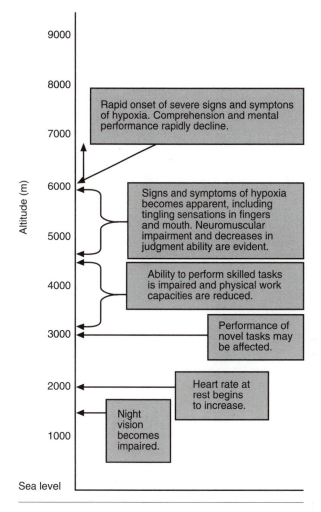

Figure 20.2 Effect of sudden exposure to various altitudes.

At elevations between 9,800 to 14,800 ft (3,000 to 4,500 m), symptoms of hypoxia may not be apparent (see table 20.2 for a list of symptoms of hypoxia), but the ability to perform skilled tasks will be impaired. In addition, physical work capacity becomes significantly diminished. If ambient temperature becomes colder than usual, symptoms of hypoxia may become apparent (Ernsting and Sharp 1978).

Between 14,800 to 19,700 ft (4,500 to 6,000 m), the signs and symptoms of hypoxia are present even at rest. Tingling sensations in the fingers and mouth may be noticed. In addition, higher mental processes and neuromuscular control are affected, with loss of critical judgment (Ernsting and Sharp 1978). Many of these symptoms that result in deterioration of performance go unnoticed by the individual, which could present a very dangerous situation. A marked change in the emotional state is frequently seen, which may manifest itself with a release of the basic personality traits of the individual, ranging from euphoria to moroseness. Physiological strain is apparent and physical exertion may increase the severity of these symptoms, often resulting in a loss of consciousness.

At elevations above 19,700 ft (6,000 m), subjects at rest rapidly experience severe signs and symptoms of hypoxia. Both comprehension and mental performance decline rapidly, and unconsciousness can occur with little to no warning (Ernsting and Sharp 1978). The time of **useful consciousness** (defined as the time elapsed between a reduction of the oxygen tension of inspired gas and the instant that there is an impairment of performance) is approximately 3.50 ± 1.36 min at 25,000 ft (7,600 m) (Ernsting and Sharp 1978). Typically, the degree of performance impairment may vary from an inability to perform psychomotor tasks to failure to respond to simple commands.

Effect of Altitude on Respiratory Response

As one ascends above sea level, the partial pressure of oxygen (PO_2) becomes reduced. To compensate for the reduced PO_2 at altitude, the breathing rate is increased. However, as breathing rate is increased (**hyperventilation**), the partial pressure of carbon dioxide (PCO_2) in the alveoli is reduced. As a result, the stimulus to maintain a high rate of ventilation may be removed because PCO_2 is an integral part of the driving force behind hyperventilation. The low PCO_2 also causes an elevation in blood pH, causing a condition known as **respiratory alkalosis.** To compensate for the increase in blood pH, the kidneys excrete more bicarbonate ions (HCO_3^-) to return the HCO_3^-/PCO_2 back to normal. The increased bicar-

bonate excretion from the kidneys appears to reduce the **buffering capacity** of the body. The importance of this will be seen later.

The relationship between ventilation and PO_2 is hyperbolic, whereas the relationship between ventilation and **arterial saturation (SaO_2)** is linear (see figure 20.3). Although variable from person to person, the increase in ventilation as one ascends does not appear to occur until inspired PO_2 reaches approximately 100 mmHg (corresponding to an alveolar PO_2 of 50 mmHg). This is equivalent to an elevation of about 9,800 ft (3,000 m) (Ward, Milledge, and West 1995). At this altitude, the saturation of oxygen to **hemoglobin** falls from 98% (normal levels seen at sea level) to less than 92%. The point at which **arterial PO_2 (PAO_2)** results in a marked increase in ventilation corresponds to the PO_2 at which the oxygen dissociation curve begins to steepen.

The decreases in SaO_2 may not be the most significant factor that limits performance at altitude. The pressure gradient between arterial PO_2 and tissue PO_2 is approximately 64 mmHg at sea level (the difference between an arterial PO_2 of 104 mmHg and a tissue PO_2 of 40 mmHg). This creates a pressure gradient, which causes oxygen to diffuse into the tissues. However, a reduction in PAO_2 at altitude results in a decrease in the pressure gradient, which decreases the diffusion capability of oxygen from the vasculature into the tissue. For instance, at an elevation of 8,200 ft (2,500 m), the PAO_2 drops to about 60 mmHg, while the tissue PO_2 remains at 40 mmHg, thus creating a pressure gradient of only 20 mmHg. This 70% reduction in the pressure gradient causes a significant reduction in the speed at which oxygen moves between the capillaries and tissues. The re-

Table 20.2   Symptoms of Hypoxia

Light	Medium	Severe
Euphoria	Anxiety	Dizziness
Loss of orientation	Desire to vomit	Delirium
Nausea	Chest pains	Coma
Headache	Apnea	Vomiting
Slightly increased blood pressure	Blood pressure is elevated	Respiration is depressed with possible Cheyne-Stokes respiration; respiration may also cease
Increased, sometimes irregular heart beat	Heart rate is reduced and irregular	
Increased breathing rate	Muscles may spasm or become rigid	Abrupt drop in breathing rate
Lack of muscular coordination		Blood pressure is weak and vanishing
Skin may be lightly cyanotic	Skin becomes deeply cyanotic with possible heavy sweating	Heart rate is low
Pupils are irregular	Pupils are alternating wideness	Muscle are relaxed, near paralysis
		Skin is gray and clammy
		Pupils are extremely dilated and rigid

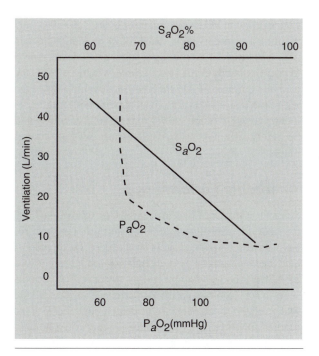

Figure 20.3 Ventilatory response to the partial pressure of alveolar oxygen (P_AO_2) and arterial oxygen saturation (S_aO_2). Adapted, by permission, from M.P. Ward, J.S. Milledge, and J.B. West, 1995, *High altitude medicine and physiology* (London: Chapman & Hall Medical Publishers), 73, 156.

duction in the pressure gradient between the vasculature and the tissues has a much greater significance on performance decrements associated with altitude than the relatively small changes (~ 6-7%) seen in arterial saturation.

Effect of Altitude on Cardiovascular Response

Acute hypoxia results in an increase in **cardiac output** both at rest and during exercise. Obviously, this is a compensation for the reduced oxygen availability caused by the lower PO_2. The primary mechanism resulting in the increased cardiac output appears to be an increase in **heart rate.** Heart rate has been shown to increase 40-50% at rest without any change in **stroke volume** (Vogel and Harris 1967). Even during exercise, the increase in cardiac output appears to be primarily the result of an increase in heart rate. In contrast to what is normally seen during exercise at sea level, stroke volume decreases during exercise at altitude (Vogel and Harris 1967). The decrease in stroke volume is apparently caused by the reduction in **plasma volume** that has been observed within a short time after arrival at altitude (Singh, Rawal, and Tyagi 1990; Wolfel et al. 1991).

During initial exposure to altitude (observed at elevations between 9,800 to 19,700 ft [3,000 to 6,000 m]),

decreases in plasma volume appear to be the result of both **diuresis** (excretion of unusually large volumes of urine) and **natriuresis** (Honig 1983). The diuresis may be explained by the large evaporative heat loss caused by ventilation of dry inspired air at altitude, which was discussed earlier. The natriuresis (increased sodium excretion in urine) appears to be the result of a neural stimulation of the kidneys to decrease the reabsorption of sodium caused by the hypoxic stimulus (Honig 1983). Even during prolonged altitude exposure, plasma volume will still remain below normal levels, and studies examining individuals that reside at altitude have reported lower plasma volumes in comparison with residents of sea-level communities (Ward, Milledge, and West 1995). Thus, acclimatization does not appear to have any significant effect on a return of blood volume to pre-exposure levels.

Effect of Altitude on Metabolic Response

During exercise at altitude, increases in lactic acid at every given work rate reflect the increase in **anaerobic** metabolism observed at altitude. Considering the limitations of oxygen availability at altitude, such a response is not unexpected. It is interesting that during exposure to altitude maximal lactic acid concentrations are generally reduced (Green et al. 1989; Sutton et al. 1988). This is likely related to the inability of the body to achieve a maximal effort because of a reduction in buffering capacity and limitations on energy production through glycolysis (Wilmore and Costill 1999). As mentioned previously, respiratory alkalosis occurring during hyperventilation at altitude causes an increase in bicarbonate ions excreted from the kidneys to compensate for the high blood pH. The mechanism that can explain the reduced **glycolytic** efficiency during exercise at altitude is not well understood.

EFFECT OF ALTITUDE ON ATHLETIC PERFORMANCE

The physiological changes seen during acute altitude exposure have significant detrimental effects on athletic performance. However, as altitude exposure is prolonged, physiological adaptations occur that not only enhance exercise performance at altitude, but may also provide significant performance benefits on return to sea level.

Endurance Performance

As one ascends from sea level, significant effects on endurance performance in athletes have been reported at relatively low to moderate levels of altitude (2,000-3,900 ft [600-1,200 m]) (Gore et al. 1996, 1997; Terrados, Mizuno, and Andersen 1985). However, the effect of altitude on endurance performance may be related to the training status of the individual. At elevations that were shown to cause a significant decline in the **$\dot{V}O_2$max** of highly trained athletes, no significant changes were seen in untrained control subjects. Although a well-trained endurance athlete may be more sensitive to changes in his or her environment than an untrained subject, subsequent studies have shown a large degree of variability, even among highly trained athletes. A study by Chapman, Emery, and Stager (1999) demonstrated that at a mild altitude (3,300 ft [1,000 m]), eight endurance athletes had significant declines in $\dot{V}O_2$max, while six other endurance athletes participating in the same study maintained their $\dot{V}O_2$max levels. The difference in the response between these subjects at altitude appeared to be related to differences in their SaO_2 at sea level. In that study, a significant correlation was seen between SaO_2 of $\dot{V}O_2$max at sea level and the decline in $\dot{V}O_2$max at altitude. Additional data from that same laboratory showed that elite distance runners with the highest SaO_2 during maximal exercise at sea level had the smallest slowing of 3,000-m run-times when performing at 6,900 ft (2,100 m) (Chapman and Levine 2000). It appears that the ability to maintain arterial oxygen saturation during heavy exercise may be more related to the ability to maintain endurance performance at altitude than training status per se (Chapman and Levine 2000).

As the ascent continues, the effect on endurance performance becomes more pronounced. At altitudes exceeding 5,200 ft (1,600 m) (approximately the elevation of Denver, CO), decreases in $\dot{V}O_2$max are commonly observed. As altitude continues to increase, a linear decline in $\dot{V}O_2$max is seen (see figure 20.4). Approximately an

11% decline in $\dot{V}O_2$max is seen for every 3,300 ft (1,000 m) increase in elevation (Buskirk et al 1967). In men with a $\dot{V}O_2$max of 50 ml · kg^{-1} · min^{-1} at sea level, a drop to approximately 5 ml · kg^{-1} · min^{-1} would be expected near the summit of Mount Everest, leaving these men with little to no ability for physical work without supplemental oxygen. In fact, a $\dot{V}O_2$max value that low would be barely sufficient to sustain resting oxygen requirements.

Anaerobic Performance

Since the 1968 Olympics in Mexico City (elevation 7,300 ft [2,240 m]), sprinting, jumping, and throwing records have been set predominately at altitude. These records have been set primarily in events that are largely driven by ATP-PC and glycolytic energy systems. The benefits seen in anaerobic exercise performance at altitude are likely related to the decrease in **drag** associated with the thinner air. Drag is the resistance that acts on a body in motion, either in air or in water. At sea level, drag is responsible for 3-9% of the energy cost of running (Pugh 1970) and for over 90% of the energy cost during cycling exercise performed at speeds > 25 mph (40 km · h^{-1}) (McCole et al. 1990). At altitude, the thinner air reduces the effects of drag, which results in faster times and a reduced energy cost. The performance benefits of reduced drag for the sprinter or cyclist is, unfortunately for the swimmer, not seen in water sports. The effects of hypobaria on the density of water are not significant enough to cause any potential performance benefit (Chapman and Levine 2000).

The effect of altitude on anaerobic sports that involve repeated high-intensity activity is not well understood. However, the decrease in buffering capacity of the muscle during acute exposure to at least a moderate altitude does suggest that prolonged anaerobic activity may be affected. In addition, the lower lactic acid concentrations seen during maximal exercise at altitude may also be suggestive of potential performance decrements. However, these potential negative effects on anaerobic performance are discussed relative to acute altitude exposure. After prolonged exposure, the acclimatization experienced by the athlete will likely result in improved performance.

ALTITUDE ACCLIMATIZATION

During prolonged altitude exposure, physiological adaptations that enhance the ability to perform at high elevations gradually occur. However, these adaptations never reach the point that fully compensates for hypobaric hypoxia. Although several physiological systems in the body are able to adapt to altitude, these systems have different time courses for adaptation.

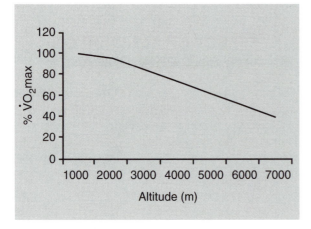

Figure 20.4 Changes in maximal aerobic power ($\dot{V}O_2$max) at differing altitudes.

Respiratory Changes After Prolonged Exposure to Altitude

During the first few days at altitude, changes in the respiratory response are seen. Initially, breathing rate increases and arterial PO_2 decreases. After a few days at altitude, arterial PO_2 begins to rise as PCO_2 values fall. However, the ventilatory rate continues to rise as a result of changes in the ventilatory response to CO_2 levels and in the sensitivity of the **carotid body.** The carotid body, situated above the bifurcation of the carotid artery, serves as a sensor of oxygen saturation in the blood. Its location is ideal for this role because it receives a large blood supply, allowing it to respond to oxygen saturation and not to oxygen content (Ward, Milledge, and West 1995). The increased sensitivity of the carotid body appears to have a biphasic response. The initial reaction may be a decrease in the hypoxic ventilatory response during the first 3-5 days of altitude exposure, but after these initial days

of exposure, an increase in the ventilatory response is seen (Ward, Milledge, and West 1995).

In addition to changes in the ventilatory response, another component of respiratory adaptation to altitude may be **diffusion capacity.** During prolonged exposure (7-10 weeks) to altitude, diffusion capacities have been reported to increase 15-20% (West 1962). Although much of this improved diffusion capacity may be accounted for by increases in hemoglobin concentrations, other studies examining highlanders (individuals permanently residing at altitude) and lowlanders (individuals residing at sea level) have also reported significant differences in diffusion capacities between these population groups (Dempsey et al. 1971). These differences may be related to the larger lung volumes developed through exposure to chronic hypoxia and the subsequent development of greater surface area for oxygen diffusion (Bartlett and Remmers 1971).

Diffusion ability appears to be affected during acute altitude exposure by a **ventilation/perfusion inequality**

©PhotoDisc

(Ward, Milledge, and West 1995). This is a limitation of the ability of oxygen to diffuse across from the alveoli to the pulmonary circulation and is reflected by a progressive decrease in arterial oxygen saturation at higher altitudes and at higher work levels. However, some evidence suggests that prolonged altitude exposure reduces the ventilation/perfusion inequality (Wagner, Saltzman, and West 1974; Wagner et al. 1987).

Cardiovascular and Hematological Changes After Prolonged Exposure to Altitude

After several weeks of altitude exposure, cardiac output remains similar to those values observed at sea level. However, similar to what occurs during acute altitude exposure, increases in cardiac output after extended stays at elevation also appear to be attributed primarily to increases in heart rate. Stroke volume continues to remain reduced (Reeves et al. 1987). A low stroke volume is also common among highlanders. When comparing highlanders to acclimatized lowlanders, similar stroke volumes between these two subject populations are commonly observed (Ward, Milledge, and West 1995). Thus, even during chronic altitude exposure, stroke volume does not improve, suggesting that other adaptations need to occur to compensate for the reduced plasma volume.

One of the best known adaptations to prolonged altitude exposure is the increase in the number of red blood cells per unit volume of blood. Exposure to hypoxic conditions results in the release of the hormone **erythropoi-**

etin. Erythropoietin is responsible for stimulating red blood cell **(erythrocyte)** production and is seen to increase within 2 h of exposure to altitude. It reaches a maximum rate of increase at about 24-48 h (Eckardt et al. 1989). After 3 weeks of altitude exposure, erythropoietin concentrations appear to return to baseline levels, but not until they have contributed to an approximate 20-25% increase in packed cell volume (Milledge and Cotes 1985). Increases in red cell mass continue even after erythropoietin returns to normal levels (Milledge and Cotes 1985). However, the mechanism that underlies this continued increase is not known.

As red cell volume increases, so does the hemoglobin concentration of the blood. This increase allows for a greater amount of oxygen to be carried per unit volume of blood. However, as red cell volume and hemoglobin concentration increase, the viscosity of the blood also increases, presenting an inherent danger associated with these physiological adaptations to altitude. The increase in hemoglobin concentration is proportional to the increase in elevation (Wilmore and Costill 1999). This is likely related to the greater diuresis and resulting loss in plasma volume associated with higher altitudes. The changes in the oxygen-carrying capacity of the blood after altitude exposure can be seen in figure 20.5.

Metabolic and Neuromuscular Changes After Prolonged Exposure to Altitude

Examinations of early studies have led to the general conclusion that prolonged exposure to moderate altitudes

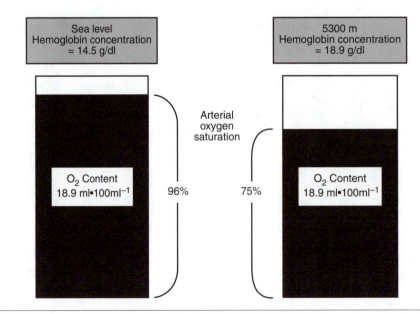

Figure 20.5   Oxygen content of arterial blood in acclimatized subjects.

Adapted, by permission, from M.P. Ward, J.S. Milledge, and J.B. West, 1995, *High altitude medicine and physiology* (London: Chapman & Hall Medical Publishers), 73, 156.

(13,100-16,400 ft [4,000-5,000 m]) results in significant increases in **oxidative** enzymes with no change in glycolytic enzyme activity (Ward, Milledge, and West 1995). This is similar to what occurs during endurance training. When challenged by oxygen deficiency either during altitude exposure or by intense endurance training, the muscles compensate by increasing their ability for oxidative metabolism.

However, when individuals ascend to more extreme altitudes, the effect on both oxidative and glycolytic enzymes appears to be different. At elevations exceeding 19,700 ft (6,000 m), several studies have observed that the oxidative enzymes decrease after prolonged exposure (Cerretelli 1987; Green et al. 1989; Howald et al. 1990). Green and colleagues (1989) reported decreases ranging from 21 to 53% in the oxidative enzymes (e.g., succinate dehydrogenase, citrate synthase, and hexokinase) during prolonged exposure. The results of these studies do not support the hypothesis that hypobaric hypoxia will result in adaptations that maximize oxidative function. The effect of extreme altitude on the glycolytic enzymes is less clear. After expeditions to Lhotse Shar or a simulated expedition to Mount Everest, glycolytic enzymes were reported to decrease (Cerretelli 1987; Green et al. 1989). In contrast, studies by Howald and colleagues (1990) after expeditions to both Mount Everest and Lhotse Shar reported increases in the glycolytic enzymes, suggesting an aerobic to anaerobic shift in muscle-energy metabolism. These contrasting results are difficult to explain and will likely be examined further in future research.

Prolonged altitude exposure also appears to affect muscle structure and **capillary density.** After 40 days of progressive decompression simulating an ascent to the summit of Mount Everest, reductions in both type I (25%, $p < 0.05$) and type II (26%, $p > 0.05$) fiber areas were seen (MacDougall et al. 1991). These results confirmed earlier reports of muscle atrophy by other investigators examining ascents to altitudes exceeding 26,200 ft (8,000 m) (Boutellier et al. 1983; Cerretelli et al. 1984). The effect of prolonged or chronic exposure to more moderate altitudes on muscle fiber size is less clear. After an 8 week expedition to the Himalayas (altitude > 16,400 ft [5,000 m]), significant reductions (–10%) in muscle cross-sectional area were observed (Hoppeler et al. 1990). However, in limited studies at lower elevations (~ 6,600 ft [2,000 m]), no significant differences in muscle fiber size have been seen (Saltin et al. 1995). In addition, endurance-trained athletes native to an altitude approximately 6,600 ft (2,000 m) above sea level did not differ from lowland endurance-trained athletes in muscle fiber size (Saltin et al. 1995). Indications from several studies suggest that the magnitude of reduction in muscle cross-sectional area depends on the elevation (see figure 20.6). However, the time frame for these adaptations to occur is less understood.

The decrease in muscle fiber area during sojourns at moderate to high elevations may be an example of a physiological adaptation to provide the best opportunity for muscle to be supplied with oxygen, thereby giving the body the maximum chance of survival. The decrease in fiber size occurs without any change in the number of capillaries within the muscle (Green et al. 1989; Hoppeler et al. 1990). As a result, the capillary density of the muscle is increased. This adaptation is similar to what is commonly seen after endurance training. The difference between the two is that increases in capillary density as a result of endurance training occur through a rise in the number of capillaries within the fiber, whereas increases in capillary density after altitude exposure are a result of muscle atrophy. Table 20.3 provides a list of potential muscle structural and metabolic changes during prolonged altitude exposure.

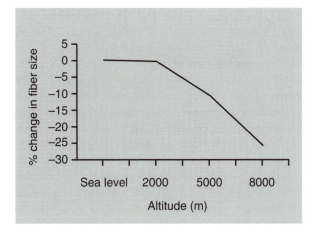

Figure 20.6 Effect of altitude on percent change in muscle fiber size.

Table 20.3 Muscle Structural and Metabolic Changes Occurring During Prolonged Exposure to Altitude

Adaptation	Low to moderate altitude (1000–5000 m)	Moderate to high altitude (> 5000 m)
Muscle fiber size	↔ or ↓	↓
Capillary number	↔	↔
Capillary density	↔ or ↑	↑
Oxidative enzymes	↑	↓
Glycolytic enzymes	↔	↑↓

TRAINING AT ALTITUDE FOR IMPROVED PERFORMANCE AT ALTITUDE

The physiological adaptations that occur as part of the acclimatization process to altitude result in improved performance in exercise at altitude. These adaptations, which include increases in ventilation, hemoglobin concentration, capillary density, and tissue myoglobin concentration, would be most beneficial for improved endurance performance. However, the reduction in muscle fiber size and body mass would have a negative effect on absolute strength and power performance. Reductions in muscle cross-sectional area may only be a concern for elevations exceeding 16,400 ft (5,000 m). In preparing to compete at altitude, a number of training strategies may be used to enhance performance. These strategies rely primarily on two options. The first option suggests that the athlete arrive within 24 h of the competition. Although this does not provide enough time for any acclimatization to occur, it helps prevent altitude sickness during competition. Symptoms of **acute mountain sickness** (discussed later) are not completely evident within the first 24 h of initial exposure to altitude. The second option suggests that the athlete arrive at least 2 weeks before the competition to begin acclimatization. However, if time permits, 4-6 weeks would allow a more complete acclimatization to occur.

Training Strategies for Competing at Altitude

Arrive at competition site within 24 h of scheduled performance.

If possible, arrive at competition site at least 2 weeks before performance for acclimatization. Ideally, for a more complete acclimatization, the athlete should arrive 4-6 weeks before competition.

Training should occur between 4,900 ft and 9,800 ft (1,500 m and 3,000 m). The lower altitude range is considered the lowest elevation at which an acclimatization effect will occur, and the upper range is thought to be the highest level for achieving a productive training session.

Training intensity should be reduced to 60-70% of sea-level intensity and gradually increased within 10-14 days.

Data from Wilmore and Costill 1999.

TRAINING AT ALTITUDE FOR IMPROVED PERFORMANCE AT SEA LEVEL

The physiological adaptations seen during acclimatization at altitude have led many scientists to hypothesize that either living or training at altitude, or a combination of the two, would enhance endurance performance at sea level. Theoretically, the physiological adaptations to altitude (caused by the hypobaric hypoxic conditions) are similar to the stimulus causing physiological adaptations to endurance training. Several studies have examined different strategies employed to elicit a training effect. These strategies have included "living high/training low," meaning that the athlete resides at a moderate altitude but trains at a lower altitude. Other strategies include "living high/training high," in which the athlete resides and trains at a moderate altitude, and "living low/training high," in which the athlete lives at sea level but trains at a moderate altitude.

Studies examining the effects of "training high" have had mixed results (Chapman and Levine 2000). Many of these studies have suffered from experimental design problems. In addition, athletes find it difficult to maintain the same exercise intensity at elevation as at sea level (Levine and Stray-Gundersen 1997). As a result, the quality of the workouts performed at elevation is reduced, and the training stimulus to enhance the physiological adaptations will thus be inferior to that occurring at sea level. In a recent well-designed study, athletes living at a moderate altitude (8,200 ft [2,500 m]) but training at a low elevation (4,100 ft [1,250 m]) had significantly greater performance improvements than athletes "living high/training high" or "living low/training low" (Levine and Stray-Gundersen 1997). In that study, 39 collegiate runners (27 men and 12 women) completed 6 weeks of preparation at sea level and were then randomly assigned to one of the three groups for 4 additional weeks of training. Both groups that lived at altitude significantly improved their $\dot{V}O_2$max (5%), and these improvements were significantly correlated to increases in red cell volume (9%, $r = 0.37$, $p < 0.05$). In addition, race times for the 5K run were significantly reduced (13.4 ± 10 s) in the high/low group only.

These same researchers then performed a retrospective examination of their studies by dividing the subjects into two groups: those they deemed responders to altitude training and those they deemed nonresponders (Chapman, Stray-Gundersen, and Levine 1998). Responders were those athletes who improved their 5K run-times by more than 14.1 s, whereas nonresponders improved less than 14.1 s. Although both groups had significant increases in erythropoietin concentrations after 30 h at 8,200 ft (2,500 m), the responders had a significantly higher erythropoietin concentration than the nonresponders. After 14 days at altitude, erythropoietin levels had declined to baseline levels in nonresponders but were still significantly elevated in the responders. As a result, the responders had a 7.9% increase in red blood cell volume and a 6.5% increase in $\dot{V}O_2$max.

Nonresponders did not have an increase in either red blood cell volume or $\dot{V}O_2$max. It is possible that erythropoietin may need to reach a threshold concentration before it is able to stimulate the production of additional red blood cells and provide an ergogenic effect for performance at sealevel (Chapman and Levine 2000).

The benefits of training at altitude for performing at sea level are still being debated. Although recent studies have suggested that there may be distinct advantages to such training strategies, several indications suggest that not all athletes will benefit from residing at altitude. Future research appears to be needed to better understand the factors that determine which athletes have the best opportunity to benefit from altitude training.

Recommendations for Training at Altitude for Sea-Level Competitions

The erythropoietin response to altitude depends on elevation above sea level. The higher the altitude, the greater the erythropoietin response (Eckardt et al. 1989). Athletes who do not increase their erythropoietin concentrations at a certain altitude may need to ascend to a higher elevation; however, the higher the altitude, the greater the chance of suffering from acute mountain sickness (AMS). Residing at an altitude greater than 6,600 ft (2,000 m) appears necessary to elicit increases in red cell volume (Chapman and Levine 2000). Training should be performed at an altitude as close to sea level as possible, thereby minimizing the reductions in training velocity and oxygen availability. It appears that 4,100 ft (1,250 m) is low enough to still provide a sufficient training stimulus for improving endurance performance while residing at higher altitudes (Levine and Stray-Gundersen 1997). Considering the biphasic time course of erythropoietin (initial increase within 24-48 h and a return to pre-altitude levels within 3-4 weeks), it is thought that residing for 3-4 weeks at altitude would be sufficient to maximize potential performance benefits (Chapman and Levine 2000).

CLINICAL PROBLEMS ASSOCIATED WITH ACUTE EXPOSURE TO ALTITUDE

On ascent to altitude, some people experience various symptoms that have been described as acute mountain sickness. Symptoms include headache, nausea, vomiting, fatigue or weakness, dizziness or lightheadedness, dyspnea (difficulty breathing), and insomnia (difficulty sleeping). Symptoms typically appear between 6 to 96 h after arrival at altitude. Symptoms start gradually and usually peak on the second or third day of exposure. Within 4 or 5 days, symptoms are usually gone and do not reoccur at that altitude. AMS, if it does not progress, is not considered a life-threatening illness.

The incidence of AMS depends on several factors, the most important being rate of ascent and altitude reached. The lowest altitude at which symptoms might appear seems to be about 8,200 ft (2,500 m) (Ward, Milledge, and West 1995), a height that may be experienced by some recreational skiers and hikers. The rate of incidence

is varied. If the stay at altitude is only 1 or 2 h, the symptoms of AMS are negligible (Ward, Milledge, and West 1995). Although fitness levels do not appear to have any significant relationship to AMS (Milledge et al. 1991), exercise itself may exacerbate the condition (Roach et al. 2000). In addition, it appears that certain individuals are more susceptible to symptoms of the illness (Forster 1984). That is, individuals who suffer from AMS during altitude exposure will likely experience similar symptoms during their next altitude exposure. The consistency of the response of individuals ascending to altitude helps predict their future response to acute altitude exposure.

The mechanism underlying AMS appears to be related to disturbances in fluid or electrolyte homeostasis caused by the hypobaric hypoxic conditions (Hackett and Rennie 1979; Hackett et al. 1981). Hypoxia causes a decrease in PO_2 and an increase in PCO_2, resulting in vasodilation and subsequent water and sodium retention. Hypoxia also causes an increase in both cerebral and pulmonary blood flow. It may also increase the microvascular permeability in these areas, resulting in a greater fluid leakage (Ward, Milledge, and West 1995). The increase of CO_2 in the tissues and the change in fluid volume, which cause a shift in fluid from intracellular to extracellular compartments, appear to be responsible for the associated symptoms. In addition, the increase in extracellular fluid may result in a severe progression of AMS to the more lethal **high altitude cerebral edema (HACE)** or **high altitude pulmonary edema (HAPE).**

Symptoms of HACE include the same symptoms associated with AMS, but the appearance of **ataxia,** irrationality, hallucinations, blurred vision, and clouding of consciousness may be a sign of the more lethal HACE. Symptoms of HAPE are again similar to those seen for AMS, but disruption of normal breathing as a result of pulmonary edema may be present, along with blueness of lips and fingernails, mental confusion, and loss of consciousness. The incidence of HAPE and HACE depends on the altitude, the rate of ascent, and the susceptibility of the individual. For example, the occurrence of HAPE increased from 2.5 to 15.5% when an 18,000-ft (5,500-m) ascent occurred rapidly by airlift versus being reached over 4-6 days by hiking (Bartsch 1999). In addition, the incidence of HAPE after a 22-h ascent to 15,100 ft (4,600 m) is reported to be 10% in mountaineers without any previous history of HAPE, but it increases to 60% in mountaineers with previously documented HAPE (Bartsch 1999). The treatment plans for both of these severe forms of AMS are similar and include administration of supplemental oxygen and removal to lower altitude.

To minimize the risk of AMS, a slow rate of ascent is recommended. In addition, only 1,000 ft (300 m) per day should be climbed above elevations of 9,800 ft (3,000 m). If symptoms of AMS are present, the individual should remain at that height until symptoms subside. Drugs are also used prophylactically to prevent symptoms. Acetazolamide and dexamethasone have been used with success to reduce the incidence of AMS and to treat it when symptoms appear.

SUMMARY

Ascent to altitude causes a number of physiological changes that limit an athlete's ability to perform. These limitations have been seen primarily in the endurance athlete, whereas the anaerobic athlete may see performance improvements at altitude because of the reduced effects of drag and energy cost on performance. Prolonged altitude exposure (between 2 to 6 weeks) results in acclimatization, which enhances the athlete's ability to perform at altitude. The physiological adaptations associated with prolonged altitude exposure (e.g., increased ventilation, increased red blood cell volume, and increased hemoglobin content) have enticed athletes who compete at sea level to reside at altitude to benefit from the ergogenic effect. Brief altitude exposure does not appear to present any clinical problems. However, as duration of exposure is prolonged, symptoms may develop that can limit an individual's performance. These symptoms indicate an illness known as acute mountain sickness. AMS is not life threatening but may progress to a more dangerous clinical situation of high altitude pulmonary edema or cerebral edema, which could indeed be life threatening if not treated.

PART V

MEDICAL AND HEALTH CONDITIONS

CHAPTER 21

OVERTRAINING

©Human Kinetics

The objective of all training programs is to optimally prepare the athlete or team for competition or, in the situation of a noncompetitive athlete, to achieve a specific training goal. The emphasis on either specific performance training (practices) or conditioning depends on the time of year and the type of athlete. An athlete who competes in a team sport (e.g., basketball, football, or baseball) has a defined season, off-season, and preseason. Although sport-specific training and conditioning occur throughout each season, the emphasis on each variable may change depending on the time of the competitive year. During the off-season, a greater emphasis is generally directed on the conditioning of the athlete, whereas during the season the primary contact time between athletes and coaches is devoted to sport-specific improvement and game or competition preparedness.

Whether emphasis is placed on an entire season of competition or on a specific meet appears to be the primary difference between athletes who participate in team sports and athletes who compete in individual sports (e.g., track and field, swimming). Athletes involved in team sports are generally brought to peak condition immediately before the season, and this level of conditioning is maintained throughout the year by sport-specific practices and competitions. In contrast, the track and field athlete or swimmer may perform early season competitions in less than peak physical condition or preparedness in order to peak for the more important competitions at the end of the season.

In both training scenarios, it is the goal of the coach to bring each athlete to peak condition at the appropriate time of the year. This is often accomplished through a periodized training program (see chapter 11) in which both training **intensity** and training **volume** are manipulated. During each phase of training, the athlete may experience a brief reduction in performance. As the athlete makes the necessary physiological **adaptations,** he or she experiences a **supercompensation** that results in improved performance (see figure 21.1). If adequate **recovery** or rest does not occur, the ability of the body to adapt to the training stimulus is affected. The performance decrements that may be experienced during each new phase of training may become further exacerbated. In addition, as athletes get closer to their genetic or performance maximum, their potential for further adaptation and subsequent performance enhancements may be limited or impossible. Further increases in training volume or intensity, or maintenance of a high training volume or intensity for a prolonged period of time, may also result in a decrease in performance. If these conditions persist, an **overtraining syndrome** may result. As can be seen in figure 21.1, there appears to be a fine line between peak performance and the potential for overtraining. It is a line that many coaches and athletes strive to reach but

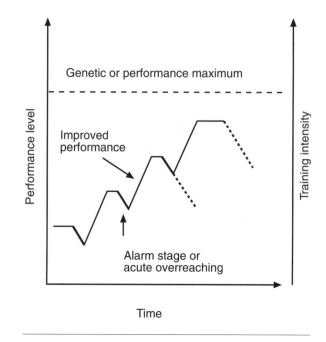

Figure 21.1 Hypothetical relationship between training adaptation, overtraining, and performance.

never cross. One of the most difficult challenges that both coaches and athletes face is determining the appropriate training stimulus (both intensity and volume) that optimizes performance without causing any undesired training responses.

This chapter focuses on understanding the different stages of overtraining and the factors that may contribute to these performance decrements. In addition, methods of identifying, monitoring, and preventing overtraining that are specific to **endurance,** anaerobic, and **strength/power** sports are discussed.

DEFINITIONS OF OVERTRAINING

The overtraining syndrome may be considered a continuum of negative adaptations to training. Symptoms of overtraining appear when the training stimulus has reached the point where either or both training intensity and training volume have become too excessive, coupled with inadequate rest and recovery. Initial stages of overtraining are generally accompanied by subjective feelings of **fatigue** and **staleness** that may or may not be accompanied by performance decrements. As the continuum proceeds, these subjective feelings of fatigue and staleness become associated with decreases in performance. When the training stimulus is excessive and recovery and adaptation do not occur within an anticipated time, the athlete is considered to be **overreaching.** With a decrease in the training stimulus and adequate rest, complete recovery usually occurs within 1-2 weeks

(Kreider, Fry, and O'Toole 1998). This recovery may also coincide with an **overcompensation** and improved performance. Overreaching is often a planned phase of many training programs.

When the imbalance between training and recovery continues for an indefinite period of time, the athlete progresses from a stage of overreaching and fatigue to the more serious problem of overtraining. Signs that are not acknowledged by an athlete or coach may be warnings of overtraining. Many times a plateau or a decrement in performance is met with frustration on the part of the athlete or coach. However, this plateau or decline may be the initial symptom of overreaching. The coach or athlete may ignore these signs, thinking that he or she needs to train harder to get past this plateau. Instead of reducing the training stimulus and resting, the training stimulus is increased. This results in a downward spiral of events that culminates in chronic fatigue and significant performance decrements associated with the overtraining syndrome. Recovery from overtraining may be a long process (possibly exceeding 6 months) (Kreider, Fry, and O'Toole 1998).

CONTRIBUTING FACTORS

A number of factors contribute to an athlete's susceptibility to overtraining. These factors are related to training program issues, training environment, psychological issues (including stress from school for the student athlete), nutritional concerns, and perhaps travel (jet lag). Although a single factor may be sufficient to cause overtraining, each additional factor adds to the total stress experienced by the athlete.

Failure to allow for adequate recovery is thought to be the primary training-related factor that causes overtraining, or the progression from overreaching to the more serious overtraining syndrome (Fry, Morton, and Keast 1991). This is especially relevant when increases in the training stimulus are performed (i.e., increases in the volume or intensity of training). Whether training intensity or training volume provides a greater stress and thereby makes the athlete more susceptible to overtraining is debatable. In addition, the impact that these training variables have on training stress may depend on the type of athlete (e.g., endurance or strength/power). This is examined further in the next section of the chapter.

During periods of intense training, the athlete must consume an adequate amount of carbohydrates and calories to prevent the catabolization of muscle for amino acids as a fuel source for exercise. In addition, prolonged exercise under hypoglycemic conditions negatively affects performance (Kuipers and Keizer 1988). An athlete's appetite may be influenced by a host of factors such as the stress of training, personal problems, sleep, and en-

vironmental conditions (Berning 1998). Athletes exercising in a hot and humid environment may have elevated body temperatures throughout the day, resulting in a decrease in hunger (Berning 1998). Regardless of the mechanism causing reduced caloric intake, the athlete will be at a greater risk of overtraining if energy demands are not met.

Psychological factors may also play a role. Monotony of training or emotional demands from family, school, or work may be a prelude to early stages of overreaching. Excessive expectations from coaches, self, or public may be another cause of emotional stress experienced by the athlete (Kuipers and Keizer 1988).

COMPARISON OF ENDURANCE AND STRENGTH/POWER ATHLETES

For both endurance and strength/power athletes, achievement of training goals is accomplished through manipulation of training volume (increase distance or increase number of sets and repetitions, respectively) and training intensity (exercise at a higher percentage of maximal oxygen uptake or 1 RM, respectively). However, the effect that each training variable has on each type of athlete in relation to increasing the susceptibility to overtraining may be different.

The results of studies designed to elicit overtraining in endurance athletes appear to be inconclusive. Several studies have suggested that increasing the volume of training for the short term does not result in any symptoms of overreaching or overtraining (Costill et al. 1988; Kirwin et al. 1988). In contrast, other studies have reported that an acute increase in training volume may be a potent stimulator of overtraining (Hooper et al. 1993, 1995; Lehmann et al. 1992). Furthermore, an increase in training intensity in a short-term study looking to elicit overreaching resulted in improved performance (Lehmann et al. 1992). Although the endurance athlete manipulates both training volume and intensity, the literature suggests that an excessive increase in the training volume without an appropriate regeneration phase places the endurance athlete at greater risk for overreaching or overtraining. However, most studies have examined overtraining over a short duration of training and not during more prolonged periods of exercise, which are common to most endurance athletes experiencing overtraining.

In strength/power athletes, overtraining has been attributed to alterations in both training volume (Stone and Fry 1998) and training intensity (Fry 1998). Apparently, changes in either variable without allowing for sufficient rest or recovery can cause overtraining in a relatively short period of time. Similar to the studies examining endurance athletes, most overtraining studies of strength/ power athletes have been of short duration and may not truly represent the influences responsible for the overtrained athlete over a prolonged season of practices. In addition, most of the research on overtraining and the strength/power athlete has examined weightlifters. Studies examining athletes participating in anaerobic sports (e.g., football or basketball) that combine several methods of training (e.g., strength, power, speed, and endurance) are sorely lacking. This is especially unfortunate considering the huge numbers of athletes participating in these sports.

Differences between endurance and strength/power athletes may become more apparent when examining symptoms of overtraining that appear to be specific to each athlete. These differences are reviewed in the section on recognition of overtraining.

SUSCEPTIBILITY TO OVERTRAINING

It is difficult to determine which athletes may be more susceptible to overtraining because all athletes at all levels of performance are at risk. However, it is the highly motivated athlete who appears to be the most susceptible (Fry, Morton, and Keast 1991). The occurrence of fatigue and other stages of overtraining depend primarily on how the individual athlete responds to the specific training stimulus. This unfortunately does not bode well for a team sport in which the training program is developed for the team as a whole and not for the individual athletes. In this situation, although most of the team may be responding well to the practice regimen, a particular athlete may be having difficulty adapting to the greater intensities or higher volume of practices. Without appreciating the possibility that this athlete may be in a particular stage of overtraining, the coach may put undue pressure on the athlete to work harder, which may cause further damage.

Track and field athletes, swimmers, and weightlifters may be in a better situation for coaches to recognize signs and symptoms of overtraining. Although it is not possible to predict who might be susceptible to overtraining, it is much easier to monitor the training performance of these athletes. Overtraining can be prevented by identifying potential markers and making appropriate adjustments to the training program. This is primarily related to the ability to objectively measure daily performance of these athletes. In contrast, athletes who participate in team sports may depend on more subjective means of evaluation (e.g., opponent's ability) to make changes in performance. This is reflected by the relatively large volume of research on overtraining in endurance athletes

(Flynn et al. 1994; Lehmann et al. 1995; Morgan et al. 1987), swimmers (Costill et al. 1988; Hooper et al. 1993; Raglin, Morgan, and O'Connor 1991), and even weightlifters (Fry, Kraemer, van Borselen, Lynch, Triplett, et al. 1994; Fry, Kraemer, van Borselen, Lynch, Marsit, et al. 1994) compared with the few studies that have examined overtraining in team sports (e.g., basketball and football) (Hoffman, Epstein, et al. 1999; Verma, Mahindroo, and Kansal 1978).

RECOGNITION OF OVERTRAINING

The primary indicator of either overreaching or overtraining is a decrement in performance. However, several reports have suggested that athletes whose performances have stagnated may also be overtrained (Hooper et al. 1995; Kuipers and Keizer 1988; Rowbottom, Keast, and Morton 1998). Much of the overtraining literature has been directed at identifying markers that would enable coaches to recognize overtraining at its earliest stages. It is thought that if recognized early, coaches could make the necessary training adjustments to prevent the more damaging and longer lasting overtraining syndrome. A host of variables, categorized according to physiological, psychological, immunological, and biochemical manifestations, have been reported in the literature as symptoms associated with overtraining. These major biological symptoms of overtraining, as indicated by their prevalence in the literature, were first compiled by Fry, Morton, and Keast (1991) and are listed in table 21.1. One or possibly several of these symptoms may be seen in athletes experiencing performance decrements. Unfortunately, these symptoms have been recorded without a single objective measure being identified as a consistent marker for overtraining. In addition, the appearance of these symptoms seems to confirm only that the athlete is in some stage of overtraining. The symptoms appear unable to indicate whether the athlete is perhaps on the threshold of overreaching or overtraining.

The following sections review some of the more prominent biological disturbances reported in athletes who are overtrained. Although there may be some cause and effect relationships between various biological disturbances, some of these imbalances may occur independently of one another.

Autonomic Nervous System Disturbances

Overtraining is thought to produce an autonomic nervous system imbalance, which results in either a **sympathetic** nervous system dominance or a **parasympathetic** nervous system dominance (Israel 1976). Sympathetic

Table 21.1   Major Symptoms of Overtraining as Indicated by Their Prevalence in the Literature

Physiological performance

Decreased performance	Inability to meet previously attained performance standards or criteria	Recovery prolonged
Reduced tolerance of loading	Decreased muscular strength	Decreased maximum work capacity
Loss of coordination	Decreased efficiency or decreased amplitude of movement	Reappearance of mistakes already corrected
Reduced capacity of differentiation and correcting technical faults	Increased difference between lying and standing heart rate	Abnormal T wave pattern in ECG
Heart discomfort on slight exertion	Changes in blood pressure	Changes in heart rate at rest, exercise, and recovery
Increased frequency of respiration	Perfuse respiration	Decreased body fat
Increased oxygen consumption	Increased ventilation and heart rate at submaximal workloads	Shift of the lactate curve towards the x-axis
Decreased evening postworkout weight	Elevated basal metabolic rate	Chronic fatigue
Insomnia with and without night sweats	Feels thirsty	Anorexia nervosa
Loss of appetite	Bulimia	Amenorrhea or oligomenorrhea
Headaches	Nausea	Increased aches and pains
Gastrointestinal disturbances	Muscle soreness or tenderness	Tendonostic complaints
Periosteal complaints	Muscle damage	Elevated C-reactive protein
Rhabdomyloysis		

Psychological/information processing

Feelings of depression	General apathy	Decreased self-esteem or worsening feelings of self
Emotional instability	Difficulty in concentrating at work and training	Sensitive to environmental and emotional stress
Fear of competition	Changes in personality	Decreased ability to narrow concentration
Increased internal and external distractability	Decreased capacity to deal with large amounts of information	Gives up when the going gets tough

Immunological

Increased susceptibility to and severity of illnesses, colds, and allergies	Flu-like illnesses	Unconfirmed glandular fever
Minor scratches heal slowly	Swelling of the lymph glands	One-day colds
Decreased functional activity of neutrophils	Decreased total lymphocyte counts	Reduced response to mitogens
Increased blood eosinophil count	Decreased proportion of null (non-Tm non-B) lymphocytes	Bacterial infection
Reactivation of herpes viral infection	Significant variations in CD4:CD8 lymphocytes	

(continued)

Table 21.1 *(continued)*

Biochemical		
Negative nitrogen balance	Hypothalamic dysfunction	Flat glucose tolerance curves
Depressed muscle glycogen concentration	Decreased bone mineral content	Delayed menarche
Decreased hemoglobin	Decreased serum iron	Decreased serum ferritin
Lowered TIBC	Mineral depletion (Zn, Co, Al, Mn, Se, Cu, etc.)	Increased urea concentrations
Elevated cortisol levels	Elevated ketosteroids in urine	Low free testosterone
Increased serum hormone binding globulin	Decreased ratio of free testosterone to cortisol of more than 30%	Increased uric acid production

Data from Fry et al. 1991.

overtraining is associated with greater sympathetic activity in the resting state, whereas parasympathetic overtraining is associated with an inhibition of sympathetic activity and a greater parasympathetic activity at both rest and during exercise.

Sympathetic overtraining is associated with restlessness and irritability, sleep disturbances, weight loss, and an elevation in resting heart rate and blood pressure. A parasympathetic imbalance is associated with fatigue and depression, a reduction in resting heart rate and blood pressure, and a suppressed heart rate, glucose, and lactate response to exercise. Neuromuscular activity also appears to be impaired (Lehmann et al. 1998). The parasympathetic imbalance is thought to be more common in endurance sports, and the hyperexcitability associated with sympathetic overtraining is more associated with explosive, power sports (Lehmann et al. 1998). However, these two forms of overtraining may also represent a continuum of varying symptoms associated with different stages of overtraining (Flynn 1998; Kuipers and Keizer 1988). Although both forms of overtraining are associated with performance decrements, the parasympathetic form may be more difficult to distinguish because the symptoms are far less alarming and, in the initial stage, are similar to the positive adaptations associated with training (Fry, Morton, and Keast 1991).

Parasympathetic overtraining is thought to be a reflection of a more advanced form of overtraining and is closely associated with exhaustion of the **neuroendocrine system.** Sympathetic overtraining reflects a fatigued state seen during the initial stages of overtraining (Kuipers and Keizer 1988). In either type of overtraining, the change in the homeostatic balance is further reflected by changes in other physiological systems.

Neuroendocrine Disturbances

Any stress, including that accompanying exercise, causes marked changes in the neuroendocrine response. These changes can result from both acute and chronic training stresses. Acute changes can be caused by manipulation of an acute program variable (e.g., training intensity). As the body adapts to this new training stimulus, the hormonal response returns to baseline levels fairly quickly. During chronic training stresses, in which recovery is inadequate or the body is unable to adapt to the greater training stress, disturbances to several hormonal axes (e.g., hypothalamic-pituitary-adrenal, hypothalamic-pituitary-gonadal) become apparent. Changes in the endocrine response to increases in training volume appear to be similar between both endurance and resistance exercise. However, these similarities do not appear to exist when overtraining is seen after increases in training intensity (Fry 1998).

The hormones **testosterone** and **cortisol** are frequently monitored when examining overtraining. Decreases in both total and free testosterone with an accompanying increase in cortisol concentrations are thought to reflect a greater catabolic state in athletes. The testosterone/cortisol ratio has been proposed as a potential monitor of training stress (Adlercreutz et al. 1986). If the ratio of free testosterone to cortisol declines more than 30% or is less than $0.35 \cdot 10^{-3}$, it is thought to represent an insufficient recovery from exercise and may be an indicator of overtraining. If such a decline occurs after 1 day of training, it is an indicator of insufficient recovery. If it occurs after 3 months of training, it may represent overtraining. A prolonged disruption in the anabolic to catabolic balance may result in decreases in body mass, primarily through a loss of lean body tissue. However, a number of studies have been unable to find any change in the testosterone/cortisol ratio during overtraining or have not seen any correlation between changes in this ratio and performance (Fry et al. 1992; Fry, Kraemer, et al. 1993; Kirwin et al. 1988; Urhausen, Gabriel, and Kindermann 1995). Thus, changes in testosterone and cortisol concentrations may not be directly responsible for perfor-

©Human Kinetics

mance decrements and may be more reflective of changes in training stimuli.

Catecholamines appear to be very responsive to training stresses. The involvement of the catecholamines with a number of physiological systems suggests that they may be potent mediators of the overtraining syndrome. In overtrained endurance athletes, a reduction in the catecholamine response to exercise, paralleling impaired glycolytic energy mobilization, may be seen (Urhausen, Gabriel, and Kindermann 1995). In addition, overtrained endurance athletes may also have reduced nocturnal levels of both **epinephrine** and **norepinephrine** (Lehmann et al. 1998). This would be consistent with parasympathetic overtraining common in this type of athlete. Furthermore, athletes suffering from parasympathetic overtraining are reported to show a 50-70% reduction in basal urinary catecholamine excretions (Lehmann et al. 1998). Decreases in catecholamine concentrations are negatively correlated to both fatigue (Lehmann et al. 1992) and latency of REM sleep (Lehmann et al. 1998), suggesting

an important role of catecholamine reduction in central fatigue.

Increases in catecholamine concentrations at rest (Hooper et al. 1995; Lehmann et al. 1992) and during exercise (Fry, Kraemer, van Borselen, Lynch, Triplett, et al. 1994; Lehmann et al. 1992) have also been reported in overtrained endurance and strength/power athletes. These increases in catecholamine concentrations may reflect a loss of sensitivity in the target organs for catecholamines, specifically norepinephrine. Decreases in **β-adrenergic receptors** have been reported after prolonged high-volume training (Jost, Weiss, and Weicker 1989). The elevated catecholamine concentrations, also seen in overtrained athletes performing high-intensity resistance training, likely also reflect a down-regulation of the β-adrenergic receptors. Regardless of the exercise stimulus, a reduction in β-adrenergic receptors in muscle causes impaired muscle performance. This is highlighted by the significant relationship reported between immediate postexercise catecholamine concentrations and maximal strength performance in resistance-trained individuals (*r* ranging from 0.79 to 0.96). However, this relationship was not seen in overtrained subjects (Fry, Kraemer, van Borselen, Lynch, Triplett, et al. 1994).

Hypothalamic-pituitary dysfunction has been shown primarily in overtrained endurance athletes. In the limited amount of research available on overtrained resistance athletes, the hypothalamic-pituitary axis does not appear to be similarly affected (Fry 1998). Barron et al. (1985) were first to show a dysfunction in the hypothalamic-pituitary axis by reporting reduced ACTH, growth hormone, and prolactin responses to insulin-induced hypoglycemia in overtrained marathon runners. Although those investigators reported that the dysfunction was seen primarily in the **hypothalamus,** they did suggest that

pituitary insensitivity was possible. Other studies have shown reduced pulsatile LH secretion (Hackney, Sinning, and Bruot 1990; MacConnie et al. 1986) and reduced β-endorphin, thyroid-stimulating hormone, and growth hormone concentrations (Fry, Morton, and Keast 1991; Keizer 1998; Urhausen, Gabriel, and Kindermann 1995) in overtrained athletes. The impaired pituitary response in overtrained athletes may cause reproductive abnormalities, such as menstrual cycle irregularities (e.g., amenorrhea) in females and reduced libido and sperm counts in males. Figure 21.2 reviews the hypothalamic-pituitary-adrenal/gonadal axes and overtraining. In addition, some of the major symptoms associated with neuroendocrine dysfunction during overtraining are described.

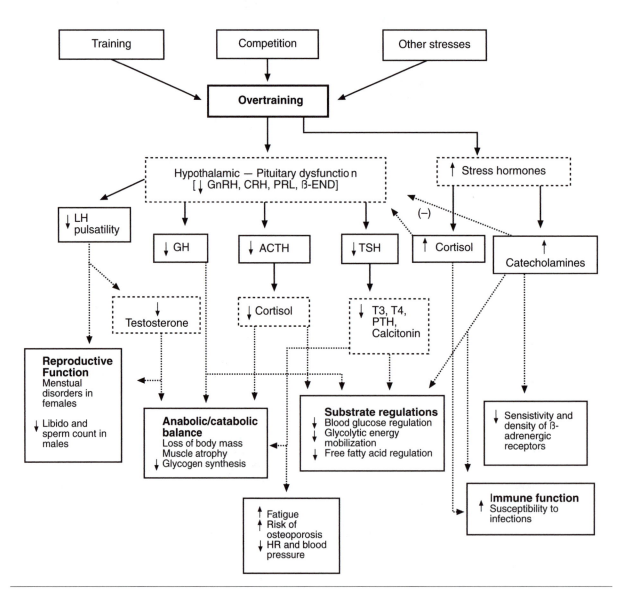

Figure 21.2   Hypothalamic-pituitary-adrenal/gonadal axes and overtraining.

Data from Fry et al. 1991; Urhausen et al. 1995.

Psychological Disturbances

Mood states have been reported to be sensitive to changes in training volume (Morgan et al. 1987; Raglin, Morgan, and Luchsinger 1990) and are thought to be useful as a potential tool for monitoring training adaptations. The **profile of mood states (POMS)** is a self-report inventory frequently used to assess the mood states of athletes (Morgan et al. 1987). Athletes tend to score lower than the normal population on scales of tension, anxiety, anger, confusion, and fatigue and score above the normal population in vigor. This mood pattern, called an **"iceberg profile"** because it has been suggested to resemble an iceberg, is typically observed at the onset of training. As a season progresses and training volume is increased, the mood profile of the athlete may become significantly altered to a pattern resembling that of the healthy nonathletic population. As training volume decreases during a taper, mood profiles return to preseason values. In contrast, a stale or overtrained athlete will not respond to the taper and his or her mood state remains significantly altered, reflected by a flattened mood profile.

Overtrained athletes may also exhibit lower confidence in their ability to succeed. A study examining overtrained resistance athletes has shown that these individuals did not exhibit any changes in their perception of the difficulty of the lifting task during 2 weeks of an overtraining protocol. However, within 8 days of the onset of this overtraining protocol, the athletes' confidence in their ability to successfully perform was significantly reduced (Fry, Fry, and Kraemer 1996). This study reflected the changes in self-efficacy that may be observed in overtrained athletes and that may contribute to the resulting impaired performance.

Immunological Disturbances

Overtrained endurance athletes appear to be at a greater risk of infectious illness, particularly upper respiratory tract infections (URTI), than their non-afflicted peers. Because most of the overtraining literature has focused on endurance athletes, it is unknown if overtrained strength/power athletes are at a similar risk. However, because immune suppression appears to be related to impaired neuroendocrine balance, it is likely that, if examined closely, overtrained strength/power athletes would also be more susceptible to infection.

Although changes in endocrine function (e.g., cortisol, catecholamines, and β-endorphins) are thought to be involved in the immune suppression reported during overtraining or during periods of intense training, the complexity of these changes suggests that more than one mechanism is responsible for alterations in immune function. As discussed in chapter 5, periods of intense training cause several changes in the immune system that may result in a greater susceptibility to illness in overtrained athletes. Reductions in resting leukocyte counts, lymphocyte counts, neutrophil activity, and immunoglobulin concentrations are likely some of the factors responsible for the increased risk of infectious disease in overreached or overtrained athletes.

Several epidemiological studies have reported a higher incidence of URTI in athletes engaged in marathon events or very heavy endurance training (Heath et al. 1991; Nieman et al. 1990; Peters and Bateman 1983). Very few studies have directly examined the incidence of illness and its relationship to overtraining. Mackinnon and Hooper (1996) compared illness rates in athletes showing symptoms of overreaching (33% of athletes examined) to those that seemed to have adapted well to the intensified training. Greater incidences of URTI were observed in those athletes who did not show symptoms of overreaching (56%) compared with the athletes who were experiencing symptoms of overreaching (12.5%). This study suggests that increases in URTI might depend more on changes in training intensity than on overreaching per se.

Biochemical Disturbances

Elevations in creatine kinase and uric acid and decreases in **glycogen** levels and **lactate** concentrations have been reported after high-intensity training periods, and each has been suggested to be a potential indicator of overtraining (Fry, Morton, and Keast 1991). However, many of these biochemical disturbances are seen as part of the normal response to an acute exercise stress and may remain impaired for several days postexercise. For example, muscle glycogen resynthesis may be impaired because of muscle damage, and full restoration of muscle glycogen after a marathon may take as long as a week (Sherman et al. 1983). Thus, caution should be used when interpreting changes in biochemical function postexercise for evaluation of overtraining. A sufficient period of recovery should be allowed to properly assess the athlete.

Performance Indicators

Several investigations have looked at the use of performance measures as an indicator of overtraining (Fry, Lawrence, et al. 1993; Fry, Kraemer, van Borselen, Lynch, Marsit, et al. 1994; Hoffman and Kaminsky 2000; Lehmann et al. 1992). Performance measures are an attractive method of monitoring training stresses. A multitude of biological variables have been identified as potential markers of overtraining; however, none have been identified as consistent predictors of overtraining. In addition, the cost associated with many of these measures makes these tests unlikely for many athletic teams.

Rowbottom, Keast, and Morton (1998) have reviewed a number of studies that have employed laboratory performance measures such as maximal oxygen uptake, running speed at 4 mmol · L^{-1} lactate, maximal effort time trials, and treadmill runs to exhaustion. It is imperative to select a performance test with established validity and reliability. Overtrained athletes may have performance decrements as much as 29%; however, performance differences are likely to be more subtle during initial stages of overreaching (Rowbottom, Keast, and Morton 1998). Endurance athletes have primarily been the focus of these performance measures. It is more complicated to select an appropriate test to monitor overreaching in strength/power athletes participating in a team sport. Hoffman and Kaminsky (2000) have reported on a testing battery used to monitor national-level youth basketball players. A major concern for the coaching staff was the number of teams that the athletes were playing for. In addition to the national team, all the players were competing for their respective high school or club team. In some cases, the athletes also played on the adult basketball team of their respective club. Thus, a number of performance tests representing the various components of fitness considered important for a basketball player (e.g., strength, speed, and agility) were used to monitor the athletes every month. The results of these tests, including the athletes' training volume and subjectively rated training intensity, can be seen in figure 21.3.

In this study, the 27-m sprint appeared to be the most sensitive test for highlighting players who were fatigued. When a "red flag" was seen (increase in sprint time greater than 0.15 s from the player's best time), a further analysis of the players' training log showed an increase in both training volume and training intensity for the 2 weeks preceding the testing period. Subsequently, the coaches of the player's club and high school teams were requested to reduce the practice volume of the player (excuse the player from practices). During the next testing session (1 month later), the player's sprint time returned to normal. This is depicted in table 21.2.

TREATMENT OF OVERTRAINING

The most important aspect of treating overtraining is simply preventing it from occurring. Obviously, this would be the primary goal of the coach. To this end, the coach will likely develop a training program based on the principles of periodization (see chapter 11) to minimize the occurrence of staleness, overreaching, or overtraining in the athlete. However, regardless of the care taken in the design of the training program, individual athletes are unique in their response to a training stimulus. This and other uncontrollable or unexpected influences (e.g., extreme environmental changes, exposure to altitude, and jet lag) may result in a poor adaptation to training stresses, causing a stagnant or decreased performance. Thus, the best way to treat overtraining would be through early detection.

During early stages of overtraining (i.e., overreaching), cessation of training for several days should be sufficient for complete recovery. After this rest period, the athlete can resume normal training. However, the cause of the staleness should be determined and appropriate adjustments should be made to the training program. During periods of competition, a decrease in the number or length of practice sessions should be considered. In addition, practice sessions during a busy competition schedule should focus primarily on technique and strategy rather than conditioning. However, a problem for the coach of a team sport is to separate the individual from the collective team. Often the stress of practices and games is not equal for all members of the team (starters versus nonstarters). Thus, reducing practice intensity and training volume may minimize the chance of staleness in the athletes playing most of the game, but it may also result in detraining the athletes who do not receive as much playing time. This highlights the importance for the coach to individualize even practices in team sports to whatever extent possible.

When an athlete is suffering from overtraining, training must be reduced drastically and competitions should be eliminated. The recovery for overtrained athletes is lengthy and may exceed 6 months in duration (Kreider, Fry, and O'Toole 1998). In such a situation, it is recommended that the athlete receive sufficient rest, sleep, relaxation, and proper nutrition (Kuipers and Keizer 1988). In addition, counseling may also help the athlete cope with the emotional conflicts and psychological demands he or she may be facing.

SUMMARY

Overtraining results when insufficient recovery accompanies an increase in the training stimulus (either an increase in training volume or an increase in training intensity). Inadequate regeneration occurs and performance can be affected. Overtraining is measured across a continuum of stages that increase in severity. During initial stages of overtraining, symptoms of fatigue may or may not be accompanied by decreased or stagnant performance. In later stages, several symptoms of overtraining become apparent, with considerable performance decrements. The mechanisms that under-

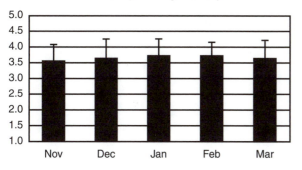

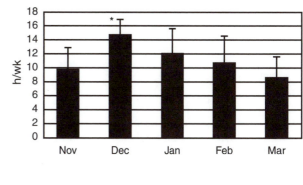

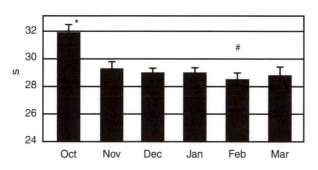

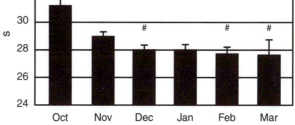

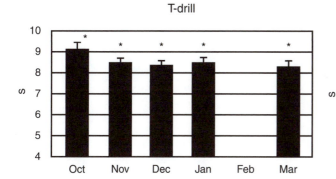

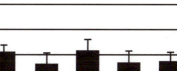

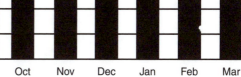

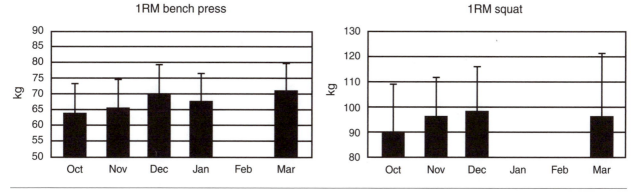

Figure 21.3 Example of athletic performance tests for monitoring a basketball team for overtraining. Intensity level based on subjective scale where 1 = very easy and 5 = very hard.

Table 21.2 Example of 'Red Flag' (in bold) and Subsequent Performance Results When Training Volume Was Reduced

Month	27-m sprint (s)	Training volume (h · wk^{-1})	Training intensity
November	4.00	15.3	3.6
December	**4.17**	22.3	4.2
January	3.98	17.3	4.0

Intensity level was subjectively rated using a 5-point scale. The rating scale included the following verbal commands: 1 = very easy, 2 = easy, 3 = average, 4 = hard, 5 = very hard. Results were averaged for the two weeks between National Team practice sessions.

lie overtraining may be different for endurance and strength/power athletes. During overreaching, several days of rest appear adequate for the athlete to recover and return to full performance. In fact, many coaches design their training programs to include periods of overreaching in hopes of overcompensation during the recovery period. In later stages of overtraining (overtraining syndrome), cessation of activity is required, and complete recovery may not be realized for more than 6 months.

CHAPTER 22

DIABETES MELLITUS

©Human Kinetics

Diabetes mellitus is the most common endocrine disorder, and it affects approximately 5% of the American population. It is possible that an equal percentage of people have this disease but are unaware of it. It is a major health problem and one of the leading causes of death by disease. The direct relationship between diabetes and mortality is difficult to assess because of the host of vascular complications common to people with diabetes, which result in heart disease and stroke. In addition, diabetes is a leading cause of visual impairment and blindness and a significant contributor to kidney disease. The incidence and prevalence of diabetes increase with age, and it is more common in minority populations. Regardless of the devastating complications that may result from diabetes, most individuals are able to live full and productive lives assuming that they are able to properly monitor and regulate their disease. With proper regulation, no limitations are imposed on people with diabetes. Many famous and successful athletes have competed at the highest levels while combating this disease. People with diabetes are playing or have played in the NBA (Chris Dudley of the New York Knicks), NFL (Jay Leeuwenburg of the Cincinnati Bengals and Michael Sinclair of the Seattle Seahawks), NHL (Bobby Clarke of the Philadelphia Flyers), and Major League Baseball (Jason Johnson of the Baltimore Orioles). In addition, Hall of Fame tennis player Bill Talbert lived with the disease for 70 years, and recent Olympic gold medalists Steve Redgrave (rowing) and Gary Hall Jr. (swimming) have also been diagnosed with diabetes. This chapter briefly reviews the physiology and the pathophysiology of the two main types of diabetes. The role of exercise in the management of diabetes is also discussed.

OVERVIEW OF DIABETES MELLITUS

Diabetes mellitus is a disease associated with the inability of the body to regulate blood **glucose** levels and **carbohydrate** metabolism. This may be caused either by an inability of the β-**cells** within the islets of Langerhans of the **pancreas** to secrete a sufficient quantity of **insulin** (the hormone responsible for glucose transport and **glycogen synthesis**) or by an inability of hepatic or peripheral tissues to respond to adequate insulin concentrations. Thus, there are two basic forms of the disease: **type 1,** or insulin-dependent diabetes mellitus **(IDDM),** and **type 2,** or non-insulin-dependent diabetes mellitus **(NIDDM).**

Type 1 Diabetes

Type 1 diabetes is characterized by an inadequate secretion of insulin by the β-cells of the pancreas. It has a sudden and dramatic onset and is diagnosed primarily in children and adolescents. IDDM accounts for approximately 5-10% of diabetes cases (Ivy, Zderic, and Fogt 1999). People with type 1 diabetes are more likely to develop **ketosis** and require exogenous insulin for survival.

An insulin deficiency has drastic effects on carbohydrate, **lipid,** and protein metabolism. Reduced or inadequate insulin concentrations cause a decrease in glucose uptake by both skeletal muscle and the liver. As a result, a reduction in glycogen synthesis occurs, which causes an increase in both **glycogenolysis** and **gluconeogenesis.** This results in greater blood glucose concentrations **(hyperglycemia).** The high glucose concentration in the blood rises to a level that exceeds the reabsorption capacity of the kidneys, causing glucose to appear in the urine **(glucosuria).** The addition of glucose to the filtrate causes **osmotic diuresis** (pulling of water into the filtrate) and creates a greater need for urination. The increased frequency of urination results in a constant urge to drink. If fluid intake is inadequate, the diabetic quickly becomes dehydrated, and it eventually leads to circulatory failure. This cycle is depicted in figure 22.1.

As a consequence of the inability to secrete insulin, a decrease in **triglyceride** synthesis is seen, which leads to an increase in circulating **free fatty acids.** This causes the liver to use a greater amount of free fatty acids as an energy substrate, causing a greater production and release of **ketone bodies** into the circulation. Subsequently, people with type 1 diabetes are at a greater risk of ketosis and **metabolic acidosis.**

Type 2 Diabetes

Type 2 diabetes is associated with varying amounts of insulin production by the pancreas. Most often, insulin concentrations are normal but at times may exceed that seen in the nondiabetic population. There is also a small subset of people with type 2 diabetes whose insulin levels are reduced because of a β-cell defect. However, the primary problem in NIDDM is a resistance to insulin action by both the liver and skeletal muscle. NIDDM accounts for approximately 90-95% of the total diabetic population (Ivy, Zderic, and Fogt 1999). Type 2 diabetes is seen primarily in an adult population (occurrence after the age of 24 in 95% of the cases). The risk of NIDDM increases with advancing age, and, although the precise etiology of the disease remains unknown, its occurrence is thought to be related to obesity and a sedentary lifestyle (Ivy, Zderic, and Fogt 1999).

Insulin resistance is the inability to respond appropriately to insulin. This can occur either by reduced insulin responsiveness or reduced insulin sensitivity. Reduced responsiveness refers to a postreceptor defect that causes a reduced biological response to a maximal

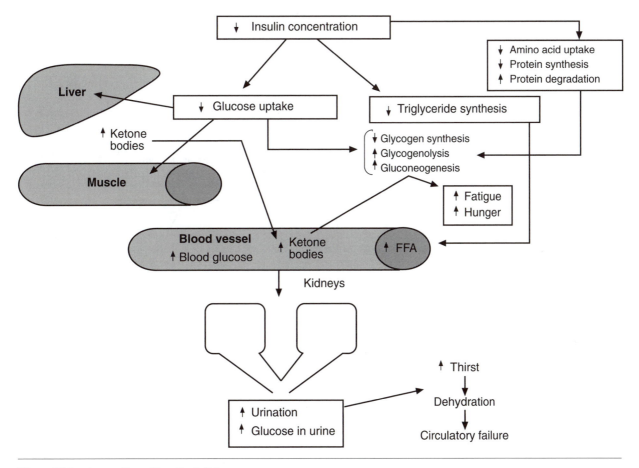

Figure 22.1 Acute effect of insulin deficiency.

stimulating insulin concentration; reduced sensitivity refers to an **insulin receptor** defect that causes a reduced biological action at a given submaximal insulin concentration (Ivy, Zderic, and Fogt 1999). Obesity is a prime cause of insulin resistance. Both **hyperinsulinemia** and **down-regulation** (reduction in number) of insulin receptors are associated with obesity. In most cases, the individual is able to secrete a greater amount of insulin, and carbohydrate homeostasis can thus be maintained. However, in individuals who have a genetic predisposition for diabetes, the stress placed on the pancreas will eventually exhaust the reserve capacity of the β-cells, causing a glucose intolerance (Hedge, Colby, and Goodman 1987).

Two mechanisms have been suggested by Ivy (1997) as a potential cause of type 2 diabetes (see figure 22.2). The first mechanism suggests that living a sedentary lifestyle results in a positive calorie balance and a subsequent increase in fat storage and **adipocyte** hypertrophy. As the adipocytes enlarge, an insulin resistance develops as a result of the reduced insulin receptor density. As free fatty acids (FFA) accumulate in the plasma, they begin to have several effects on blood glucose, including

stimulating gluconeogenesis and hepatic glucose output and inhibiting insulin-stimulated muscle glucose clearance. FFA may also accumulate within muscle tissue, causing insulin resistance and a compensatory increase in β-cell insulin production. This cycle continues until the β-cells become impaired and insulin production becomes reduced. This intensifies the insulin-resistant state and the reduced FFA clearance, and it accelerates hepatic glucose output, resulting in type 2 diabetes.

The second possible mechanism suggests that a sedentary lifestyle exposes a genetic defect in skeletal muscle, which results in muscle insulin resistance. Similar to the first mechanism, a hyperinsulinemia results in response to the elevated blood glucose. However, in this scenario, the hyperinsulinemia suppresses FFA oxidation and increases triglyceride storage and adipocyte hypertrophy. As a result, the adipocytes become insulin resistant and an increase in FFA concentration is observed. Eventually, this cycle will result in β-cell impairment and the development of type 2 diabetes.

A number of complications are associated with diabetes, especially NIDDM. These complications include an increased risk of coronary heart disease because of

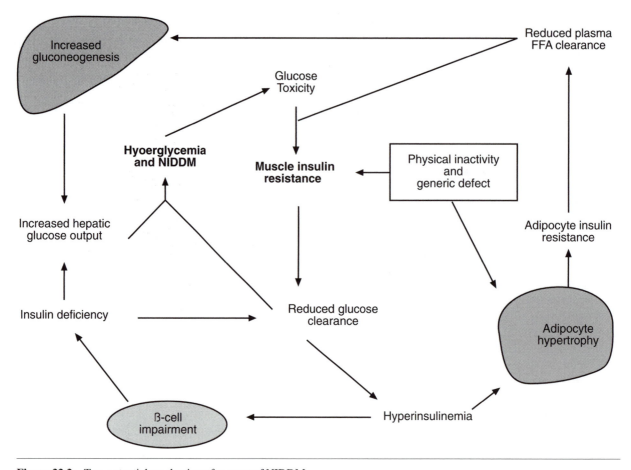

Figure 22.2 Two potential mechanisms for onset of NIDDM.

Adapted, by permission, from J.L. Ivy, T.W. Zderic, D.L. Fogt, 1999, "Prevention and treatment of non-insulin-dependent diabetes mellitus," *Exercise and Sport Science Reviews* 27:1-35.

atherosclerotic changes to the coronary vasculature. The atherosclerotic changes common in people with diabetes are not limited to the coronary vasculature but are also seen in the cerebral and peripheral vascular beds. The high insulin concentrations associated with this disease cause an increase in lipid synthesis and deposition to arterial walls. In addition, microvascular lesions in the kidneys and retinas of people with diabetes are commonly seen, leading to a high prevalence of kidney disease, visual impairment, and blindness. Multiple neuropathies are also frequently found in people with diabetes, resulting in dysfunction of peripheral nerves, the spinal cord, and the brain and causing disruption to most physiological systems in the body. The mechanisms directly responsible for these complications are not well understood but are largely believed to be the result of the inadequate insulin and the metabolic disturbances associated with diabetes. In addition, there is some debate whether proper management of diabetes will reduce the incidence of these complications. Some theories suggest that the vascular lesions seen in the people with diabetes are genetic in origin, and the occurrence of diabetes is associated with

a genetic predisposition to the disease (Hedge, Colby, and Goodman 1987).

Treatment and Management of Diabetes

For individuals with type 1 diabetes, insulin is the only treatment available and in some cases may be the best therapeutic option for type 2 (NIDDM) diabetes. Insulin is available in many different formulations that differ in their onset of action, maximal activity, and duration of action. Depending on what chemical insulin is conjugated with, its action may become evident within 30 min or up to 24 h after administration. The large variability in the onset of insulin action is related to several factors, including site of injection, volume of insulin injected, and physical activity of the patient.

In NIDDM, the use of oral hypoglycemic agents may be the primary pharmacological means to control this type of diabetes. The use of this medication is limited to those individuals who are capable of secreting insulin. It is thought that these hypoglycemic agents enhance the binding of insulin to peripheral receptors, causing an increase in glucose utilization and a decrease in hepatic glucose production.

EXERCISE AND DIABETES

The benefits of exercise as a treatment modality for diabetes have been known for centuries (Wallberg-Henriksson 1992). Even when insulin was not yet available, exercise was a generally prescribed course of treatment for the people with diabetes. The benefit of exercise has been related primarily to enhanced glucose uptake and insulin sensitivity, resulting in improved **glycemic control.** However, exercise may have even added importance for the person with type 2 diabetes by reducing the risk of coronary heart disease, which is common in the diabetic population.

Exercise and Carbohydrate Metabolism in Nondiabetic Individuals

Exercise places a large demand on carbohydrate metabolism. Although exercise and metabolism have been covered in greater detail in chapter 3, it is important to briefly review carbohydrate metabolism in the nondiabetic individual to appreciate the demands placed on the diabetic person during exercise.

In people without diabetes, glucose uptake and utilization may increase more than 10-fold during exercise (Kanj, Schneider, and Ruderman 1988). However, the body is able to regulate blood glucose concentrations effectively so that only minimal changes are seen. During exercise, the increase in glucose demand is met by an accompanying increase in hepatic glucose production, which prevents a state of **hypoglycemia** (low blood sugar). Hepatic production of glucose is stimulated by an increase in the circulating concentrations of **catecholamines** and **glucagon** as well as a decrease in insulin concentration. Insulin and glucagon are both released from the pancreas and have opposite effects. As insulin concentrations in the circulation decrease, glycogen **phosphorylase** activity (the enzyme responsible for the breakdown of glycogen) in the liver increases, thereby enhancing glycogenolysis (the process of breaking down the glycogen molecule to its simplest form, glucose). At the same time, glucagon levels tend to elevate, which also has a glycogenolytic effect. When exercise ceases and the depleted glycogen stores need to be replaced, the reverse actions of insulin and glucagon on the liver take place. The interaction of these hormones on the liver is critical for regulating both glucose production and glycogen synthesis.

After exercise, a heightened sensitivity to insulin makes replenishing depleted glucose stores easily accomplished at the next meal. The enhanced insulin sensitivity appears to last for several hours after exercise and is also a hallmark of metabolic adaptation to endurance training (see figure 22.3).

Effect of Exercise on Type 1 (Insulin-Dependent) Diabetes

The primary concern for people with insulin-dependent diabetes is avoiding either hypo- or hyperglycemic conditions. The hypoglycemic state is the most frequent disturbance associated with exercise in this population. As previously mentioned, the person with type 1 diabetes lacks the ability to perform normal glucose regulation. If insulin levels are too high at the onset of exercise (possibly from too great an insulin injection dose or from an accelerated absorption of insulin from the injection site), insulin levels will not decrease in a normal physiological fashion during exercise. This prevents the liver from producing sufficient glucose to meet peripheral glucose demand, causing hypoglycemia. Exercise intensity and duration also appear to be determining factors in the magnitude of hypoglycemia (Wallberg-Henriksson 1992). As exercise intensity or duration increases, the risk of developing hypoglycemia increases as well. Hypoglycemia may not be a problem only during exercise but may also occur 4-6 h after an exercise session (Campaign, Wallberg-Henriksson, and Gunnarsson 1987).

Hyperglycemia in people with type 1 diabetes is rarely seen during exercise and may occur if blood glucose levels are initially high when exercise begins (Wahren, Hagenfeldt, and Felig 1975). The lack of insulin impairs glucose transport into the exercising muscles, forcing a greater reliance on free fatty acids for fuel. An increase in ketone levels may result from a glucose–fatty acid cycle, which may be further accelerated by increases in the counter-regulatory hormones

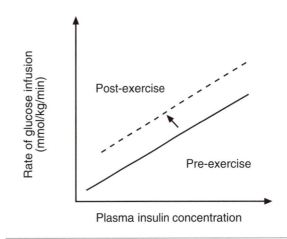

Figure 22.3 Enhanced insulin sensitivity after exercise.

©Human Kinetics

(e.g., glucagon, catecholamines, and growth hormone). This in turn may exacerbate the hyperglycemic state and possibly lead to the development of ketosis. For this reason, people with diabetes must be under adequate control before beginning an exercise program. The American College of Sports Medicine, American Dietetic Association, and Dietitians of Canada (2000) have suggested a number of precautions that a diabetic individual should follow before exercise to minimize the risk of unwanted reactions during exercise.

- Measure blood glucose before, during, and after exercise.
- Avoid exercise during periods of peak insulin activity.
- Unplanned exercise should be preceded by extra carbohydrates (e.g., 20-30 g per 30 min of exercise); insulin may have to be decreased after exercise.
- If exercise is planned, insulin dosages must be decreased before and after exercise according to the exercise intensity and duration as well as the personal experience of the patient; insulin dosage re-

ductions may amount to 50-90% of daily insulin requirements.
- During exercise, easily absorbable carbohydrates may have to be consumed.
- After exercise, an extra carbohydrate-rich snack may be necessary.
- Be knowledgeable of the signs and symptoms of hypoglycemia.
- Exercise with a partner.

Acute exercise in people with type 1 diabetes has been shown to reduce blood glucose concentrations (Wallberg-Henriksson 1992). The greater glucose uptake by the exercising muscles exerts an insulin-like effect. However, in long-term controlled studies, investigators have been unable to demonstrate that exercise training is capable of improving glycemic control in people with type 1 diabetes (Wallberg-Henriksson et al. 1982, 1986; Zinman, Zuniga-Guajardo, and Kelly 1984). Studies examining children and adolescents have met with conflicting results. An improved glycemic control has been reported in some studies (Campaign et al. 1984; Dahl-Jorgensen et al. 1980), whereas other studies have been unable to see any improvement in glycemic control (Hansen et al. 1989; Huttunen et al. 1989). These contrasting results may be related to the level of pretraining metabolic control (Wallberg-Henriksson 1992). Exercise programs for children with initially poor glycemic control appeared to be of benefit by bringing blood glucose levels under better control. However, in children who had good metabolic control before the onset of the exercise program, no further changes in metabolic control were seen.

Although exercise may not improve glycemic control, people with type 1 diabetes appear to benefit from long-term exercise by improving insulin sensitivity. Insulin sensitivity refers to a reduced concentration of insulin required to stimulate transport of a similar concentration of glucose into the muscle. Insulin sensitivity may increase 20% in the type 1 diabetic after 16 weeks of exercise (Wallberg-Henriksson 1992). Practically speaking, the person with type 1 diabetes will be able to reduce the dose concentration in the insulin injection.

People with type 1 diabetes appear able to achieve similar improvements in maximal aerobic capacity as those without diabetes (Wallberg-Henriksson 1992). In addition, physiological adaptations associated with endurance training, such as increases in mitochondrial enzyme concentrations, are also seen (Costill, Cleary, et al. 1979; Wallberg-Henriksson et al. 1984). However, several physiological adaptations associated with training may be compromised in the diabetic population with long-standing IDDM. For instance, the magnitude of the capillary density increase appears to be blunted in people with type 1 diabetes who have had the disease for more than 15 years in comparison with healthy individuals or even those with type 1 diabetes who have had IDDM for less than 15 years (Wallberg-Henriksson et al. 1984). In addition, decreases in some glycolytic enzymes have also been reported in people with type 1 diabetes (Wallberg-Henriksson 1992). The potential for physiological adaptation and improvement in exercise capacity may be impaired in those with type 1 diabetes who have complications associated with diabetes, such as autonomic neuropathies or nephropathy.

Effect of Exercise on Type 2 (Non-Insulin-Dependent) Diabetes

Several studies examining the effect of exercise training programs on people with type 2 diabetes have demonstrated consistent improvements in both glycemic control and insulin sensitivity (Dela et al. 1995; Holloszy et al. 1986; Reitman et al. 1984; Schneider et al. 1984). However, the magnitude of glycemic control, similar to what was seen in those with in type 1 diabetes, appeared to be related to the pretraining status (individual's initial

level before the start of the exercise program). In subjects whose blood glucose levels were under proper control, the extent of improvement was minimal, but in subjects whose preexercise glycemic control was impaired, exercise appeared to provide substantial benefits for better control of blood glucose concentrations (Holloszy et al. 1986).

As previously stated, NIDDM is associated with obesity and a sedentary lifestyle. Individuals who are 20-30% overweight are at a higher risk of developing type 2 diabetes (Wallberg-Henriksson 1992). Therefore, in addition to the benefits of exercise on glucose tolerance and insulin sensitivity, further benefits of exercise for the person with type 2 diabetes include reducing body fat, improving blood lipid profiles, and decreasing hypertension. These training adaptations reduce the inherently high risk of developing cardiovascular disease in this subject population as well as improve well-being and quality of life. In addition, the physiological adaptations resulting from chronic exercise training may also prevent or delay the cellular changes associated with the development of NIDDM (Ivy, Zderic, and Fogt 1999). The mechanisms that have been suggested to improve insulin action and glucose control during exercise training are depicted in figure 22.4.

A decrease in body fat, especially abdominal fat, has a significant role in reducing insulin resistance of obese individuals (Despres, Nadeau, and Bouchard 1988). Ivy, Zderic, and Fogt (1999) have suggested several potential explanations, although the exact mechanism is not known. Obese individuals with NIDDM are known to be resistant to insulin suppression of plasma FFA (Ivy, Zderic, and Fogt 1999). As FFA accumulate in the blood, an increase in both gluconeogenesis and hepatic glucose output is seen. The inability to properly regulate glucose transport into tissue is further diminished in these individuals, and this negative chain of events leads to the hyperglycemic state common in NIDDM. The size of the adipocytes is also related to the degree of insulin resistance. An inverse relationship has been demonstrated between adipocyte size and insulin receptor density (Craig et al. 1981). A down-regulation in insulin receptors reduces the chances of insulin interacting with its receptor and reduces insulin-stimulated glucose uptake. In addition, increases in plasma FFA may also result in an increase in triglyceride formation and accumulation in muscle tissue. As muscle triglyceride concentration increases, an inverse relationship is seen with insulin-stimulated glucose uptake (Goodpaster et al. 1997; Phillips et al. 1996). This may result in peripheral insulin resistance and subsequent development of NIDDM (Ivy, Zderic, and Fogt 1999). Thus, the importance placed on reducing body fat through an exercise and diet program is easily understood considering the

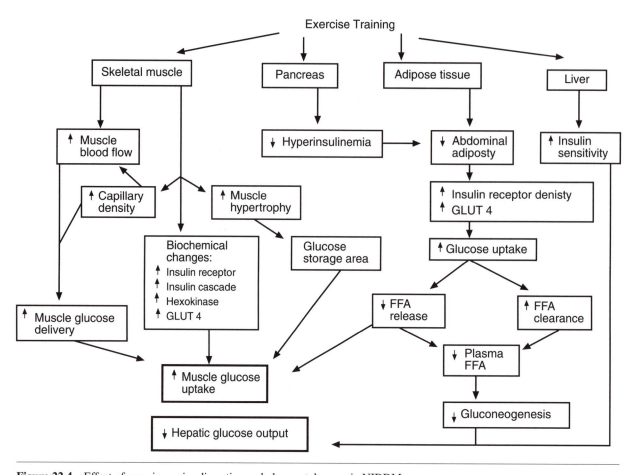

Figure 22.4   Effect of exercise on insulin action and glucose tolerance in NIDDM.

Adapted, by permission, from J.L. Ivy, T.W. Zderic, D.L. Fogt, 1999, "Prevention and treatment of non-insulin-dependent diabetes mellitus," *Exercise and Sport Science Reviews* 27:1-35.

possible consequences when a sedentary lifestyle leads to obesity. The benefits of exercise training are reflected in either preventing the disease or limiting its progression and improving its management.

Exercise training causes a number of skeletal muscle adaptations that were reviewed in chapter 1. These physiological adaptations depend on the type of training program employed. Endurance training may result in an increase in skeletal muscle blood flow because of the increased capillary density associated with such training. Improved skeletal muscle blood flow would be beneficial for correcting any possible vascular deficiencies reported in NIDDM patients (Laakso et al. 1992). This may also be a potential mechanism of the improved insulin sensitivity and glucose uptake seen in people with type 2 diabetes after prolonged endurance training (Dela et al. 1995).

People with type 2 diabetes are also reported to have a reduced number of insulin receptors located on the cell membrane (Caro et al. 1987; Olefsky 1976). This down-regulation reduces the opportunity for interaction between circulating insulin and its receptor. Down-regulation is also seen in obese individuals (Caro et al. 1987; Olefsky 1976) and may possibly be reversed during exercise training (Dohm, Sinha, and Caro 1987). However, research is still ongoing in this important area and, as yet, **up-regulation** (increase in receptor number) has not been demonstrated in human NIDDM patients.

The insulin resistance seen in people with type 2 diabetes may not necessarily be associated with a receptor-hormone interaction defect. For glucose to move across the cell membrane, a transporter protein is needed. This transporter protein **(GLUT 4)** is located intracellularly and is translocated to the cell membrane by insulin action. As the concentration of GLUT 4 increases, the rate of glucose transport across the cell membrane increases as well. A deficiency in GLUT 4 is not typically seen in those with type 2 diabetes (Lund et al. 1993). More likely, the problem involves a reduced ability of insulin to translocate GLUT 4 from its intracellular storage site to the cell membrane (Ivy, Zderic, and Fogt 1999). Since GLUT 4 has been shown to increase as a result of training

(Houmard et al. 1991), the benefit for the those with type 2 diabetes may involve increasing GLUT 4 concentrations within the cell. This possibly compensates for the defect in translocation by somehow positioning GLUT 4 closer to the cell membrane and enhancing glucose transport (Ivy, Zderic, and Fogt 1999).

EXERCISE PRESCRIPTION FOR PEOPLE WITH DIABETES

There are no physical limitations placed on young, active people who have type 1 diabetes who do not have any complications and have good blood glucose control. Those with type 1 diabetes participate in all sports at all levels, including elite and professional. A primary concern for these athletes is the proper adjustment of insulin dosage and diet to ensure safe participation and maximum performance. It is also important that people with type 1 diabetes and their training partners or coaches be aware of management techniques and treatment of hypoglycemia.

For those with type 1 diabetes who are sedentary and have had IDDM for several years, and for those with type 2 diabetes, it is highly recommended that a complete cardiovascular evaluation be performed before the person begins an exercise regimen. Many of these patients have asymptomatic coronary artery disease and need to be thoroughly evaluated before participating in any physical exercise program. In addition, the mode of exercise selected will depend on any complicating diseases. Certain exercises will be contraindicated according to the medical limitation. For example, in diabetics with peripheral neuropathy, jogging may cause trauma to the lower extremities and would not be recommended for the exercise program.

Relatively few studies have examined the optimal intensity and duration of exercise for the diabetic patient. However, the American College of Sports Medicine in collaboration with the American Diabetes Association (1997) reported, based on several long-term training studies, that exercise at an intensity ranging from 50 to 80% of $\dot{V}O_2$ max, performed three to four times per week for 30-60 min per session, appears to have the greatest potential for eliciting the desired metabolic adaptations. Aerobic exercise (e.g., brisk walking, jogging, cycling, or swimming) appears to be the most desirable; however, resistance training has also been reported to improve glucose tolerance in the diabetic patient (Miller, Sherman, and Ivy 1984). Nevertheless, for older people with diabetes and for those with other complications such as retinopathy, some concern has been raised about the advisability of performing high-intensity anaerobic exercise (Kanj, Schneider, and Ruderman 1988).

SUMMARY

This chapter demonstrated the importance of exercise as a treatment for both type 1 and type 2 diabetes. For those with type 1 diabetes, there do not appear to be any limitations to the exercise regimens they can participate in. The primary concern for these patients is adjusting their insulin dosage and diet so that they can exercise without increasing the risk of hypoglycemia.

Exercise is of vital importance in improving the metabolic dysfunction associated with the disease. However, exercise may play an even greater role in preventing the disease or in limiting its progression. Large muscle-mass exercises such as brisk walking, jogging, swimming, or cycling appear most able to elicit the desired metabolic adaptations. Resistance training has also been shown to be of benefit as part of the exercise training regimen for the person with type 2 diabetes.

CHAPTER 23

EXERCISE-INDUCED ASTHMA

©Susan Hindman

Exercise-induced asthma (EIA) is a condition in which vigorous exercise stimulates an acute narrowing of the airway in individuals with heightened airway sensitivity. Exercise-induced **bronchospasm** has also been used to describe this condition and may even provide a more accurate description (McFadden Jr. and Gilbert 1994). Symptoms of EIA include cough, wheezing, and chest tightness during or after exercise (Storms 1999). EIA often occurs in otherwise healthy individuals who experience these symptoms only during exercise. However, up to 90% of people with asthma may experience EIA when exercising at a relatively high **intensity** (Godfrey 1988). The prevalence of EIA is reported to be two to three times greater than that of asthma and is thought to reflect the large subset of individuals whose airways are sensitive to an exercise stress (Smith and LaBotz 1998).

Asthma is one of the most common conditions seen in the United States. The number of individuals diagnosed with asthma has more than doubled in the past 15 years (Centers for Disease Control and Prevention 1998). Approximately 15 million people have asthma in the United States and 25% of these people are under the age of 14. Athletes appear to be as susceptible to this disease as any other population group. However, asthma does not limit athletes from achieving their maximum potential. This was especially highlighted during the 1984 Summer Olympics in which 67 of the 597 American athletes (11.2%) were reported to have EIA (Voy 1986). Of these athletes, 41 won medals (15 gold, 20 silver, and 6 bronze) in 14 different sports. During the 1996 Summer Olympics, 20% of the participating American athletes were reported to have EIA (Weiler, Layton, and Hunt 1998). The American delegation is not the only group reporting a high incidence of EIA among its athletes. Australian teams have reported that greater than 9% of their athletes participating in the Olympic Games between 1976 and 1992 had asthma (Morton 1995). The occurrence of EIA in Winter Olympic athletes may be even greater. Wilber et al. (2000) reported that 23% of the athletes on the 1998 United States Winter Olympic team had EIA. This is consistent with other reports of a higher incidence of EIA among winter sport athletes (Heir and Oseid 1994; Larsson, Hemmingsson, and Boethius 1994).

WHAT IS EXERCISE-INDUCED ASTHMA?

Asthma is characterized by hypersensitive airways that overreact to air pollutants, **allergens** (e.g., pollen, dust, specific foods), psychological stress, and other triggers such as temperature changes or physical exertion. The response to these stimuli results in a reversible airway obstruction caused by **bronchoconstriction** or inflammation of the mucosal linings. As a result, the afflicted individual is unable to exhale completely and both **residual volume** and **vital capacity** are reduced. The feeling that one might experience during a mild or even a moderate asthma attack at rest may be described as nothing more than an inconvenience. However, if such an attack occurs during moderate physical activity, the individual may experience severe respiratory difficulty (Morton 1995).

When physical activity is the trigger for inducing an asthmatic reaction, it is referred to as EIA. Exercise can be the primary and only stimulus that results in an asthmatic response. Often when intense exercise begins, the **peak expiratory flow rate (PEFR), forced expiratory volume in the first second (FEV$_1$),** and **forced expiratory volume percent (FEV%)** are elevated. This is common in all exercising individuals, asthmatic or not. However, in the individual with EIA, these values become reduced to below preexercise levels as the exercise bout continues. Specific values considered to be an indicator of EIA are further discussed in the section on the diagnosis of EIA.

The onset of EIA appears to occur only when a relatively high intensity of exercise is reached. Bronchospasms are most often reported when exercise intensity is greater than 65% of the individual's **maximal oxygen consumption** (Morton 1995). As exercise intensity elevates, an increase in **catecholamine** concentrations is seen (see chapter 2). The rise in catecholamine concentrations reflects the increase in **bronchodilation** in both asthmatic and nonasthmatic subjects at the onset of high-intensity exercise. The elevated catecholamine concentrations may serve a protective function for people with bronchoconstriction

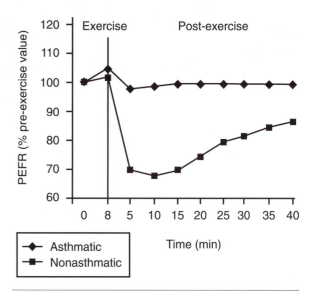

Figure 23.1   Changes in PEFR after exercise in asthmatic and nonasthmatic subjects.

during exercise. This may possibly explain the relatively large incidence of EIA commonly seen after exercise when catecholamine concentrations return to baseline levels. It is important to note, however, that much of this information has come as a result of traditional clinical testing for EIA in which most protocols call for 6-8 min of strenuous exercise. Most athletic events require exercise durations that far exceed 6-8 min of intense activity and that often result in the appearance of symptoms of EIA.

At cessation of exercise, FEV$_1$ and PEFR in the person with asthma will drop by at least 15% of their preexercise values and reach their lowest points at about 3-10 min postexercise (Morton 1995). Recovery is gradual, with a return to baseline levels in about 60 min (see figure 23.1). In some populations, primarily children, a second or late reaction may appear 3-4 h after exercise (Morton 1995).

When exercise produces an asthmatic reaction, there appears to be a period of time in which any further exercise results in a reduced bronchoconstrictive response in comparison with the initial episode of EIA. This period of time may last for up to several hours and is known as the **refractory period.** A possible mechanism responsible for the refractory period is an increase in **prosta-**
glandins, which results in an increase in bronchodilation (Manning, Watson, and O'Byrne 1993).

PATHOPHYSIOLOGY OF EIA

The pathogenesis of EIA is not completely understood, although it is generally believed that EIA is closely associated with changes in both heat and water exchange in the **tracheobronchial tree** (McFadden Jr. and Gilbert 1994). **Hyperventilation** associated with intense exercise results in an inability to properly warm and humidify the inhaled air. As the air travels down the tracheobronchial tree, it remains cool and dry, which may cause an alteration in the airway fluid-layer homeostasis (Smith and LaBotz 1998). There is some debate about the relative contribution of bronchial heat and water loss in EIA, and this has resulted in two separate hypotheses governing the pathophysiology of EIA.

The first of these hypotheses is the **hyperosmolarity theory.** During exercise, the ventilation rate is dramatically increased. This rapid breathing may result in an increased evaporation of mucosal surface water as the air travels from the upper to lower airways. As the water loss increases, changes in intracellular osmolarity and temperature occur, although the

mechanisms responsible are not completely understood. The hyperosmolarity of the airway surface water, or **airway drying,** appears to result in **mast cell** degranulation, **histamine** release, and airway smooth muscle constriction (McFadden Jr. and Gilbert 1994). Support for this theory has been generated primarily by studies showing bronchoconstriction after inhalation of hyperosmolar saline at rest (Storms 1999). However, no direct evidence demonstrates that airway drying develops, and some studies suggest that it may not (Gilbert, Fouke, and McFadden Jr. 1987, 1988).

The second hypothesis, referred to as the **thermal expenditure theory,** proposes that rapid rewarming of the airways after exercise results in bronchoconstriction (Storms 1999). This is believed to be caused by **hyperemia** (increased blood flow) in the vasculature of the airway, with a resulting edema, and does not involve constriction of the bronchial smooth muscle itself. During exercise, heat is transferred from the bronchiolar vasculature in an attempt to warm the inhaled air. After exercise, there is a rapid rewarming of the airways. As a result, the bronchiolar vessels become dilated (from the influx of blood for rewarming), causing the airways to narrow and, hence, the bronchoconstriction. The plausibility of this hypothesis has gained support based on studies demonstrating that the capillary bed is more permeable in the asthmatic than in the nonasthmatic population (McFadden Jr. and Gilbert 1994; Persson 1986). In addition, changes to bronchiolar blood vessels have been shown to influence the cooling and heating of the airways as well as affect pulmonary function (Gilbert and McFadden Jr. 1992; McFadden Jr. and Gilbert 1994). As these vessels dilate, some fluid may leak into the tissue, leading to an inflammatory mediator release and resulting in bronchospasm (Storms 1999). Figure 23.2 depicts the two hypotheses that have been suggested as mechanisms leading to EIA.

DIAGNOSIS OF EIA

The diagnosis of EIA is frequently based on self-reported symptoms (chest tightness, **dyspnea** out of proportion to the exercise intensity, coughing, wheezing, and excess sputum) without pulmonary function testing. However, Rundell and colleagues (2000) have highlighted the inadequacy of relying only on symptoms of EIA for diagnosis. In their study, conducted on United States national-level cold-weather athletes, they reported that 45% of the athletes who had normal pulmonary function exams reported symptoms of EIA, and only 61% of the athletes with positive pulmonary function tests reported these symptoms. The researchers concluded that the use

of symptoms alone appears to be unreliable for the diagnosis of EIA.

A general guideline for assessment of EIA includes 6-8 min of exercise at an intensity approximately 85% of the subject's maximal predicted heart rate under standard laboratory conditions (e.g., normal room temperature). However, some athletes (especially cold-weather athletes) with EIA may not experience any symptoms until exercise intensity reaches their race pace (90-100% HRmax), and the ambient temperature during exercise is cold (Rundell et al. 2000). Exercise protocols requiring the individual to exercise at a percentage of maximal oxygen consumption have also been recommended (McFadden Jr. and Gilbert 1994). When possible, the individual should be tested in the mode of activity and environment that induces the asthmatic symptoms. The diagnosis of EIA is confirmed with pulmonary function tests. A fall in PEFR or FEV_1 greater than 15% is indicative of airway flow obstruction (McFadden Jr. and Gilbert 1994; Rupp et al. 1992), although some reports have suggested decrements as low as 10% to be of clinical significance (Tan and Spector 1998). A recent study by Rundell and colleagues (2000) has shown that field tests performed on cold-weather athletes may be more sensitive than laboratory-based exercise challenges in the diagnosis of EIA. Field tests used were the actual competition or simulated competition of the athlete. Exercise duration varied from approximately 1 min and 20 s for speedskaters to over 1 h for cross-country skiers. Results of this study suggested that not only were the field tests more sensitive than laboratory measures of these athletes but that accepted criteria for EIA diagnosis (discussed previously) may be too restrictive. The authors concluded that EIA may be diagnosed in cold-weather athletes using a sport/environment field test based on postexercise decreases of –8.3% for forced vital capacity, –6.5% for FEV_1, and –12% for PEFR.

FACTORS MODIFYING THE ASTHMATIC RESPONSE TO EXERCISE

A number of variables influence the asthmatic response to a bout of exercise. As previously mentioned, if the goal is to induce an asthmatic response, it would be most productive to use the mode of activity that the individual primarily performs. However, in general, running is the most likely mode of exercise to induce an asthmatic response, whereas swimming and walking are the least likely activities reported to induce EIA.

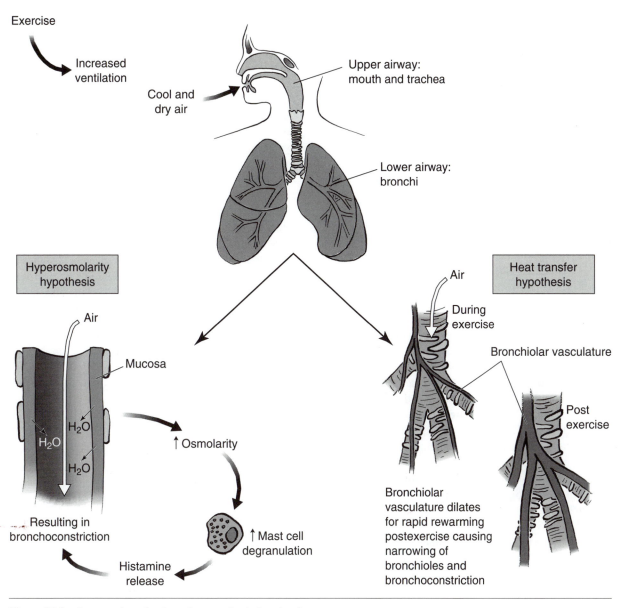

Figure 23.2 Suggested mechanisms for exercise-induced asthma.

Both the duration and intensity of exercise are important variables for triggering an asthmatic response. When exercise is of short duration (e.g., 8 min, as might be the case in an exercise challenge performed to induce EIA), bronchodilation is generally seen during the exercise period. However, at the conclusion of exercise, bronchoconstriction often occurs during the recovery period. When exercise is of longer duration, bronchoconstriction may be evident after 15 min of activity and may remain constant until the end of the exercise session (Beck 1999). During intermittent types of exercise, similar to most sporting events, bronchodilation is frequently seen during the periods of high intensity, and bronchoconstriction is seen when the exercise intensity is reduced (Beck, Offord, and Scanlon 1994). Postexercise bronchoconstriction is related to the intensity of exercise. The maximum postexercise asthmatic response occurs when exercise intensity is 65-75% of maximal oxygen consumption or 75-85% of the predicted maximal heart rate (Morton 1995). However, as mentioned earlier, this may be higher in elite-level athletes (Rundell et al. 2000).

The environment also has a significant influence on EIA. A cold, dry climate appears to be the most likely to induce an asthmatic response during exercise. This is reflected by the higher incidence of EIA reported in Olympic winter-sport athletes than in Olympic summer-sport athletes (Wilber et al. 2000). Other environmental factors that have been shown to influence the severity of EIA include the level of air pollution and allergens.

TREATMENT OF EIA

The primary goal of treating individuals with EIA is prophylaxis (preventing an asthmatic attack), but therapy may also be based on symptoms. The treatment plans for individuals with EIA can be of either pharmacological or nonpharmacological means and are critical for helping both the recreational and competitive athlete perform. However, the competitive athlete and his or her physician need to be aware of the restrictions placed on certain medications used to treat asthma. Some of these therapies have proven ergogenic effects and are banned by major sport governing bodies.

Inhalation therapy is the most common form of treatment. β_2-agonists are the first line of therapy in the prevention and treatment of EIA (Smith and LaBotz 1998). These drugs bind to β_2-adrenergic receptors, causing bronchodilation. The binding properties of a β_2-agonist are similar to those of catechlolamines. The major advantage of this particular pharmacological therapy is the longer duration of effect of the β_2-agonists in comparison with the endogenous catecholamine response. Although side effects (tachycardia and muscle tremors) caused by cross-reactivity with α- and β_1-receptors were a concern when this therapy was initially introduced, most of the β_2-agonists used today have a high β_2-receptor selectivity. Albuterol is one of the most common β_2-agonists currently being used. It is of short duration and is generally given prophylactically. Protection against EIA lasts for 1 to 2 h after administration (Smith and LaBotz 1998). If exercise is of longer duration, selection of a β_2-agonist with longer-lasting protection (e.g., bitolterol or salmeterol in Canada and in European nations) could be considered (McFadden Jr. and Gilbert 1994; Smith and LaBotz 1998). β_2-agonists may also be used during exercise when symptoms are slow to resolve.

Other standard pharmacological therapies used to combat EIA or asthma include the **khellin derivatives** (cromolyn sodium and nedocromil), **anticholinergics,** and **glucocorticoids.** The khellin derivatives are often used as a second-line therapy for prophylaxis. Cromolyn inhibits mast cell degranulation, and nedocromil has anti-inflammatory properties. However, the relatively high degree of efficacy with minimal side effects of these medications may make them the first choice for prophy-

laxis in some situations (Smith and LaBotz 1998). The use of anticholinergic agents may be limited because of their slow onset of action in the treatment of EIA. Glucocorticoid treatment is common and effective for managing asthma, but its efficacy in treating strictly EIA is relatively unknown. However, the use of inhaled glucocorticoids in combination with a β_2-agonist is thought to have possible benefits for the individual with severe EIA symptoms (Smith and LaBotz 1998). Other therapies that have been suggested or are in experimental stages (e.g., antihistamines and leukotriene antagonists) may have potential use in the treatment of EIA.

Bronchodilator agents effectively improve the ventilatory capacity of those with asthma. However, this improvement is generally reflected as a return to baseline; bronchodilators do not have the ability to increase ventilatory capacity beyond normal levels. β_2-agonists do have a stimulatory effect, and some have been suggested to have anabolic effects (primarily clenbuterol). Because of the stimulatory effect associated with the use of β_2-agonists, the National Collegiate Athletic Association (NCAA) and the United States Olympic Committee (USOC) have banned the systemic use of these agents. Although the NCAA permits the use of most inhaled prescription therapies, the USOC has banned all β_2-agonists except for albuterol, terbutaline, and salmeterol. The USOC requires written notification of the use of these β_2-agonists. However, khellin derivatives do not require written notification and are permitted to be used by the athletes.

HOW TO EXERCISE WITH ASTHMA (NONPHARMACOLOGICAL THERAPY)

As seen in the introduction to this chapter, EIA should not pose any limitations for athletes trying to reach their maximum performance potential. In addition, although pharmacological therapy is a staple of many treatment plans, EIA can be significantly modified or prevented through nonpharmacological interventions (Smith and LaBotz 1998).

If an athlete is aware that exercise induces an asthmatic response, it is recommended that he or she precede all exercise sessions with a warm-up. The warm-up should consist of low-intensity activity for a period of time sufficient to raise body temperature (see chapter 13 for further details concerning the warm-up). Although the warm-up's efficacy in producing a normal bronchial response to exercise in the asthmatic person is not clear, it is thought to induce a refractory period in which bronchoconstriction during the exercise session is reduced. This refractory period may last between 1 to 4 h

(Weiler 1998) and is specific to exercise only. The airways will still be sensitive to other stimuli that can easily cause bronchospasm.

It may also be possible to limit EIA by controlling the activity or the environment. For most recreational athletes, this may be easily accomplished by selecting an appropriate activity (e.g., swimming and walking are less likely to trigger an asthmatic response than running) or deciding to exercise indoors versus outdoors. However, this may be more difficult to control for competitive athletes, because such variables may not be of their choice.

Table 23.1   High and Low Asthmogenic Activities (adapted from Storm, 1999)

High Asthmogenic Activities	Low Asthmogenic Activities
Basketball	Baseball
Cross-country snow skiing	Boxing
Cycling	Football
Ice hockey	Golf
Ice skating	Gymnastics
Long-distance running	Karate
Rugby	Racquet sports (i.e., tennis, racquetball)
Soccer	
	Sprinting
	Swimming and water sports
	Wrestling
	Weightlifting

As previously mentioned, exercise in a cold, dry climate is the most potent stimulator of EIA. However, if the individual cannot exercise indoors, it is recommended that he or she warm the inspired air by wearing a scarf or a mask (Smith and LaBotz 1998). Running sports have the greatest **asthmogenicity** (ability to cause an asthmatic response) of all activities. Table 23.1 lists both high and low asthmogenic sports. If possible, an activity with a minimal asthmogenic response should be selected. In addition, low-intensity exercise performed intermittently may provide the greatest protection against EIA.

Conditioning does appear to be beneficial for the person with asthma. Ram, Robinson, and Black (2000) have shown that aerobic conditioning programs improve the aerobic capacity of people with asthma. However, even more important for the athlete with EIA, a higher level of aerobic fitness may increase the tolerance and threshold levels of those with asthma so that a greater stimulus is necessary to elicit an asthmatic response (Morton 1995).

SUMMARY

The prevalence of EIA is high, even in Olympic athletes, with a greater occurrence seen in winter-sport athletes. EIA does not prevent an athlete from reaching peak athletic potential. The proper use of both pharmacological and nonpharmacological treatments minimizes the discomfort and enhances performance during athletic competition. For competitive athletes, consideration should be given to the type of medication so that it falls within the confines of the governing body of the respective athletic association.

REFERENCES

Adamovich, D.R. 1984. *The heart: Fundamentals of electrocardiography, exercise physiology and exercise stress testing.* Freeport, NY: Sports Medicine Books.

Adams, D., J.P. O'Shea, K.L. O'Shea, and M. Climstein. 1992. The effect of six weeks of squat, plyometric, and squat-plyometric training on power production. *Journal of Applied Sport Science Research* 6:36-41.

Adams, G.R., B.M. Hather, K.M. Baldwin, and G.A. Dudley. 1993. Skeletal muscle myosin heavy chain composition and resistance training *Journal of Applied Physiology* 74:911-15.

Adams, W.C., R.H. Fox, A.J. Fry, and I.C. MacDonald. 1975. Thermoregulation during marathon running in cold, moderate and hot environments. *Journal of Applied Physiology* 38:1030-37.

Adlercreutz, H., M. Harkonen, K. Kuoppasalmi, I. Huhtaniemi, H. Tikanen, K. Remes, A. Dessypris, and J. Karvonen. 1986. Effect of training on plasma anabolic and catabolic steroid hormones and their response during physical exercise. *International Journal of Sports Medicine* 7(supp):227-8.

Ahlborg, G., P. Felig, L. Hagenfeldt, R. Hendler, and J. Wahren. 1974. Substrate turnover during prolonged exercise in man: Splanchnic and leg metabolism of glucose, free fatty acids, and amino acids. *Journal of Clinical Investigations* 53:1080-90.

Akil, H., S.J. Watson, E. Young, M.E. Lewis, H. Khachaturian, and J.M. Walker. 1984. Endogenous opioids: Biology and function. *Annual Review of Neuroscience* 7:223-55.

Alberici, J.C., P.A. Farrell, P.M. Kris-Etherton, and C.A. Shively. 1993. Effects of pre-exercise candy bar ingestion on glycemic response, substrate utilization, and performance. *International Journal of Sports Nutrition* 3:323-33.

Alen, M. 1985. Androgenic steroid effects on liver and red cells. *British Journal of Sports Medicine* 19:15-20.

Alen, M., and K. Hakkinen. 1985. Physical health and fitness of an elite bodybuilder during 1 year of self-administration of testosterone and anabolic steroids: A case study. *International Journal of Sports Medicine* 6:24-9.

Alen, M., K. Hakkinen, and P.V. Komi. 1984. Changes in neuromuscular performance and muscle fibre characteristics of elite power athletes self-administering androgenic and anabolic steroids. *Acta Physiologica Scandinavica* 122:535-44.

Alen, M., A. Pakarinen, and K. Hakkinen. 1993. Effects of prolonged training on serum thyrotropin and thyroid hormones in elite strength athletes. *Journal of Sports Sciences* 11:493-7.

Alen, M., and P. Rahkila. 1988. Anabolic-androgenic steroid effects on endocrinology and lipid metabolism in athletes. *Sports Medicine* 6:327-32.

Alen, M., M. Reinila, and R. Vihko. 1985. Response of serum hormones to androgen administration in power athletes. *Medicine and Science in Sports and Exercise* 17:354-9.

Alexander, M.J.L. 1989. The relationship between muscle strength and sprint kinematics in elite sprinters. *Canadian Journal of Sport Science* 14:148-57.

Allerheiligen, W.B. 1994a. Speed development and plyometric training. In *Essentials of strength training and conditioning,* ed. T. Baechle, 314-43. Champaign, IL: Human Kinetics.

————. 1994b. Stretching and warm-up. In *Essentials of strength training and conditioning,* ed. T. Baechle, 289-313. Champaign, IL: Human Kinetics.

Alter, M. 1996. *Science of flexibility and stretching.* Champaign, IL: Human Kinetics.

Alway, S.E., W.H. Grumbt, W.J. Gonyea, and J. Stray-Gunderson. 1989. Contrasts in muscle and myofibers of elite male and female body builders. *Journal of Applied Physiology* 67:24-31.

Alway, S.E., W.H. Grumbt, J. Stray-Gunderson, and W.J. Gonyea. 1992. Effects of resistance training on elbow flexors of highly competitive bodybuilders. *Journal of Applied Physiology* 72:1512-21.

Alway, S.E., J.D. MacDougall, and D.G. Sale. 1989. Contractile adaptations in human triceps surae after isometric exercise. *Journal of Applied Physiology* 66:2725-32.

Alway, S.E., P.K. Winchester, M.E. Davis, and W.J. Gonyea. 1989. Regionalized adaptations and fiber proliferation in stretch-induced muscle enlargement. *Journal of Applied Physiology* 66:771-81.

American College of Sports Medicine (ACSM). 2000a. *Guidelines for exercise testing and prescription.* Edited by B.A. Franklin. Philadelphia: Lippincott, Williams, and Wilkins.

————. 2000b. The physiological and health effects of oral creatine supplementation. Consensus Statement. *Medicine and Science in Sports and Exercise* 32:706-17.

American College of Sports Medicine and American Diabetes Association Joint Position Statement. 1997. Diabetes mellitus and exercise. *Medicine and Science in Sports and Exercise* 29(12):i-vi.

American College of Sports Medicine, American Dietetic Association, and Dietitians of Canada Joint Position Statement. 2000. Nutrition and athletic performance. *Medicine and Science in Sports and Exercise* 32:2130-45.

Anderson, M.A., J.B. Gieck, D. Perrin, A. Weltman, R. Rutt, and C. Denegar. 1991. The relationships among isometric, isotonic and isokinetic quadriceps and hamstring force and three components of athletic performance. *Journal of Orthopedic Sports Physical Therapy* 14:114-20.

Anderson, T., and J.T. Kearney. 1982. Effects of three resistance training programs on muscular strength and absolute and relative endurance. *Research Quarterly for Exercise and Sport* 2:27-30.

Andrew, G.M., C.A. Guzman, and M.R. Becklake. 1966. Effect of athletic training on exercise cardiac output. *Journal of Applied Physiology* 21:603-8.

Anselme, F., K. Collump, B. Mercier, S. Ahmaidi, and C. Prefaut. 1992. Caffeine increases maximal anaerobic power and blood lactate concentrations. *European Journal of Applied Physiology* 65:188-91.

Antal, L., and C. Good. 1980. Effects of oxprenolol on pistol shooting under stress. *Practitioner* 224:755-60.

Antonio, J., and W.J. Gonyea. 1993. Skeletal muscle fiber hyperplasia. *Medicine and Science in Sports and Exercise* 25:1333-45.

————. 1994. Muscle fiber splitting in stretch-enlarged avian muscle. *Medicine and Science in Sports and Exercise* 26:973-77.

Appell, H.J., S. Forsberg, and W. Hollmann. 1988. Satellite cell activation in human skeletal muscle after training: Evidence for muscle fiber neoformation. *International Journal of Sports Medicine* 9:297-9.

Armstrong, L.E. 1988. Research update: Fluid replacement and athlete hydration. *National Strength and Conditioning Journal* 10:69-71.

————. 2000. *Performing in extreme environments.* Champaign, IL: Human Kinetics.

Armstrong, L.E., D.L. Costill, and W.J. Fink. 1985. Influence of diuretic-induced dehydration on competitive running performance. *Medicine and Science in Sports and Exercise* 17:456-61.

Armstrong, L.E., A.E. Crago, R. Adams, W.O. Roberts, and C.M. Maresh. 1996. Whole-body cooling of hyperthermic runners: Comparison of two field therapies. *American Journal of Emergency Medicine* 14:355-58.

Armstrong, L.E., J.P. Deluca, and R.W. Hubbard. 1990. Time course of recovery and heat acclimation ability of prior exertional heatstroke patients. *Medicine and Science in Sports and Exercise* 22:36-48.

Armstrong, L.E., and J.E. Dziados. 1986. Effects of heat exposure on the exercising adult. In *Sports Physical Therapy,* ed. D.B. Bernhardt, 197-214. New York: Churchill Livingstone, Inc.

Armstrong, L.E., R.W. Hubbard, B.H. Jones, and J.T. Daniels. 1986. Preparing Alberto Salazar for the heat of the 1984 Olympic marathon. *Physician and Sportsmedicine* 14:73-81.

Armstrong, L.E., R.W. Hubbard, W.J. Kraemer, J.P. Deluca, and E. Christensen. 1987. Signs and symptoms of heat exhaustion during strenuous exercise. *Annals of Sports Medicine* 3:182-9.

Armstrong, L.E., and C.M. Maresh. 1991. The induction and decay of heat acclimatization in trained athletes. *Sports Medicine* 12:302-12.

————. 1993. The exertional heat illnesses: A risk of athletic participation. *Medicine, Exercise, Nutrition, and Health* 2:125-34.

Armstrong, L.E., C.M. Maresh, M. Whittlesey, M.F. Bergeron, C. Gabaree, and J.R. Hoffman. 1994. Longitudinal exercise-heat tolerance and running economy of collegiate distance runners. *Journal of Strength and Conditioning Research* 8:192-197.

Armstrong, L.E., and K.B. Pandolf. 1988. Physical training, cardiorespiratory physical fitness and exercise-heat tolerance. In *Human performance physiology and environmental medicine at terrestrial extremes,* ed. K.B. Pandolf, M.N. Sawka, and R.R. Gonzalez, 199-226. Indianapolis: Benchmark Press.

Asmussen, E., and F. Bonde-Peterson, and K. Jorgensen. 1976. Mechano-elastic properties of human muscles at different temperatures. *Acta Physiologica Scandinavica* 96:86-93.

Astrand, P.O. 1965. *Work tests with the bicycle ergometer.* Varberg, Sweden: AB Cykelfabriken Monark.

Atha, J. 1981. Strengthening muscle. *Exercise and Sport Sciences Reviews* 9:1-73.

Ayalon, A., O. Inbar, and O. Bar-Or. 1974. Relationships among measurements of explosive strength and anaerobic power. In *International series on sport sciences.* Vol. 1. Biomechanics IV, ed. Nelson R.C. and C.A. Morehouse , 527-32. Baltimore: University Park Press.

Ayers, J.W.T., Y. Komesu, T. Romani, and R. Ansbacher. 1985. Anthropometric, hormonal, and psychological correlates of semen quality in endurance trained male athletes. *Fertility and Sterility* 43:917-21.

Bahrke, M. 1993. Psychological effects of endogenous testosterone and anabolic-androgenic steroids. In *Anabolic steroids in sport and exercise,* ed. C.E. Yesalis, 161-92. Champaign, IL: Human Kinetics.

Baker, D., and S. Nance. 1999. The relation between running speed and measures of strength and power in professional rugby players. *Journal of Strength and Conditioning Research* 13:230-35.

Baker, D., G. Wilson, and R. Carlyon. 1994. Periodization: The effect on strength of manipulating volume and intensity. *Journal of Strength and Conditioning Research* 8:235-42.

Ball, T.C., S.A. Headley, P.M. Vanderburgh, and J.C. Smith. 1995. Periodic carbohydrate replacement during 50-min of high-intensity cycling improves subsequent sprint performance. *International Journal of Sports Nutrition* 5:151-8.

Ballor, D.L., and E.T. Poehlman. 1992. Resting metabolic rate and coronary-heart-disease risk factors in aerobically and resistance-trained women. *American Journal of Clinical Nutrition* 56:968-74.

Balsam, A., and L.E. Leppo. 1975. Effect of physical training on the metabolism of thyroid hormones in man. *Journal of Applied Physiology* 38:212-5.

Balsom, P.D., B. Ekblom, K. Soderlund, B. Sjoden, and E. Hultman. 1993. Creatine supplementation and dynamic high-intensity intermittent exercise. *Scandinavian Journal of Medicine and Science in Sports* 3:143-9.

Balsom, P.D., K. Soderlund, and B. Ekblom. 1994. Creatine in humans with special reference to creatine supplementation. *Sports Medicine* 3:143-9.

Bar-Or, O. 1987. The Wingate anaerobic test: An update on methodology, reliability and validity. *Sports Medicine* 4:381-94.

Bar-Or, O., R. Dotan, O. Inbar, A. Rotstein, J. Karlsson, and P. Tesch. 1980. Anaerobic capacity and muscle fiber type distribution in man. *International Journal of Sports Medicine* 1:89-92.

Barron, J.L., T.D. Noakes, W. Levy, C. Smith, and R.P. Millar. 1985. Hypothalamic dysfunction in overtrained athletes. *Journal of Clinical Endocrinology and Metabolism* 60:803-6.

Bartlett, D., and J.E. Remmers. 1971. Effects of high altitude exposure on the lungs of young rats. *Respiratory Physiology* 13:116-25.

Bartsch, P. 1999. High altitude pulmonary edema. *Medicine and Science in Sports and Exercise* 31:S23-7.

Bassett, D.R., and E.T. Howley. 1997. Maximal oxygen uptake: "Classical" versus "contemporary" viewpoints. *Medicine and Science in Sports and Exercise* 29:591-603.

———. 2000. Limiting factors for maximum oxygen uptake and determinants of endurance performance. *Medicine and Science in Sports and Exercise* 32:70-84.

Baum, M., H. Liesen, and J. Enneper. 1994. Leucocytes, lymphocytes, activation parameters and cell adhesion molecules in middle-distance runners under different training conditions. *International Journal of Sports Medicine* 15:S122-6.

Beck, K.C. 1999. Control of airway function during and after exercise in asthmatics. *Medicine and Science in Sports and Exercise* 31:S4-11.

Beck, K.C., K.P. Offord, and P.D. Scanlon. 1994. Bronchoconstriction occurring during exercise in asthmatic subjects. *American Journal of Respiratory Critical Care Medicine* 149:352-7.

Becque, M.D., J.D. Lochmann, and D.R. Melrose. 2000. Effects of oral creatine supplementation on muscular strength and body composition. *Medicine and Science in Sports and Exercise* 32:654-8.

Beelen, A., and A.J. Sargeant. 1991. Effect of lowered muscle temperature on the physiological response to exercise in men. *European Journal of Applied Physiology* 63:387-92.

Bell, D.G., and I. Jacobs. 1999. Combined caffeine and ephedrine ingestion improves run times of Canadian forces warrior test. *Aviation, Space and Environmental Medicine* 70:325-9.

Bell, D.G., I. Jacobs, T.M. McLellan, and J. Zamecnik. 2000. Reducing the dose of combined caffeine and ephedrine preserves the ergogenic effect. *Aviation, Space and Environmental Medicine* 71:415-519.

Bell, D.G., I. Jacobs, and J. Zamecnik. 1998. Effects of caffeine, ephedrine and their combination on time to exhaustion during high-intensity exercise. *European Journal of Applied Physiology and Occupational Physiology* 77:427-33.

Bell, G.J., S.R. Petersen, J. Wessel, K. Bagnall, and H.A. Quinney. 1991. Physiological adaptations to concurrent endurance training and low velocity resistance training. *International Journal of Sports Medicine* 12:384-90.

Bell, G., D. Syrotuik, T.P. Martin, R. Burnham, and H.A. Quinney. 2000. Effect of concurrent strength and endurance training on skeletal muscle properties and hormone concentrations in humans. *European Journal of Applied Physiology* 81:418-27.

Bell, G., D. Syrotuik, T. Socha, I. Maclean, and H.A. Quinney. 1997. Effect of strength training and concurrent strength and endurance training on strength, testosterone and cortisol. *Journal of Strength and Conditioning Research* 11:57-64.

Bendich, A. 1989. Carotenoids and the immune response. *Journal of Nutrition* 119:112-5.

Benoni, G., P. Bellavite, A. Adami, S. Chirumbolo, G. Lippi, G. Brocco, G.M. Guilini, and L. Cuzzolin. 1995. Changes in several neutrophil functions in basketball players before, during and after the sports season. *International Journal of Sports Medicine* 16:34-7.

Berger, R.A. 1962. Effect of varied weight training programs on strength. *Research Quarterly for Exercise and Sport* 33:168-81.

———. 1963. Comparative effects of three weight training programs. *Research Quarterly for Exercise and Sport* 34:396-8.

Bergeron, M.F. 1996. Heat cramps during tennis: A case report. *International Journal of Sports Nutrition* 6:62-8.

Bergh, U., and B. Ekblom. 1979. Influence of muscle temperature on maximal muscle strength and power output in human muscle. *Acta Physiologica Scandinavica.* 107:332-7.

Bergstrom, J., L. Hermansen, E. Hultman, and B. Saltin. 1967. Diet, muscle glycogen and physical performance. *Acta Physiologica Scandinavica* 71:140-50.

Bergstrom, M., and E. Hultman. 1988. Energy cost and fatigue during intermittent electrical stimulation of human skeletal muscle. *Journal of Applied Physiology* 65:1500-5.

Berk, L.S., D.C. Nieman, W.S. Youngberg, K. Arabatzis, M. Simpson-Westerberg, J.W. Lee, S.A. Tan, and W.C. Eby. 1990. The effect of long endurance running on natural killer cells in marathoners. *Medicine and Science in Sports and Exercise* 22:207-12.

Benardot, D. 2000. *Nutrition for serious athletes.* Champaign, IL: Human Kinetics.

Berning, J.R. 1998. Energy intake, diet, and muscle wasting. In *Overtraining in sport,* ed. R.B. Kreider, A.C. Fry, and M.L. O'Toole, 275-88. Champaign, IL: Human Kinetics.

Bessman, S.P., and F. Savabi. 1990. The role of the phosphocreatine energy shuttle in exercise and muscle hypertrophy. In *Biochemistry of exercise VII,* ed. A.W. Taylor, P.D. Gollnick, H.J. Green, C.D. Ianuzzo, E.G.

Noble, G. Metivier, and J.R. Sutton, 167-77. Champaign, IL: Human Kinetics.

Binkhorst, R.A., L. Hoofd, and A.C.A. Vissers. 1977. Temperature and force-velocity relationship of human muscles. *Journal of Applied Physiology: Respiratory, Environmental, Exercise Physiology* 42:471-5.

Bishop, D., D.G. Jenkins, L.T. Mackinnon, M. McEniery, and M.F. Carey. 1999. The effects of strength training on endurance performance and muscle characteristics. *Medicine and Science in Sports and Exercise* 31:886-91.

Bjorneboe, A., G.A. Bjorneboe, and C.A. Drevon. 1990. Absorption, transport and distribution of vitamin E. *Journal of Nutrition* 120:233-42.

Bjorntorp, P. 1991. Importance of fat as a support nutrient for energy: Metabolism of athletes. *Journal of Sports Sciences* 9:71-6.

Black, W., and E. Roundy. 1994. Comparisons of size, strength, speed and power in NCAA division I-A football players. *Journal of Strength and Conditioning Research* 8:80-5.

Blannin, A.K., L.J. Chatwin, R. Cave, and M. Gleeson. 1996. Effects of submaximal cycling and long-term endurance training on neutrophil phagocytic activity in middle aged men. *British Journal of Sports Medicine* 30:125-9.

Blomqvist, C.G., and B. Saltin. 1983. Cardiovascular adaptations to physical training. *Annual Review Physiology* 45:169-89.

Bloom, S., R. Johnson, D. Park, M. Rennie, and W. Sulaiman. 1976. Differences in the metabolic and hormonal response to exercise between racing cyclists and untrained individuals. *Journal of Physiology* 258:1-18.

Blume, F.D., S.J. Boyer, L.E. Braverman, A. Cohen, J. Dirkse, and J.P. Mordes. 1984. Impaired osmoregulation at high altitude: Studies on Mt. Everest. *Journal of the American Medical Association* 252:524-6.

Bobbert, M.F., K.G.M. Gerritsen, M.C.A. Litjens, and A.J. Van Soest. 1996. Why is countermovement jump height greater than squat jump height? *Medicine and Science in Sports and Exercise* 28:1402-12.

Bogdanis, G.C., M.E. Nevill, L.H. Boobis, and H.K.A. Lakomy. 1996. Contribution of phosphocreatine and aerobic metabolism to energy supply during repeated sprint exercise. *Journal of Applied Physiology* 80:876-84.

Bompa, T.O. 1999. *Periodization: Theory and methodology of training.* Champaign, IL: Human Kinetics.

Bonow, R.O. 1994. Left ventricular response to exercise. In *Cardiovascular response to exercise*, ed. G.F. Fletcher, 31-48. Mount Kisco, NY: Futura.

Boobis, L.H. 1987. Metabolic aspects of fatigue during sprinting. In *Exercise: Benefits, limitations and adaptations,* ed. D.R. Macleod, R. Maughan, M. Nimmo, T. Reilly, and C. Williams, 116-43. London: E&FN Spon.

Borg, G.A.V. 1982. Psychophysical bases of perceived exertion. *Medicine and Science in Sports and Exercise* 14:377-81.

Bosco, C., and P.V. Komi. 1979. Mechanical characteristics and fiber composition of human leg extensor muscles. *European Journal of Applied Physiology* 24:21-32.

Bosco, C., P. Mognoni, and P. Luhtanen. 1983. Relationship between isokinetic performance and ballistic movement. *European Journal of Applied Physiology* 51:357-64.

Bosco, C., J.T. Viitalsalo, P.V. Komi, and P. Luhtanen. 1982. Combined effect of elastic energy and myoelectric potentiation during stretch-shortening cycle exercise. *Acta Physiologica Scandinavica* 114:557-65.

Bottecchia, D., D. Bordin, and R. Martino. 1987. Effect of different kinds of physical exercise on the plasmatic testosterone level of normal adult males. *Journal of Sports Medicine and Physical Fitness* 27:1-5.

Boutellier, U., H. Howald, P.E. di Prampero, D. Giezendanner, and P. Cerretelli. 1983. Human muscle adaptations to chronic hypoxia. *Progress in Clinical Biology Research* 136:273-85.

Brandenberger, G., V. Candas, M. Follenius, J.P. Libert, and J.M. Kahn. 1986. Vascular fluid shifts and endocrine responses to exercise in the heat. *European Journal of Applied Physiology* 55:123-9.

Brener, W., T.R. Hendrix, and P.R. McHugh. 1983. Regulation of the gastric emptying of glucose. *Gastroenterology* 85:76-82.

Brenner, M., J. Walberg-Rankin, and D. Sebolt. 2000. The effect of creatine supplementation during resistance training in women. *Journal of Strength and Conditioning Research* 14:207-13.

Brooks, S., J. Burrin, M.E. Cheetham, G.M. Hall, T. Yeo, and C. Williams. 1988. The responses of the catecholamines and β-endorphin to brief maximal exercise in man. *European Journal of Applied Physiology* 57:230-4.

Brooks, S., M.E. Nevill, L. Meleagros, H.K.A. Lakomy, G.M. Hall, S.R. Bloom, and C. Williams. 1990. The hormonal responses to repetitive brief maximal exercise in humans. *European Journal of Applied Physiology* 60:144-8.

Brown, C.H., and J.H. Wilmore. 1974. The effects of maximal resistance training on the strength and body composition of women athletes. *Medicine and Science in Sports and Exercise* 6:174-7.

Brown, G.A., M.D. Vukovich, R.L. Sharp, T.A. Reifenrath, K.A. Parsons, and D.S. King. 1999. Effect of oral DHEA on serum testosterone and adaptations to resistance training in young men. *Journal of Applied Physiology* 87:2274-83.

Brown, M.E., J.L. Mayhew, and L.W. Boleach. 1986. Effect of plyometric training on vertical jump performance in high school basketball players. *Journal Sports Medicine Physical Fitness Quarterly Review* 26:1-4.

Bruce, C.R., M.E. Anderson, S.F. Fraser, N.K. Stepto, R. Klein, W.G. Hopkins, and J.A. Hawley. 2000. Enhancement of 2000-m rowing performance after caffeine ingestion. *Medicine and Science in Sports and Exercise* 32:1958-63.

Bruce, R.A., F. Kusumi, and D. Hosmer. 1973. Maximal oxygen uptake and nomographic assessment of functional aerobic impairment in cardiovascular disease. *American Heart Journal* 85:546-62.

Buckley, W.E., C.E. Yesalis, K.E. Friedl, W.A. Anderson, A.L. Streit, and J.E. Wright. 1988. Estimated prevalence of anabolic steroid use among male high school seniors. *Journal of the American Medical Association* 260:3441-5.

Buick, F.J., N. Gledhill, A.B. Froese, L. Spriet, and E.C. Meyers. 1980. Effect of induced erythrocythemia on aerobic work capacity. *Journal of Applied Physiology* 48:636-42.

Bunt, J.C., R.A. Boileau, J.M. Bahr, and R.A. Nelson. 1986. Sex and training differences in human growth hormone during prolonged exercise. *Journal of Applied Physiology* 61:1796-801.

Buono, M.J., J.E. Yeager, and J.A. Hodgdon. 1986. Plasma adrenocorticotropin and cortisol responses to brief high-intensity exercise in humans. *Journal of Applied Physiology* 61:1337-9.

Burke, L.M., G.R. Collier, and M. Hargreaves. 1993. Muscle glycogen storage after prolonged exercise: Effect of the glycemic index of carbohydrate feeding. *Journal of Applied Physiology* 75:1019-23.

Burkett, L.N. 1970. Causative factors in hamstring strains. *Medicine and Science in Sports and Exercise* 2:39-42.

Buskirk, E.R., P.F. Iampietro, and D.E. Bass. 1958. Work performance after dehydration: Effects of physical conditioning and heat acclimatization. *Journal of Applied Physiology* 12:189-94.

Buskirk, E.R., J. Kollias, R.F. Akers, E.K. Prokop, and E.P. Reategui. 1967. Maximal performance at altitude and return from altitude in conditioned runners. *Journal of Applied Physiology* 23:259-66.

Caizzo, V.J., J.J. Perrine, and V.R. Edgerton. 1981. Training-induced alterations of the in vivo force-velocity relationship of human muscle**.** *Journal of Applied Physiology: Respiratory, Environmental, Exercise Physiology* 51:750-4.

Caldwell, J.E., E. Ahonen, and U. Nousiainen. 1984. Differential effects of sauna-, diuretic-, and exercise-induced hypohydration. *Journal of Applied Physiology: Respiratory, Environmental, Exercise Physiology* 57:1018-23.

Campaign, B.N., T.B. Gilliam, M.L. Spencer, R.M. Lampman, and M.A. Schork. 1984. Effects of a physical activity program on metabolic control and cardiovascular fitness in children with insulin dependent diabetes mellitus. *Diabetes Care* 7:57-62.

Campaign, B.N., H. Wallberg-Henriksson, and R. Gunnarsson. 1987. Glucose and insulin responses in relation to insulin dose and caloric intake 12 h after acute physical exercise in men with IDDM. *Diabetes Care* 10:716-21.

Cannon, J.G., S.N. Meydani, R.A. Fielding, M.A. Fiatarone, M. Meydani, N. Farhangmeh, S.F. Orencole, J.B. Blumberg, and W.J. Evans. 1991. Acute phase response in exercise II: Association with vitamin E, cytokines and muscle proteolysis. *American Journal of Physiology* 260:R1235-40.

Cannon, J.G., S.F. Orencole, R.A. Fielding, M. Meydani, S.N. Meydani, M.A. Fiatarone, J.B. Blumberg, and W.J. Evans. 1990. Acute phase response in exercise: Interaction of age and vitamin E on neutrophils and muscle enzyme release. *American Journal of Physiology* 259:R1214-19.

Cappon, J., J.A. Brasel, S. Mohan, and D.M. Cooper. 1994. Effect of brief exercise on circulating insulin-like growth factor I. *Journal of Applied Physiology* 76:2490-6.

Caro, J.F., M.K. Sinha, S.M. Raju, O. Ittoop, W.J. Pories, E.G. Flickinger, D. Meelheim, and D. Dohm. 1987. Insulin receptor kinase in human skeletal muscle from obese subjects with and without noninsulin dependent diabetes. *Journal of Clinical Investigations* 79:1330-7.

Carr, G. 1999. *Fundamentals of track and field.* Champaign, IL: Human Kinetics.

Carroll, J.F., V.A. Convertino, C.E. Wood, J.E. Graves, D.T. Lowenthal, and M.L. Pollack. 1995. Effect of training on blood volume and plasma hormone concentrations in the elderly. *Medicine and Science in Sports and Exercise* 27:79-84.

Castell, L.M., J.R. Poortmans, R. Leclercq, M. Brasseur, J. Duchateau, and E.A. Newsholme. 1997. Some aspects of the acute phase response after a marathon race, and the effects of glutamine supplementation. *International Journal of Sports Medicine* 75:47-53.

Castellani, J.W., A.J. Young, J.E. Kain, A. Rouse, and M. Sawka. 1999. Thermoregulation during cold exposure: Effects of prior exercise. *Journal of Applied Physiology* 87:247-52.

Cavanagh, P.R., and K.R. Williams. 1982. The effect of stride length variation on oxygen uptake during distance running. *Medicine and Science in Sports and Exercise* 14:30-5.

Centers for Disease Control and Prevention. 1983. Current trends hypothermia: United States. *Morbidity and Mortality Weekly Report* 32:46-48.

———. 1998. Surveillance for Asthma: United States, 1960-1995. *Morbidity and Mortality Weekly Report* 47:1-28.

Cerretelli, P. 1987. Extreme hypoxia in air breathers. In *Comparative physiology of environmental adaptations,* ed. P. Dejours. Basel, Switzerland: Karger.

Cerretelli, P., C. Marconi, O. Deriaz, and D. Giezendanner. 1984. After-effects of chronic hypoxia on cardiac output and muscle blood flow at rest and exercise. *European Journal of Applied Physiology* 53:92-6.

Chang, F.E., W.G. Dodds, M. Sullivan, M.H. Kim, and W.B. Malarkey. 1986. The acute effects on prolactin and growth hormone secretion: Comparison between sedentary women and women runners with normal and abnormal menstrual cycles. *Journal of Clinical Endocrinology and Metabolism* 62:551-6.

Chapman, R.F., M. Emery, and J.M. Stager. 1999. Degree of arterial desaturation in normoxia influences the decline in $\dot{V}O_2$max in mild hypoxia. *Medicine and Science in Sports and Exercise* 31:658-63.

Chapman, R.F., and B.D. Levine. 2000. The effects of hypo- and hyperbaria on performance. In *Exercise and sport science,* ed. W.E. Garrett and D.T. Kirkendall, 447-58. Philadelphia: Lippincott, Williams, and Wilkins.

Chapman, R.F., J. Stray-Gundersen, and B.D. Levine. 1998. Individual variation in response to altitude training. *Journal of Applied Physiology* 85:1448-56.

Charlton, G.A., and M.H. Crawford. 1997. Physiological consequences of training. *Cardiology Clinics* 15:345-54.

Chattoraj, S.C., and N.B. Watts. 1987. Endocrinology. In *Fundamentals of clinical chemistry,* ed. N.W. Tietz, 175-80. Philadelphia: W.B. Saunders.

Cheetham, M.E., L.H. Boobis, S. Brooks, and C. Williams. 1986. Human muscle metabolism during sprint running. *Journal of Applied Physiology* 61:54-60.

Chu, D. 1992. *Jumping into plyometrics.* Champaign, IL: Human Kinetics.

Claremont, A.D., F. Nagle, W.D. Reddan, and G.A. Brooks. 1975. Comparison of metabolic, temperature, heart rate and ventilatory responses to exercise at extreme ambient temperatures (0° and 35°C). *Medicine and Science in Sports and Exercise* 7:150-4.

Clarkson, P.M. 1991. Minerals, exercise performance and supplementation in athletes. *Journal of Sports Sciences* 9:91-116.

———. 1997. Eccentric exercise and muscle damage. *International Journal of Sports Medicine* 18(supp):S314-6.

Coggan, A.R., and E.F. Coyle. 1989. Metabolism and performance following carbohydrate ingestion late in exercise. *Medicine and Science in Sports and Exercise* 21:59-65.

————. 1991. Carbohydrate ingestion during prolonged exercise: Effects on metabolism and performance. *Exercise and Sport Sciences Reviews* 19:1-40.

Collins, M.A., and T.K. Snow. 1993. Are adaptations to combined endurance and strength training affected by the sequence of training? *Journal of Sports Sciences* 11:485-91.

Collomp, K., S. Ahmaidi, M. Audran, J.L. Chanal, and C. Prefaut. 1991. Effects of caffeine ingestion on performance and anaerobic metabolism during the Wingate test. *International Journal of Sports Medicine* 12:439-43.

Collomp, K., S. Ahmaidi, J.C. Chatard, M. Audran, and C. Prefaut. 1992. Benefits of caffeine ingestion on sprint performance in trained and untrained swimmers. *European Journal of Applied Physiology* 64:377-80.

Conley, D.L., and G. Krahenbuhl. 1980. Running economy and distance running performance of highly trained athletes. *Medicine and Science in Sports and Exercise* 12:357-60.

Convertino, V.A. 1991. Blood volume: Its adaptation to endurance training. *Medicine and Science in Sports and Exercise* 23:1338-48.

Convertino, V.A., L.C. Keil, E.M. Bernauer, and J.E. Greenleaf. 1981. Plasma volume, osmolality, vasopressin, and renin activity during graded exercise in man. *Journal of Applied Physiology: Respiratory, Environmental, Exercise Physiology* 50:123-8.

Convertino, V.A., L.C. Keil, and J.E. Greenleaf. 1983. Plasma volume, renin and vasopressin responses to graded exercise after training. *Journal of Applied Physiology: Respiratory, Environmental, Exercise Physiology* 54:508-14.

Cook, E.E., V.L. Gray, E. Savinar-Nogue, and J. Medeiros. 1987. Shoulder antagonistic strength ratios: A comparison between college-level baseball pitchers and nonpitchers. *Journal of Orthopedic and Sports Physical Therapy* 8:451-60.

Cooke, W.H., P.W. Grandjean, and W.S. Barnes. 1995. Effect of oral creatine supplementation on power output and fatigue during bicycle ergometry. *Journal of Applied Physiology* 78:670-3.

Costill, D.L. 1988. Carbohydrates for exercise: Dietary demands for optimal performance. *International Journal of Sports Medicine* 9:1-18.

Costill, D.L., P. Cleary, W.J. Fink, C. Foster, J.L. Ivy, and F. Witzmann. 1979. Training adaptations in skeletal muscle of juvenile diabetics. *Diabetes* 28:818-22.

Costill, D.L., R. Cote, and W.J. Fink. 1976. Muscle water and electrolytes following varied levels of dehydration in man. *Journal of Applied Physiology* 40:6-11.

Costill, D.L., R. Cote, W.J. Fink, and P. Van Handel. 1981. Muscle water and electrolyte distribution during prolonged exercise. *International Journal of Sports Medicine* 2:130-4.

Costill, D.L., E. Coyle, G. Dalsky, W. Evans, W.J. Fink, and D. Hoopes. 1977. Effects of elevated plasma FFA and insulin on muscle glycogen usage during exercise. *Journal of Applied Physiology* 43:695-9.

Costill, D.L., E.F. Coyle, W.J. Fink, G.R. Lesmes, and F.A. Witzmann. 1979. Adaptations in skeletal muscle following strength training. *Journal of Applied Physiology* 46:96-9.

Costill, D.L., G.P. Dalsky, and W.J. Fink. 1978. Effects of caffeine ingestion on metabolism and exercise performance. *Medicine and Science in Sports and Exercise* 10:155-8.

Costill, D.L., W.J. Fink, and M.L. Pollock. 1976. Muscle fiber composition and enzyme activities of elite distance runners. *Medicine and Science in Sports and Exercise* 8:96-100.

Costill, D.L., M.G. Flynn, J.P. Kirwin, J.A. Houmard, J.B. Mitchell, R. Thomas, and S.H. Park. 1988. Effects of repeated days of intensified training on muscle glycogen and swimming performance. *Medicine and Science in Sports and Exercise* 20:249-54.

Costill, D.L., and E.L. Fox. 1969. Energetics of marathon running. *Medicine and Science in Sports and Exercise* 1:81-6.

Costill, D.L., and B. Saltin. 1974. Factors limiting gastric emptying during rest and exercise. *Journal of Applied Physiology* 37:679-83.

Costill, D.L., R. Thomas, R.A. Roberts, D.D. Pascoe, C.P. Lampert, S.I. Barr, and W.J. Fink. 1991. Adaptations to swimming training: Influence of training volume. *Medicine and Science in Sports and Exercise* 23:371-7.

Costill, D.L., F. Verstappen, H. Kuipers, E. Jansson, and W. Fink. 1984. Acid-base balance during repeated bouts of exercise: Influence of HCO_3^-. *International Journal of Sports Medicine* 5:228-31.

Costrini, A. 1990. Emergency treatment of exertional heatstroke and comparison of whole body cooling techniques. *Medicine and Science in Sports and Exercise* 22:15-8.

Coyle, E.F., M.K. Hemmert, and A.R. Coggan. 1986. Effects of detraining on cardiovascular response to exercise: Role of blood volume. *Journal of Applied Physiology* 60:95-9.

Coyle, E.F., W.H. Martin III, S.A. Bloomfield, O.H. Lowry, and J.O. Holloszy. 1985. Effects of detraining on responses to submaximal exercise. *Journal of Applied Physiology* 59:853-9.

Coyle, E.F., W.H. Martin III, D.R. Sinacore, M.J. Joyner, J.M. Hagberg, and J.O. Holloszy. 1984. Time course for loss of adaptation after stopping prolonged intense endurance training. *Journal of Applied Physiology* 57:1857-64.

Craig, B.W., and H.Y. Kang. 1994. Growth hormone release following single versus multiple sets of back squats: Total work versus power. *Journal of Strength and Conditioning Research* 8:270-5.

Craig, B.W., G.T. Hammons, S.M. Garthwite, L. Jarett, and J.O. Holloszy. 1981. Adaptations of fat cells to exercise: Response of glucose uptake and oxidation to insulin. *Journal of Applied Physiology* 51:1500-6.

Craig, B.W., K. Thompson, and J.O. Holloszy. 1983. Effect of stopping training on size and response to insulin of fat cells in female rats. *Journal of Applied Physiology* 54:571-5.

Craig, F.N., and E.G. Cummings. 1966. Dehydration and muscular work. *Journal of Applied Physiology* 21:670-4.

Crist, D.M., P.J. Stackpole, and G.T. Peake. 1983. Effects of androgenic-anabolic steroids on neuromuscular power and body composition. *Journal of Applied Physiology* 54:366-70.

Cunningham, D.A., and J.A. Faulkner. 1969. The effect of training on aerobic and anaerobic metabolism during a short exhaustive run. *Medicine and Science in Sports and Exercise* 1:65-9.

Cunningham, D.A., D.H. Paterson, C.J. Blimkie, and A.P. Donner. 1984. Development of cardiorespiratory function in circumbertal boys: A longitudinal study. *Journal of Applied Physiology*. 56:302-7.

Dahl-Jorgensen, K., H.D. Meen, K.F. Hanssen, and O. Asgenaes. 1980. The effect of exercise on diabetic control and hemoglobin A_1 (HbA_1) in children. *Acta Paediatics Scandinavica* 283(supp):53-6.

Daniels, J.T. 1985. A physiologist's view of running economy. *Medicine and Science in Sports and Exercise* 17:332-8.

Danielsson, U. 1996. Windchill and the risk of tissue freezing. *Journal of Applied Physiology* 81:2666-73.

Danzl, D.F., R.S. Pozos, and M.P. Hamlet. 1995. Accidental hypothermia. In *Wilderness medicine: Management of wilderness and environmental emergencies,* ed. P.S. Auerbach, 51-103. St. Louis: Mosby.

Davies, C.T.M., I.K. Mecrow, and M.J. White. 1982. Contractile properties of the human triceps surae with some observations on the effects of temperature and exercise. *European Journal of Applied Physiology* 49:255-69.

Davies, C.T.M., and K. Young. 1983a. Effect of temperature on contractile properties and muscle power of triceps surae in humans. *Journal of Applied Physiology: Respiratory, Environmental, Exercise Physiology* 55:191-5.

———. 1983b. Effects of training at 30 and 100% maximal isometric force on the contractile properties of the triceps surae of man. *Journal of Physiology* 336:22-3.

Davis, J.A., M.H. Frank, B.J. Whipp, and K. Wasserman. 1979. Anaerobic threshold alterations caused by endurance training in middle-aged men. *Journal of Applied Physiology* 46:1039-46.

Davis, J.M., D.A. Jackson, M.S. Broadwell, J.L. Query, and C.L. Lambert. 1997. Carbohydrate drinks delay fatigue during intermittent, high-intensity cycling in active men and women. *International Journal of Sports Medicine* 7:261-73.

Dawson, B., M. Cutler, A. Moody, S. Lawrence, C. Goodman, and N. Randall. 1995. Effects of oral creatine loading on single and repeated maximal short sprints. *Australian Journal of Science and Medicine in Sport* 27:56-61.

Dela, F., J.J., Larson, K.J. Mikines, and H. Galbo. 1995. Normal effect of insulin to stimulate leg blood flow in NIDDM. *Diabetes* 44:221-6.

Delecluse, C. 1997. Influence of strength training on sprint running performance: Current findings and implications for training. *Sports Medicine* 24:147-56.

Delecluse, C., H.V. Coppenolle, E. Willems, M.V. Leemputte, R. Diels, and M. Goris. 1995. Influence of high-resistance and high-velocity training on sprint performance. *Medicine and Science in Sports and Exercise* 27:1203-9.

Deligiannis, A., E. Zahopoulou, and K. Mandroukas. 1988. Echocardiographic study of cardiac dimensions and function in weight lifters and body builders. *Journal of Sports Cardiology* 5:24-32.

Dempsey, J.A., W.G. Reddan, M.L. Birnbaum, H.V. Forster, J.S. Thoden, R.F. Grover, and J. Rankin. 1971. Effects of acute though life-long hypoxic exposure on exercise pulmonary gas exchange. *Respiration Physiology* 13:62-89.

Dengel, D.R., P.G. Weyand, D.M. Black, and K.J. Cureton. 1992. Effect of varying levels of hypohydration on responses during submaximal cycling. *Medicine and Science in Sports and Exercise* 24:1096-101.

Deschenes, M.R., W.J. Kraemer, C.M. Maresh, and J.F. Crivello. 1991. Exercise-induced hormonal changes and their effects upon skeletal tissue. *Sports Medicine* 12:80-93.

DeSouza, M.J., C.M. Maresh, M.S. Maguire, W.J. Kraemer, G. Flora-Ginter, and K.L. Goetz. 1989. Menstrual status and plasma vasopressin, renin activity, and aldosterone exercise responses. *Journal of Applied Physiology* 67:736-43.

Despres, J.P., A. Nadeau, and C. Bouchard. 1988. Physical training and changes in regional adipose tissue distribution. *Acta Medical Scandinavica* 723(supp):205-12.

Dessypris, A., K. Kuoppasalmi, and H. Adlercreutz. 1976. Plasma cortisol, testosterone, androstenedione and luteinizing hormone (LH) in a non-competitive marathon run. *Journal of Steroid Biochemistry* 7:33-7.

Devlin, J.T., J. Calles-Escandon, and E.S. Horton. 1986. Effects of preexercise snack feeding on endurance cycle exercise. *Journal of Applied Physiology* 60:980-5.

Dickerman, R.D., R.M. Pertusi, N.Y. Zachariah, D.R. Dufour, and W.J. McConathy. 1999. Anabolic steroid-induced hepatotoxicity: Is it overstated? *Clinical Journal of Sports Medicine* 9:34-9.

Dimri, G.P., M.S. Malhotra, J. Sen Gupta, T.S. Kumar, and B.S. Arora. 1980. Alterations in aerobic-anaerobic proportions of metabolism during work in heat. *European Journal of Applied Physiology* 45:43-50.

Dohm, G.L., M.K. Sinha, and J.F. Caro. 1987. Insulin receptor binding and protein kinase activity in muscle of trained rats. *American Journal of Physiology* 252:E170-5.

Dolezal, B.A., and J.A. Potteiger. 1998. Concurrent resistance and endurance training influence basal metabolic rate in nondieting individuals. *Journal of Applied Physiology* 85:695-700.

Donevan, R.H., and G.M. Andrew. 1987. Plasma β-endorphin immunoreactivity during graded cycle ergometry. *Medicine and Science in Sports and Exercise* 19:229-33.

Dotan, R., and O. Bar-Or. 1980. Climatic heat stress and performance in the Wingate anaerobic test. *European Journal of Applied Physiology* 44:237-43.

Dotan, R., A. Rotstein, R. Dlin, O. Inbar, H. Kofman, and Y. Kaplansky. 1983. Relationships of marathon running to physiological, anthropometric and training indices. *European Journal of Applied Physiology* 51:281-93.

Drinkwater, B.L., and S.M. Horvath. 1972. Detraining effects in young women. *Medicine and Science in Sports and Exercise* 4:91-5.

Dudley, G.A., W.M. Abraham, and R.L. Terjung. 1982. Influence of exercise intensity and duration on biochemical adaptations in skeletal muscle. *Journal of Applied Physiology* 53:844-50.

Dudley, G.A., and R. Djamil. 1985. Incompatibility of endurance and strength training modes of exercise. *Journal of Applied Physiology* 59:1446-51.

Dufaux, B., U. Order, and H. Liesen. 1991. Effect of a short maximal physical exercise on coagulation, fibrinolysis, and complement system. *International Journal of Sports Medicine* 12:S38-42.

Durnin, J.V.G.A., and J. Womersley. 1974. Body fat assessment from total body density and its estimation from skinfold thickness: Measurements on 481 men and women aged 16-72 years. *British Journal of Nutrition* 32:77-97.

Ebbeling, C.B., A. Ward, E.M. Puleo, J. Widrick, and J.M. Rippe. 1991. Development of a single-stage submaximal treadmill walking test. *Medicine and Science in Sports and Exercise* 23:966-73.

Eckardt, K., U. Boutellier, A. Kurtz, M. Schopen, E.A. Koller, and C. Bauer. 1989. Rate of erythropoietin formation in humans in response to acute hypobaric hypoxia. *Journal of Applied Physiology* 66:1785-8.

Eckert, H.M. 1968. Angular velocity and range of motion in the vertical and standing broad jump. *Research Quarterly for Exercise and Sport* 39:937-42.

Ekblom, B., and B. Berglund. 1991. Effect of erythropoietin administration on maximal aerobic power. *Scandinavian Journal of Medicine and Science in Sports* 1:88-93.

Eldridge, F.L. 1994. Central integration of mechanisms in exercise hyperpnea. *Medicine and Science in Sports and Exercise* 26:319-27.

Ellenbecker, T.S. 1991. A total arm strength isokinetic profile of highly skilled tennis players. *Isokinetic Exercise Science* 1:9-21.

Enoka, R.M. 1994. *Neuromechanical basis of kinesiology.* 2nd ed. Champaign, IL: Human Kinetics.

Epstein, Y. 1990. Heat intolerance: Predisposing factor or residual injury. *Medicine and Science in Sports and Exercise* 22:29-35.

Ernsting, J., and G.R. Sharp. 1978. Prevention of hypoxia at altitudes below 40,000 feet. In *Aviation Medicine, Physiology and Human Factors,* ed. J. Ernsting, 84-127. London: Tri-Med Books.

Esperson, G.T., A. Elbaek, E. Ernst, E. Toft, S. Kaalund, C. Jersild, and N. Grunnet. 1990. Effect of physical exercise on cytokines and lymphocyte transformation and antibody formation. *APMIS* 98:395-400.

Esperson, G.T., E. Toft, E. Ernst, S. Kaalund, and N. Grunnet. 1991. Changes of polymorphonuclear granulocyte migration and lymphocyte subpopulations in human peripheral blood. *Scandinavian Journal of Medicine and Science in Sports* 1:158-62.

Essig, D., D.L. Costill, and P.J. Van Handel. 1980. Effects of caffeine ingestion on utilization of muscle glycogen and lipid during leg ergometer cycling. *International Journal of Sports Medicine* 1:86-90.

Ettema, G.J.C., A.J. Van Soest, and P.A. Huijing. 1990. The role of series elastic structures in prestretch-induced work enhancement during isotonic and isokinetic contractions. *Journal of Experimental Biology* 154:121-36.

Evans, W.J., and J.G. Cannon. 1991. The metabolic effects of exercise-induced muscle damage. *Exercise and Sport Sciences Reviews* 19:99-126.

Fagard, R.H. 2001. Exercise characteristics and the blood pressure response to dynamic physical training. *Medicine and Science in Sports and Exercise* 33(6)(supp):S484-92.

Fahey, T.D., and C.H. Brown. 1973. The effects of an anabolic steroid on the strength, body composition, and endurance of college males when accompanied by a weight training program. *Medicine and Science in Sports and Exercise* 5:272-6.

Fahey, T.D., R. Rolph, P. Moungmee, J. Nadel, and S. Martara. 1976. Serum testosterone, body composition, and strength of young adults. *Medicine and Science in Sports and Exercise* 8:31-4.

Falk, B., S. Radom-Isaac, J.R. Hoffman, Y. Wang, Y. Yarom, A. Magazanik, and Y. Weinstein. 1998. The effect of heat exposure on performance of and recovery from high-intensity, intermittent exercise. *International Journal of Sports Medicine* 19:1-6.

Falk, B., Y. Weinstein, R. Dotan, D.R. Abramson, D. Mann-Segal, and J.R. Hoffman. 1996. A treadmill test of sprint running. *Scandinavian Journal of Medicine and Science in Sports* 6:259-64.

Farrel, M., and J.G. Richards. 1986. Analysis of the reliability and validity of the kinetic communicator exercise device. *Medicine and Science in Sports and Exercise* 18:44-9.

Farrell, P.A., T.L. Garthwaite, and A.B. Gustafson. 1983. Plasma adrenocorticotropin and cortisol responses to submaximal and exhaustive exercise. *Journal of Applied Physiology: Respiratory, Environmental, Exercise Physiology* 55:1441-4.

Farrell, P.A., A. Gustafson, T. Garthwaite, R. Kalkhoff, A. Cowley Jr., and W.P. Morgan. 1986. Influence of endogenous opioids on the response of selected hormones to exercise in humans. *Journal of Applied Physiology* 61:1051-7.

Farrell, P.A., M. Kjaer, F.W. Bach, and H. Galbo. 1987. Beta-endorphin and adrenocorticotropin response to supramaximal treadmill exercise in trained and untrained males. *Acta Physiologica Scandinavica* 130:619-25.

Farrell, P.A., J.H. Wilmore, E.F. Coyle, J.E. Billing, and D.L. Costill. 1979. Plasma lactate accumulation and distance running performance. *Medicine and Science in Sports and Exercise* 11:338-44.

Febbraio, M.A., T.R. Flanagan, R.J. Snow, S. Zhao, and M.F. Carey. 1995. Effect of creatine supplementation on intramuscular TCr, metabolism and performance during intermittent, supramaximal exercise in humans. *Acta Physiologica Scandinavica* 155:387-95.

Fehr, H.G., H. Lotzerich, and H. Michna. 1988. The influence of physical exercise on peritoneal macrophage functions: Histochemical and phagocytic studies. *International Journal of Sports Medicine* 9:77-81.

———. 1989. Human macrophage function and physical exercise: Phagocytic and histochemical studies. *European Journal of Applied Physiology* 58:613-17.

Felig, P., and J. Wahren. 1971. Amino acid metabolism in exercising man. *Journal of Clinical Investigations* 50:2703-14.

Felsing, N.E., J.A. Brasel, and D.M. Cooper. 1992. Effect of low and high intensity exercise on circulating growth hormone in men. *Journal of Clinical Endocrinology and Metabolism* 75:157-62.

Fern, E.B., R.N. Bielinski, and Y. Schutz. 1991. Effects of exaggerated amino acid and protein supply in man. *Experientia* 47:168-72.

Ferretti, G. 1992. Cold and muscle performance. *International Journal of Sports Medicine* 13(supp):S185-7.

Ferry, A., F. Picard, A. Duvallet, B. Weill, and M. Rieu. 1990. Changes in blood leukocyte populations induced by acute maximal and chronic submaximal exercise. *European Journal of Applied Physiology* 59:435-42.

Few, J.D. 1974. Effect of exercise on the secretion and metabolism of cortisol in man. *Journal of Endocrinology* 62:341-53.

Fielding, R.A., T.J. Manfredi, W. Ding, M.A. Fiatarone, W.J. Evans, and J.G. Cannon. 1993. Acute phase response in exercise III: Neutrophil and IL-1β accumulation in skeletal muscle. *American Journal of Physiology: Regulatory, Integrative, Comparative Physiology* 34: R166-72.

Fisher, A.G., and C.R. Jensen. 1990. *Scientific basis of athletic conditioning.* Malvern, PA: Lea & Febiger.

Fitts, R.H. 1992. Substrate supply and energy metabolism during brief high intensity exercise: Importance in limiting performance. In *Energy metabolism in exercise and sport,* ed. D.R. Lamb and C.V. Gisolfi, 53-105. Carmel, IN: Brown & Benchmark.

Fleck, S.J. 1988. Cardiovascular adaptations to resistance training. *Medicine and Science in Sports and Exercise* 20:S146-51.

———. 1992. Cardiovascular responses to strength training. In *Strength and power in sport,* ed. P.V. Kovi, 305-15. Oxford, England: Blackwell Scientific.

———. 1999. Periodized strength training: A critical review. *Journal of Strength and Conditioning Research* 13:82-9.

Fleck, S.J., and J.E. Falkel. 1986. Value of resistance training for the reduction of sports injuries. *Sports Medicine* 3:61-8.

Fleck, S.J., C. Henke, and W. Wilson. 1989. Cardiac MRI of elite junior Olympic weight lifters. *International Journal of Sports Medicine* 10:329-33.

Fleck, S.J., and W.J. Kraemer. 1997. *Designing resistance training programs.* Champaign, IL: Human Kinetics.

Fleck, S.J., and R.C. Schutt. 1985. Types of strength training. *Clinics in Sports Medicine* 4:159-68.

Flora, G. 1985. Secondary treatment of frostbite. In *High altitude deterioration,* ed. J.P. Rivolier, P. Cerretelli, J. Foray, and P. Segantini, 159-69. Basel, Switzerland: Karger.

Florini, J. 1985. Hormonal control of muscle growth. *Journal of Animal Science* 61(supp):21-37.

Flynn, M.G. 1998. Future research needs and discussion. In *Overtraining in sport,* ed. R.B. Kreider, A.C. Fry, and M.L. O'Toole, 373-84. Champaign, IL: Human Kinetics.

Flynn, M.G., F.X. Pizza, J.B. Boone, F.F. Andres, T.A. Michaud, and J.R. Rodriguez-Zayas. 1994. Indices of training stress during competitive running and swimming seasons. *International Journal of Sports Medicine* 15:21-6.

Foray, J. 1992. Mountain frostbite. *International Journal of Sports Medicine* 13(supp):S193-6.

Forbes, G.B. 1985. The effect of anabolic steroids on lean body mass: The dose response curve. *Metabolism* 34:571-3.

Ford, L.E. 1976. Heart size. *Circulatory Research* 39:299-303.

Ford Jr., J.F., J.R. Puckett, J.P. Drummond, K. Sawyer, K. Gantt, and C. Fussell. 1983. Effects of three combinations of plyometric and weight training programs on selected physical fitness test items. *Perceptual and Motor Skills* 56:59-61.

Forster, P. 1984. Reproducibility of individual response to exposure to high altitude. *British Medical Journal* 289:1269.

Fortney, S.M., C.B. Wenger, J.R. Bove, and E.R. Nadel. 1984. Effect of hyperosmolality on control of blood flow and sweating. *Journal of Applied Physiology* 57:1688-95.

Foster, C., D.L. Costill, and W.J. Fink. 1979. Effects of preexercise feedings on endurance performance. *Medicine and Science in Sports and Exercise* 11:1-5.

———. 1980. Gastric emptying characteristics of glucose and glucose polymers. *Research Quarterly for Exercise and Sport* 51:299-305.

Foster, C., L.L. Hector, R. Welsh, M. Schrager, M.A. Green, and A.C. Snyder. 1995. Effects of specific versus cross-training on running performance. *European Journal of Applied Physiology* 70:367-72.

Foster, N.K., J.B. Martyn, R.E. Rangno, J.C. Hogg, and R.L. Pardy. 1986. Leukocytosis of exercise: Role of cardiac output and catecholamines. *Journal of Applied Physiology* 61:2218-23.

Fowler Jr., W.M., G.W. Gardner, and G.H. Egstrom. 1965. Effect of an anabolic steroid on physical performance in young men. *Journal of Applied Physiology* 20:1038-40.

Fox, E.L., R.L. Bartels, C.E. Billings, D.K. Mathews, R. Bason, and W.M. Webb. 1973. Intensity and distance of interval training programs and changes in aerobic power. *Medicine and Science in Sports and Exercise* 5:18-22.

Fraioli, F., C. Moretti, D. Paolucci, E. Alicicco, F. Crescenzi, and G. Fortunio. 1980. Physical exercise stimulates marked concomitant release of β-endorphin and ACTH in peripheral blood in man. *Experientia* 36:987-9.

Francesconi, R.P., M.N. Sawka, K.B. Pandolf, R.W. Hubbard, A.J. Young, and S. Muza. 1985. Plasma hormonal responses at graded hypohydration levels during exercise-heat stress. *Journal of Applied Physiology* 59:1855-60.

Freund, B.J., E.M. Shizuru, G.M. Hashiro, and J.R. Claybaugh. 1991. Hormonal, electrolyte and renal responses to exercise are intensity dependent. *Journal of Applied Physiology* 70:900-6.

Friedl, K.E. 1993. Effects of anabolic steroids on physical health. In *Anabolic steroids in sport and exercise,* ed. C.E. Yesalis, 107-50. Champaign, IL: Human Kinetics.

Friedman, J.E., and P.W.R. Lemon. 1989. Effect of chronic endurance exercise on the retention of dietary protein. *International Journal of Sports Medicine* 10:118-23.

Froberg, S.O. 1971. Effect of training and of acute exercise in trained rats. *Metabolism* 20:1044-51.

Fry, A.C. 1998. The role of training intensity in resistance exercise overtraining and overreaching. In *Overtraining in sport,* ed. R.B. Kreider, A.C. Fry, and M.L. O'Toole, 107-30. Champaign, IL: Human Kinetics.

Fry, A.C., C.A. Allemeier, and R.S. Staron. 1994. Correlation between percentage fiber type area and myosin heavy chain content in human skeletal muscle. *European Journal of Applied Physiology* 68:246-51.

Fry, A.C., and W.J. Kraemer. 1991. Physical performance characteristics of American collegiate football players. *Journal of Applied Sport Science Research* 5:126-38.

Fry, A.C., W.J. Kraemer, M.H. Stone, J.T. Kearney, S.J. Fleck, and C.A. Weseman. 1993. Endocrine and performance responses to high volume training and amino acid supplementation in elite junior weightlifters. *International Journal of Sports Nutrition* 3:306-22.

Fry, A.C., W.J. Kraemer, F. van Borselen, J.M. Lynch, J.L. Marsit, E.P. Roy, N.T. Triplett, and H.G. Knuttgen. 1994. Performance decrements with high-intensity resistance exercise overtraining. *Medicine and Science in Sports and Exercise* 26:1165-73.

Fry, A.C., W.J. Kraemer, F. van Borselen, J.M. Lynch, N.T. Triplett, L.P. Koziris, and S.J. Fleck. 1994. Catecholamine responses to short-term high-intensity resistance exercise overtraining. *Journal of Applied Physiology* 77:941-6.

Fry, A.C., W.J. Kraemer, C.A. Weseman, B.P. Conroy, S.E. Gordon, J.R. Hoffman, and C.M. Maresh. 1991. The effects of an off-season strength and conditioning program on starters and non-starters in women's intercollegiate volleyball. *Journal of Applied Sport Science Research* 5:174-81.

Fry, A.C., and D.R. Powell. 1987. Hamstring/quadricep parity with three different weight training methods. *Journal of Sports Medicine and Physical Fitness.* 27:362-7.

Fry, M.D., A.C. Fry, and W.J. Kraemer. 1996. Self-efficacy responses to short-term high-intensity resistance exercise overtraining. *International Conference on Overtraining and Overreaching in Sport: Physiological, Psychological, and Biomedical Considerations.* Memphis, TN.

Fry, R.W., S.R. Lawrence, A.R. Morton, A.B. Schreiner, T.D. Polglaze, and D. Keast. 1993. Monitoring training stress in endurance sports using biological parameters. *Clinical Journal of Sports Medicine* 3:6-13.

Fry, R.W., A.R. Morton, P. Garcia-Webb, G.P.M. Crawford, and D. Keast. 1992. Biological responses to overload training in endurance sports. *European Journal of Applied Physiology* 64:335-44.

Fry, R.W., A.R. Morton, and D. Keast. 1991. Overtraining in athletes. *Sports Medicine* 12:32-65.

Gabriel, H., H.J. Miller, A. Urhausen, and W. Kindermann. 1994. Suppressed PMA-induced oxidative burst and unimpaired phagocytosis of circulating granulocytes one week after a long endurance exercise. *International Journal of Sports Medicine* 15:441-5.

Gabriel, H., L. Schwarz, P. Bonn, and W. Kindermann. 1992. Differential mobilization of leucocyte and lymphocyte subpopulations into the circulation during endurance exercise. *European Journal of Applied Physiology* 65:529-34.

Gabriel, H., A. Urhausen, and W. Kindermann. 1992. Mobilization of circulating leucocyte and lymphocyte subpopulations during and after short, anaerobic exercise. *European Journal of Applied Physiology* 65:164-70.

Gaitanos, G.C., M.E. Nevill, S. Brooks, and C. Williams. 1991. Repeated bouts of sprint running after induced alkalosis. *Journal of Sports Science* 9:355-70.

Gaitanos, G.C., C. Williams, L. Boobis, and S. Brooks. 1993. Human muscle metabolism during intermittent maximal exercise. *Journal of Applied Physiology* 75:712-9.

Galbo, H. 1981. Endocrinology and metabolism in exercise. *International Journal of Sports Medicine* 2:2203-11.

———. 1985. The hormonal response to exercise. *Proceedings of the Nutrition Society* 44:257-66.

Galbo, H., L. Hammer, I.B. Peterson, N.J. Christensen, and N. Bic. 1977. Thyroid and testicular hormone responses to gradual and prolonged exercise in man. *Journal of Applied Physiology* 36:101-6.

Gallagher, P.M., J.A. Carrithers, M.P. Godard, K.E. Schulze, and S. Trappe. 2000a. β-hydroxy-β-methylbutyrate ingestion, part I: Effects on strength and fat free mass. *Medicine and Science in Sports and Exercise* 32:2109-15.

———. 2000b. β-hydroxy-β-methylbutyrate ingestion, part II: Effects on hematology, hepatic and renal function. *Medicine and Science in Sports and Exercise.* 32:2116-19.

Galloway, S.D.R., and R.J. Maughan. 1997. Effects of ambient temperature on the capacity to perform prolonged cycle exercise in man. *Medicine and Science in Sports and Exercise* 29:1240-9.

Garagioloa, U., M. Buzzetti, E. Cardella, F. Confaloneieri, E. Giani, V. Polini, P. Ferrante, R. Mancuso, M. Montanari, E. Grossi, and A. Pecori. 1995. Immunological patterns during regular intensive training in athletes: Quantification and evaluation of a preventive pharmacological approach. *Journal of International Medical Research* 23:85-95.

Gareau, R., M. Audran, R.D. Baynes, C.H. Flowers, A. Duvallet, L. Senecal, and G.R. Brisson. 1996. Erythropoietin abuse in athletes. *Nature* 380:113.

Garhammer, J., and R. Gregor. 1992. Propulsion forces as a function of intensity for weightlifting and vertical jumping. *Journal of Applied Sport Science Research* 6:129-34.

Gergley, T.J., W.D. McArdle, P. DeJesus, M.M. Toner, S. Jacobowitz, and R.J. Spina. 1984. Specificity of arm training on aerobic power during swimming and running. *Medicine and Science in Sports and Exercise* 16:349-54.

Gettman, L.R., and M.L. Pollock. 1981. Circuit weight training: Critical review of its physiological benefits. *Physician and Sportsmedicine* 9:45-57.

Gilbert, I.A., J.M. Fouke, and E.R. McFadden Jr. 1987. Heat and water flux in the intrathoracic airways and exercise-induced asthma. *Journal of Applied Physiology* 63:1681-91.

———. 1988. Intra-airway thermodynamics during exercise and hyperventilation in asthmatics. *Journal of Applied Physiology* 64:2167-74.

Gilbert, I.A., and E.R. McFadden Jr. 1992. Airway cooling and rewarming: The second reaction sequence in exercise-induced asthma. *Journal of Clinical Investigations* 90:699-704.

Gillam, G.M. 1981. Effects of frequency of weight training on muscular strength. *Journal of Sports Medicine* 21:432-6.

Girandola, R.N., and F.L. Katch. 1976. Effects of physical training on ventilatory equivalent and respiratory exchange ratio during weight supported, steady-state exercise. *European Journal of Applied Physiology and Occupational Physiology* 21:119-25.

Girouard, C.K., and B.F. Hurley. 1995. Does strength training inhibit gains in range of motion from flexibility training in older adults? *Medicine and Science in Sports and Exercise* 27:1444-9.

Gisolfi, C.V. 1973. Work-heat tolerance derived from interval training. *Journal of Applied Physiology* 35:349-54.

Gisolfi, C.V., and J.S. Cohen. 1979. Relationships among training, heat acclimation, and heat tolerance in men and women: The controversy revisited. *Medicine and Science in Sports and Exercise* 11:56-9.

Gisolfi, C.V., and S. Robinson. 1969. Relations between physical training, acclimatization, and heat tolerance. *Journal of Applied Physiology* 26:530-4.

Gleeson, M., W.A. McDonald, A.W. Cripps, D.B. Pyne, R.L. Clancy, and P.A. Fricker. 1995. The effect on immunity of long-term intensive training in elite swimmers. *Clinical and Experimental Immunology* 102:210-16.

Gleim, G.W., P.A. Witman, and J.A. Nicholas. 1984. Indirect assessment of cardiovascular "demands" using telemetry on professional football players. *American Journal of Sports Medicine* 9:178-83.

Gmunder, F.K., P.W. Joller, H.I. Joller-Jemelka, B. Bechler, M. Cogoli, W.H. Ziegler, J. Muller, R.E. Aeppli, and A. Cogoli. 1990. Effect of herbal yeast food supplements and long-distance running on immunological parameters. *British Journal of Sports Medicine* 24:103-12.

Godfrey, S. 1988. Exercise-induced asthma. In *Allergic diseases from infancy to adulthood*. 2nd ed., ed. W.C. Bierman and D.S. Pearlman, 597. Philadelphia: Saunders.

Godt, R.E., and B.D. Lindly. 1982. Influence of temperature upon contractile activation and isometric force production in mechanically skinned muscle fibers of the frog. *Journal of General Physiology* 80:279-97.

Goforth, H.W., A.N. Campbell, J.A. Hodgdon, and A.A. Sucec. 1982. Hematologic parameters of trained distance runners following induced erythrocythemia. *Medicine and Science in Sports and Exercise* 14:174.

Goldberg, L., D.L. Elliot, and K.S. Kuehl. 1994. A comparison of the cardiovascular effects of running and weight training. *Journal of Strength and Conditioning Research* 8:219-24.

Golden, F.S.C., I.F.G. Hampton, G.R. Hervery, and A.V. Knibbs. 1979. Shivering intensity in humans during immersion in cold water. *Journal of Physiology* 277:48.

Goldfarb, A.H., B.D. Hatfield, D. Armstrong, and J. Potts. 1990. Plasma beta-endorphin concentration: Response to intensity and duration of exercise. *Medicine and Science in Sports and Exercise* 22:241-68.

Goldfinch, J., L. McNaughton, and P. Davies. 1988. Induced metabolic alkalosis and its effects on 400-m racing time. *European Journal of Applied Physiology* 57:45-8.

Golding, L.A., J.E. Freydinger, and S.S. Fishel. 1974. The effect of an androgenic-anabolic steroid and a protein supplement on size, strength, weight and body composition in athletes. *Physician and Sportsmedicine* 2:39-45.

Goldman, R. 1984. *Death in the locker room*. South Bend, IN: Icarus Press, Inc.

Goldspink, G. 1970. The proliferation of myofibrils during muscle fibre growth. *Journal of Cell Science* 6:593-603.

Gollnick, P.D., R.B. Armstrong, C.W. Saubert, K. Piehl, and B. Saltin. 1972. Enzyme activity and fiber composition in skeletal muscle of untrained and trained men. *Journal of Applied Physiology* 33:312-9.

Gollnick, P.D., D. Parsons, M. Riedy, and R.L. Moore. 1983. Fiber number and size in overloaded chicken anterior latissimus dorsi muscle. *Journal of Applied Physiology: Respiratory, Environmental, Exercise Physiology* 54:1292-7.

Gollnick, P.D., and B. Saltin. 1982. Significance of skeletal muscle oxidative enzyme enhancement with endurance training. *Clinical Physiology* 2:1-12.

———. 1988. Fuel for muscular exercise: Role of fat. In *Exercise, nutrition, and energy metabolism,* ed. E.S. Horton and R.L. Terjung, 72-88. New York: Macmillan.

Gollnick, P.D., B.F. Timson, R.L. Moore, and M. Riedy. 1981. Muscular enlargement and number of fibers in skeletal muscle of rats. *Journal of Applied Physiology: Respiratory, Environmental, Exercise Physiology* 50:936-43.

Gonyea, W.J. 1980a. Muscle fiber splitting in trained and untrained animals. *Exercise and Sport Sciences Reviews* 8:19-39.

———. 1980b. Role of exercise in inducing increases in skeletal muscle fiber number. *Journal of Applied Physiology: Respiratory, Environmental, Exercise Physiology* 48:421-6.

Gonyea, W.J., D.G. Sale, F. Gonyea, and A. Mikesky. 1986. Exercise induced increases in muscle fiber number. *European Journal of Applied Physiology* 55:137-41.

Goodman, H.M. 1988. *Basic medical endocrinology*. New York: Raven Press Publishers.

Goodman, H.M., and J.C.S. Fray. 1988. Regulation of sodium and water balance. In *Basic medical endocrinology,* ed. H.M. Goodman, 153-74. New York: Raven Press Publishers.

Goodpaster, B.H., F.L. Thaete, J.A. Simoneau, and D.E. Kelly. 1997. Subcutaneous abdominal fat and thigh muscle composition predict insulin sensitivity independently of visceral fat. *Diabetes* 46:1579-85.

Gore, C.J., A.G. Hahn, G.C. Scroop, D.B. Watson, K.I. Norton, R.J. Wood, D.P. Campbell, and D.L. Emonson. 1996. Increased arterial desaturation in trained cyclists during maximal exercise at 580 m altitude. *Journal of Applied Physiology* 80:2204-10.

Gore, C.J., S.C. Little, A.G. Hahn, G.C. Scroop, K.I. Norton, P.C. Bourdon, S.M. Woolford, J.D. Buckley, T. Stanef, D.P. Campbell, D.B. Watson, and D.L. Emonson. 1997. Reduced performance of male and female athletes at 580 m altitude. *European Journal of Applied Physiology* 75:136-43.

Gorostiaga, E.M., C.B. Walter, C. Foster, and R.C. Hickson. 1991. Uniqueness of interval and continuous training at the same maintained exercise intensity. *European Journal of Applied Physiology and Occupational Physiology* 63:101-7.

Graham, T.E., E. Hibbert, and P. Sathasivam. 1998. Metabolic and exercise endurance effects of coffee and caffeine ingestion. *Journal of Applied Physiology* 85:883-9.

Graham, T.E., and L.L. Spriet. 1995. Metabolic, catecholamine and exercise performance responses to varying doses of caffeine. *Journal of Applied Physiology* 78:867-74.

Gravelle, B.L., and D.L. Blessing. 2000. Physiological adaptation in women concurrently training for strength and endurance. *Journal of Strength and Conditioning Research* 14:5-13.

Gray, A.B., R.D. Telford, M. Collins, and M.J. Weidemann. 1993. The response of leukocyte subsets and plasma hormones to interval exercise. *Medicine and Science in Sports and Exercise* 25:1252-8.

Green, H.J., S. Jones, M. Ball-Burnett, B. Farrance, and D. Ranney. 1995. Adaptations in muscle metabolism to prolonged voluntary exercise and training. *Journal of Applied Physiology* 78:138-45.

Green, H.J., J.R. Sutton, G. Coates, M. Ali, and S. Jones. 1991. Response of red cells and plasma volume to prolonged training in humans. *Journal of Applied Physiology* 70:1810-15.

Green, H.J., J. Sutton, P. Young, A. Cymerman, and C.S. Houston. 1989. Operation Everest II: Muscle energetics during maximal exhaustive exercise. *Journal of Applied Physiology* 66:142-50.

Green, R.L., S.S. Kaplan, B.S. Rabin, C.L. Stanitski, and U. Zdiarski. 1981. Immune function in marathon runners. *Annals of Allergy* 47:73-5.

Greenhaff, P.L. 1995. Creatine and its application as an ergogenic aid. *International Journal of Sports Nutrition* 5:S100-10.

Greer, F., C. McLean, and T.E. Graham. 1998. Caffeine, performance, and metabolism during repeated Wingate exercise tests. *Journal of Applied Physiology* 85:1502-8.

Griggs, R.C., W. Kingston, R.F. Jozefowicz, B.E. Herr, G. Forbers, and D. Halliday. 1989. Effect of testosterone on muscle mass and muscle protein synthesis. *Journal of Applied Physiology* 66:498-503.

Gruber, A.J., and H.G. Pope. 1998. Ephedrine abuse among 36 female weightlifters. *American Journal of Addiction* 7:256-61.

Guglielmini, C., A.R. Paolini, and F. Conconi. 1984. Variations of serum testosterone concentrations after physical exercise of different duration. *International Journal of Sports Medicine* 5:246-9.

Hack, B., G. Strobel, M. Weiss, and H. Weicker. 1994. PMN cell counts and phagocytic activity of highly trained athletes depend on training period. *Journal of Applied Physiology* 77:1731-5.

Hackett, P.H., and D. Rennie. 1979. Rales, peripheral edema, retinal hemorrhage and acute mountain sickness. *American Journal of Medicine* 67:214-8.

Hackett, P.H., D. Rennie, R.F. Glover, and J.T. Reeves. 1981. Acute mountain sickness and the edemas of high altitude: A common pathogenesis? *Respiratory Physiology* 46:383-90.

Hackney, A.C., W.E. Sinning, and B.C. Bruot. 1988. Reproductive hormonal profiles of endurance-trained and untrained males. *Medicine and Science in Sport and Exercise* 20:60-5.

———. 1990. Hypothalamic-pituitary-testicular axis function in endurance-trained males. *International Journal of Sports Medicine* 11:298-303.

Hafe, G.G., K.B. Kirksey, M.H. Stone, B.J. Warren, R.L. Johnson, M. Stone, H. O'Bryant, and C. Proulx. 2000. The effect of 6 weeks of creatine monohydrate supplementation on dynamic rate of force development. *Journal of Strength and Conditioning Research* 14:426-33.

Hakkinen, K., M. Alen, and P.V. Komi. 1985. Changes in isometric force- and relaxation-time, electromyographic and muscle fibre characteristics of human skeletal muscle during strength training and detraining. *Acta Physiologica Scandinavica* 125:573-85.

Hakkinen, K., and P.V. Komi. 1983. Electromyographic changes during strength training and detraining. *Medicine and Science in Sports and Exercise* 15:455-60.

———. 1985a. Changes in electrical and mechanical behavior of leg extensor muscles during heavy resistance strength training. *Scandinavian Journal of Sports Sciences* 7:55-64.

———. 1985b. The effect of explosive type strength training on electromyographic and force production characteristics of leg extensor muscles during concentric and various stretch-shortening cycle exercises. *Scandinavian Journal of Sports Sciences* 7:65-76.

———. 1986. Training induced changes in neuromuscular performance under voluntary and reflex conditions. *European Journal of Applied Physiology* 55:147-55.

Hakkinen, K., P.V. Komi, and M. Alen. 1985. Effect of explosive type strength training on isometric force- and relaxation-time, electromyographic and muscle fibre characteristics of leg extensor muscles*Acta Physiologica Scandinavica* 125:587-600.

Hakkinen, K., P.V. Komi, M. Alen, and H. Kauhanen. 1987a. EMG, muscle fibre and force production characteristics during a 1 year training period in highly competitive weightlifters. *European Journal of Applied Physiology and Occupational Physiology* 56:419-27.

Hakkinen, K., and A. Pakarinen. 1993. Acute hormonal responses to two different fatiguing heavy-resistance protocols in male athletes. *Journal of Applied Physiology* 74:882-7.

Hakkinen, K., A. Pakarinen, M. Alen, H. Kauhanen, and P.V. Komi. 1987b. Relationships between training volume, physical performance capacity, and serum hormone concentration during prolonged training in elite weightlifters. *International Journal of Sports Medicine* 8(supp):61-5.

———. 1988. Neuromuscular and hormonal adaptations in athletes to strength training in two years. *Journal of Applied Physiology* 65:2406-12.

Hakkinen, K., A. Pakarinen, M. Alen, and P.V. Komi. 1985. Serum hormones during prolonged training of neuromuscular performance. *European Journal of Applied Physiology* 53:287-93.

Hakkinen, K., A. Pakarinen, P.V. Komi, T. Ryushi, and H. Kauhanen. 1989. Neuromuscular adaptations and hormone balance in strength athletes, physically active males and females during intensive strength training. In *Proceedings of XII International Congress of Biomechanics.* no. 8, ed. R.J. Gregor, R.F. Zernicke, and W.C. Whiting, 889-98. Champaign, IL: Human Kinetics.

Hamlet, M.P. 1988. Human cold injuries. In *Human performance physiology and environmental medicine at terrestrial extremes,* ed. K.B. Pandolf, M.N. Sawka, and R.R. Gonzalez, 435-66. Indianapolis: Benchmark Press.

Hansen, A.A.P. 1973. Serum growth hormone response to exercise in non-obese and obese normal subjects. *Scandinavian Journal of Clinical Investigations* 31:175-8.

Hansen, J.B., and D.K. Flaherty. 1981. Immunological responses to training in conditioned runners. *Clinical Science* 60:225-8.

Hansen, J.B., L. Wilsgard, and B. Osterud. 1991. Biphasic changes in leukocytes induced by strenuous exercise. *European Journal of Applied Physiology* 62:157-61.

Hansen, L.P., B.B. Jacobsen, P.E.L. Kofeod, M.L. Larsen, T. Tougaard, and I. Johansen. 1989. Serum fructosamine and HbA1c in diabetic children before and after attending a winter camp. *Acta Paediatics Scandinavica* 78:451-2.

Harmon, E., J. Garhammer, and C. Pandorf. 2000. Administration, scoring, and interpretation of selected tests. In *Essentials of strength and conditioning,* ed. T. Baechle and R. Earle, 287-318. Champaign, IL: Human Kinetics.

Harris, R.C., R.H.T. Edwards, E. Hultman, L.O. Nordesjo, B. Nylind, and K. Sahlin. 1976. The time course of phosphorylcreatine resynthesis during recovery of the quadriceps muscle in man. *Pflugers Archives* 367:137-42.

Harris, R.C., K. Soderlund, and E. Hultman. 1992. Elevation of creatine in resting and exercised muscle of normal subjects by creatine supplementation. *Clinical Science* 83:367-74.

Hartley, L.H., J.W. Mason, R.P. Hogan, L.G. Jones, T.A. Kotchen, E.H. Mougey, F.E. Wherry, L.L. Pennington, and P.T. Ricketts. 1972. Multiple hormonal responses to prolonged exercise in relation to physical training. *Journal of Applied Physiology* 33:607-10.

Hather, B.M., P.A. Tesch, P. Buchanan, and G.A. Dudley. 1991. Influence of eccentric actions on skeletal muscle adaptations to resistance training. *Acta Physiologica Scandinavica* 143: 177-85.

Hayward, J.S., and J.D. Eckerson. 1984. Physiological responses and survival time prediction for humans in ice-water. *Aviation, Space and Environmental Medicine* 55:206-12.

Hayward, J.S., J.D. Eckerson, and M.L. Collis. 1975. Effect of behavioral variables on cooling rate of man in cold water. *Journal of Applied Physiology* 38:1073-7.

Heath, G.W., E.S. Ford, T.E. Craven, C.A. Macera, K.L. Jackson, and R.R. Pate. 1991. Exercise and the incidence of upper respiratory tract infections. *Medicine and Science in Sports and Exercise* 25:186-90.

Heath, G.W., C.A. Macera, and D.C. Nieman. 1992. Exercise and upper respiratory tract infections: Is there a relationship? *Sports Medicine* 14:353-65.

Heck, H., A. Mader, G. Hess, S. Mucke, R. Muller, and W. Hollmann. 1985. Justification of the 4 mmol/L lactate threshold. *International Journal of Sports Medicine* 6:117-30.

Hedge, G.A., H.D. Colby, and R.L. Goodman. 1987. *Clinical Endocrine Physiology* Philadelphia: W.B. Saunders.

Heigenhauser, G.J.F., and N.L. Jones. 1991. Bicarbonate loading. In *Perspectives in exercise science and sports medicine.* Vol. 4. Ergogenics, ed. D.R. Lamb and M.H. Williams, 183-212. Carmel, IN: Benchmark Press.

Heir, T., and S. Oseid. 1994. Self-reported asthma and exercise-induced asthma symptoms in high-level competitive cross-country skiers. *Scandinavian Journal of Medicine and Science in Sports* 4:128-33.

Hennessy, L.C., and A.W.S. Watson. 1994. The interference effects of training for strength and endurance simultaneously. *Journal of Strength and Conditioning Research* 8:12-9.

Hermansen, L., and B. Saltin. 1969. Oxygen uptake during maximal treadmill and bicycle exercise. *Journal of Applied Physiology.* 26:31-37.

Hermansen, L., and M. Wachtlova. 1971. Capillary density of skeletal muscle in well-trained and untrained men. *Journal of Applied Physiology* 30:860-3.

Hervey, G.R., A.V. Knibbs, L. Burkinshaw, D.B. Morgan, P.R.M. Jones, D.R. Chettle, and D. Vartsky. 1981.

Effects of methandienone on the performance and body composition of men undergoing athletic training. *Clinical Science* 60:457-61.

Heymsfield, S.B., C. Arteaga, C. McManus, J. Smith, and S. Moffitt. 1983. Measurement of muscle mass in humans: Validity of the 24-hour urinary creatinine method. *American Journal of Clinical Nutrition* 37:478-94.

Heyward, V.H. 1997. *Advanced fitness assessment & exercise prescription,*. Champaign, IL: Human Kinetics.

Heyward, V.H., and L.M. Stolarczyk. 1996. *Applied body composition assessment.* Champaign, IL: Human Kinetics.

Hickner, R.C., C.A. Horswill, J. Welker, J.R. Scott, J.N. Roemmich, and D.L. Costill. 1991. Test development for study of physical performance in wrestlers following weight loss. *International Journal of Sports Medicine* 12:557-62.

Hickson, R.C. 1981. Skeletal muscle cyctochrome cand myoglobin, endurance, frequency of training. *Journal of Applied Physiology* 51:746-9.

Hickson, R.C., B.A. Dvorak, E.M. Gorostiaga, T.T. Kurowski, and C. Foster. 1988. Potential for strength and endurance training to amplify endurance performance. *Journal of Applied Physiology* 65:2285-90.

Hickson, R.C., K. Hidaka, C. Foster, M.T. Falduto, and R.T. Chatterton Jr. 1994. Successive time course of strength development and steroid hormone responses to heavy resistance training. *Journal of Applied Physiology* 76:663-70.

Hickson, R.C., M.A. Rosenkoetter, and M.M. Brown. 1980. Strength training effects on aerobic power and short-term endurance. *Medicine and Science in Sports and Exercise* 12:336-9.

Hirvonen, J., A. Nummela, H. Rusko, S. Rehunen, and M. Harkonen. 1992. Fatigue and changes of ATP, creatine phosphate, and lactate during the 400-m sprint. *Canadian Journal of Sport Science* 17:141-4.

Hirvonen, J., S. Rehunen, H. Rusko, and M. Harkonen. 1987. Breakdown of high-energy phosphate compounds and lactate accumulation during short supramaximal exercise. *European Journal of Applied Physiology* 56:253-9.

Ho, K.W., R.R. Roy, C.D. Tweedle, W.W. Heusner, W.D. Van Huss, and R.E. Carrow. 1980. Skeletal muscle fiber splitting with weight-lifting exercise. *American Journal of Anatomy* 116:57-65.

Hoeger, W.W.K., S.L. Barette, D.F. Hale, and D.R. Hopkins. 1987. Relationship between repetitions and selected percentages of one repetition maximum. *Journal of Applied Sports Science Research* 1:11-3.

Hoeger, W.W.K., and S.A. Hoeger. 2000. *Lifetime physical fitness & wellness.* Englewood, CO: Morton.

Hoeger, W.W.K., D.R. Hopkins, S.L. Barette, and D.F. Hale. 1990a. Relationship between repetitions and selected percentages of one repetition maximum: A com-

parison between untrained and trained males and females. *Journal of Applied Sports Science Research* 4:47-54.

Hoeger, W.W.K., D.R. Hopkins, S. Button, and T.A. Palmer. 1990b. Comparing the sit and reach with the modified sit and reach in measuring flexibility in adolescents. *Pediatric Exercise Science*, 2:156-162.

Hoffman, J.R. 1997. The relationship between aerobic fitness and recovery from high-intensity exercise in infantry soldiers. *Military Medicine* 162:484-8.

Hoffman, J.R., S. Epstein, M. Einbinder, and Y. Weinstein. 1999. The influence of aerobic capacity on anaerobic performance and recovery indices in basketball players. *Journal of Strength and Conditioning Research* 13:407-11.

———. 2000. A comparison between the Wingate anaerobic power test to both vertical jump and line drill tests in basketball players. *Journal of Strength and Conditioning Research* 14:261-4.

Hoffman, J.R., S. Epstein, Y. Yarom, L. Zigel, and M. Einbinder. 1999. Hormonal and biochemical changes in elite basketball players during a 4-week training camp. *Journal of Strength and Conditioning Research* 13:280-5.

Hoffman, J.R., A.C. Fry, R. Howard, C.M. Maresh, and W.J. Kraemer. 1991a. Strength, speed and endurance changes during the course of a division I basketball season. *Journal of Applied Sport Science Research* 5:144-9.

Hoffman, J.R., and M. Kaminsky. 2000. Use of performance testing for monitoring overtraining in elite youth basketball players. *Strength and Conditioning* 22:54-62.

Hoffman, J.R., and S. Klafeld. 1998. The effect of resistance training on injury rate and performance in a self-defense course for females. *Journal of Strength and Conditioning Research* 12:52-6.

Hoffman, J.R., W.J. Kraemer, A.C. Fry, M. Deschenes, and M. Kemp. 1990. The effect of self-selection for frequency of training in a winter conditioning program for football. *Journal of Applied Sport Science Research* 3:76-82.

Hoffman, J.R., and C.M. Maresh. 2000. Physiology of basketball. In *Exercise and sport science,* ed. W.E. Garrett and D.T. Kirkendall, 733-44. Philadelphia: Lippincott, Williams, and Wilkins.

Hoffman, J.R., C.M. Maresh, and L.E. Armstrong. 1992. Isokinetic and dynamic constant resistance strength testing: Implications for sport. *Physical Therapy Practice* 2:42-53.

Hoffman, J.R., C.M. Maresh, L.E. Armstrong, C.L. Gabaree, M.F. Bergeron, R.W. Kenefick, J.W. Castellani, L.E. Ahlquist, and A. Ward. 1994. The effects of hydration status on plasma testosterone, cortisol, and catecholamine concentrations before and during mild exercise at elevated temperature. *European Journal of Applied Physiology* 69:294-300.

Hoffman, J.R., C.M. Maresh, L.E. Armstrong, and W.J. Kraemer. 1991b. Effects of off-season and in-season resistance training programs on a collegiate male basketball team. *Journal of Human Muscle Performance* 1:48-55.

Hoffman, J.R., H. Stavsky, and B. Falk. 1995. The effect of water restriction on anaerobic power and vertical jumping height in basketball players. *International Journal of Sports Medicine* 16:214-8.

Hoffman, J.R., G. Tenenbaum, C.M. Maresh, and W.J. Kraemer. 1996. Relationship between athletic performance tests and playing time in elite college basketball players. *Journal of Strength and Conditioning Research* 10:67-71.

Hoffman, T., R.W. Stauffer, and A.S. Jackson. 1979. Sex differences in strength. *American Journal of Sports Medicine* 7:265-7.

Holloszy, J.O. 1975. Adaptation of skeletal muscle to endurance exercise. *Medicine and Science in Sports and Exercise* 7:155-64.

———. 1988. Metabolic consequences of endurance exercise training. In *Exercise, nutrition, and energy metabolism,* ed. E.S. Horton and R.L. Terjung, 116-31. New York: Macmillan.

Holloszy, J.O., and F.W. Booth. 1976. Biochemical adaptations to endurance exercise in muscle. *Annual Review of Physiology* 38:273-91.

Holloszy, J.O., and E.F. Coyle. 1984. Adaptations of skeletal muscle to endurance exercise and their metabolic consequences. *Journal of Applied Physiology* 56:831-8.

Holloszy, J.O., L.B. Oscai, I.J. Don, and P.A. Mole. 1970. Mitochondrial citric acid cycle and related enzymes: Adapted response to exercise. *Biochemical Biophysical Research Communications* 40:1368-73.

Holloszy, J.O., J. Schultz, J. Kusnierkiewicz, J.M. Hagberg, and A.A. Eshani. 1986. Effects of exercise on glucose tolerance and insulin resistance: A brief review and some preliminary results. *Acta Medical Scandinavica* 711(supp):55-65.

Holt, L.E., T.M. Travis, and T. Okita. 1970. Comparative study of three stretching techniques. *Perceptual and Motor Skills* 31:611-6.

Honig, A. 1983. Role of arterial chemoreceptors in the reflex control of renal function and body fluid volumes in acute arterial hypoxia. In *Physiology of the peripheral arterial chemoreceptors,* ed. H. Acher and R.G. O'Regan, 395-429. Amsterdam: Elsevier.

Hooper, S., L.T. Mackinnon, R.D. Gordon, and A.W. Bachmann. 1993. Hormonal responses of elite swimmers to overtraining. *Medicine and Science in Sports and Exercise* 25:741-7.

Hooper, S., L.T. Mackinnon, A. Howard, R.D. Gordon, and A.W. Bachmann. 1995. Markers for monitoring overtraining and recovery in elite swimmers. *Medicine and Science in Sports and Exercise* 27:106-12.

Hoppeler, H., E. Kleinert, C. Schlegel, H. Claassen, H. Howald, S.R. Kayar, and P. Cerretelli. 1990. Morphological adaptations of human skeletal muscle to chronic hypoxia. *International Journal of Sports Medicine* 11:S3-9.

Horowitz, J.F., and E.F. Coyle. 1993. Metabolic responses to pre-exercise meals containing various carbohydrates and fat. *American Journal of Clinical Nutrition* 58:235-41.

Horswill, C.A., D.L. Costill, W.J. Fink, M.G. Flynn, J.P. Kirwin, J.B. Mitchell, and J.A. Houmard. 1988. Influence of sodium bicarbonate on sprint performance: Relationship to dosage. *Medicine and Science in Sports and Exercise* 20:566-9.

Horswill, C.A., R.C. Hickner, J.R. Scott, D.L. Costill, and K. Gould. 1990. Weight loss, dietary carbohydrate modifications and high intensity physical performance. *Medicine and Science in Sports and Exercise* 22:470-6.

Horvath, S.M. 1981. Exercise in a cold environment. *Exercise and Sport Sciences Reviews* 9:221-63.

Houmard, J.A., D.L. Costill, J.A. Davis, J.B. Mitchell, D.D. Pascoe, and R.A. Robergs. 1990. The influence of exercise intensity on heat acclimation in trained subjects. *Medicine and Science in Sports and Exercise* 22:615-20.

Houmard, J.A., P.C. Egan, P.D. Neufer, J.E. Friedman, W.S. Wheeler, R.G. Israel, and G.L. Dohm. 1991. Elevated skeletal muscle glucose transporter levels in exercise-trained middle-aged men. *American Journal of Physiology* 261:E437-43.

Housh, T.J., G.O. Johnson, L. Marty, G. Eichen, C. Eishen, and D. Housh. 1988. Isokinetic leg flexion and extension strength of university football players. *Journal of Orthopedic and Sports Physical Therapy* 9:365-9.

Houston, M.E., H. Bentzen, and H. Larsen. 1979. Interrelationships between skeletal muscle adaptations and performance as studied by detraining and retraining. *Acta Physiologica Scandinavica* 105:163-70.

Houston, M.E., D.A. Marin, H.J. Green, and J.A. Thomson. 1981. The effect of rapid weight loss on physiological function in wrestlers. *Physician and Sportsmedicine* 9:73-8.

Houston, M.E., D.M. Wilson, H.J. Green, J.A. Thomson, and D.A. Ranney. 1981. Physiological and muscle enzyme adaptations to two different intensities of swim training. *European Journal of Applied Physiology* 46:283-91.

Howald, H., H. Hoppeler, H. Claassen, O. Mathieu, and R. Staub. 1985. Influence of endurance training on the ultrastructural composition of the different muscle fiber types in humans. *Pflugers Archives* 403:369-76.

Howald, H., D. Pette, J.A. Simoneau, A. Uber, H. Hoppeler, and P. Cerretelli. 1990. Effect of chronic hypoxia on muscle enzyme activities. *International Journal of Sports Medicine* 11:S10-4.

Howard, R.L., W.J. Kraemer, D.C. Stanley, L.E. Armstrong, and C.M. Maresh. 1994. The effects of cold immersion on muscle strength. *Journal of Strength and Conditioning Research* 8:129-33.

Hubbard, R.W. 1990. Heatstroke pathophysiology: The energy depletion model. *Medicine and Science in Sports and Exercise* 22:19-28.

Hubbard, R.W., and L.E. Armstrong. 1988. The heat illnesses: Biochemical, ultrastuctural, and fluid-electrolyte considerations. In *Human performance physiology and environmental medicine at terrestrial extremes,* ed. K.B. Pandolf, M.N. Sawka, and R.R. Gonzalez, 305-60. Indianapolis: Benchmark Press.

———. Hyperthermia: New thoughts on an old problem. *Physician and Sportsmedicine* 17:97-113.

Hubbard, R.W., B.L. Sandick, W.T. Matthews, R.P. Francesconi, J.B. Sampson, M.J. Durkot, O. Maller, and D.B. Engell. 1984. Voluntary dehydration and alliesthesia for water. *Journal of Applied Physiology: Respiratory, Environmental, Exercise Physiology* 57:868-75.

Hubbard, R.W., P.C. Szlyk, and L.E. Armstrong. 1990. Influence of thirst and fluid palatability on fluid ingestion during exercise. In *Perspectives in exercise science and sports medicine.* Vol. 3. Fluid homeostasis during exercise, ed. C.V. Gisolfi and D.R. Lamb, 39-96. Carmel, IN: Benchmark Press.

Hubinger, L.M, L.T. Mackinnon, L. Barber, J. McCosker, A. Howard, and F. Lepre. 1997. The acute effects of treadmill running on lipoprotein (a) levels in males and females. *Medicine and Science in Sports and Exercise* 29:436-42.

Hughes, R.J., G.O. Johnson, T.J. Housh, J.P. Weir, and J.E. Kinder. 1996. The effect of submaximal treadmill running on serum testosterone levels. *Journal of Strength and Conditioning Research* 10:224-7.

Hultman, E., K. Soderlund, J.A. Timmons, G. Cederblad, and P.L. Greenhaff. 1996. Muscle creatine loading in man. *Journal of Applied Physiology* 81:232-7.

Hunter, G.R. 1985. Changes in body composition, body build and performance associated with different weight training frequencies in males and females. *National Strength and Conditioning Association Journal* 7:26-8.

Hunter, G., R. Demment, and D. Miller. 1987. Development of strength and maximum oxygen uptake during simultaneous training for strength and endurance. *Journal of Sports Medicine and Physical Fitness* 27:269-75.

Hunter, G.R., J. Hilyer, and M. Forster. 1993. Changes in fitness during 4 years of intercollegiate basketball. *Journal of Strength and Conditioning Research* 7:26-9.

Hurley, B.F., J.M. Hagberg, W.K. Allen, D.R. Seals, J.C. Young, R.W. Cuddihee, and J.O. Holloszy. 1984.

Effect of training on blood lactate levels during submaximal exercise. *Journal of Applied Physiology* 56:1260-4.

Huttunen, N.P., S.L. Lankela, M. Knip, P. Lautala, M.L. Kaar, K. Laasonen, and P. Puuka. 1989. Effect of once-a-week training program on physical fitness and metabolic control in children with IDDM. *Diabetes Care* 12:737-40.

Huttunen, P., J. Hirvonen, and V. Kinnula. 1981. The occurrence of brown adipose tissue in outdoor workers. *European Journal of Applied Physiology* 46:339-45.

Huxley, H. 1969. The mechanism of muscular contraction. *Science* 164:1356-66.

Idstrom, J.P., C.B. Subramanian, B. Chance, T. Schersten, and A.C. Bylund-Fellenius. 1985. Oxygen dependence of energy metabolism in contracting and recovery rat skeletal muscle. *American Journal of Physiology* 248:H40-8.

Ingjer, F. 1979. Capillary supply and mitochondrial content of different skeletal muscle fiber types in untrained and endurance trained men: A histochemical and ultra structural study. *European Journal of Applied Physiology* 40:197-209.

Israel, S. 1976. Problems of overtraining from an internal medical and performance physiological standpoint. *Medizin Sport* 16:1-12.

Iverson, P.O., B.L. Arvesen, and H.B. Benestad. 1994. No mandatory role for the spleen in the exercise-induced leucocytosis in man. *Clinical Science* 86:505-10.

Ivy, J.L. 1997. Role of exercise training in the prevention and treatment of insulin resistance and non-insulin-dependent diabetes mellitus. *Sports Medicine* 24:3221-36.

Ivy, J.L., D.L. Costill, W.J. Fink, and R.W. Lower. 1979. Influence of caffeine and carbohydrate feedings on endurance performance. *Medicine and Science in Sports and Exercise* 11:6-11.

Ivy, J.L., A.L. Katz, C.L. Cutler, W.M. Sherman, and E.F. Coyle. 1988. Muscle glycogen synthesis after exercise: Effect of time of carbohydrate ingestion. *Journal of Applied Physiology* 64:1480-5.

Ivy, J.L., T.W. Zderic, D.L. Fogt. 1999. Prevention and treatment of non-insulin-dependent diabetes mellitus. *Exercise and Sport Sciences Reviews* 27:1-35.

Jackman, M., P. Wendling, D. Friars, and T.E. Graham. 1996. Metabolic, catecholamine, and endurance responses to caffeine during intense exercise. *Journal of Applied Physiology* 81:1658-63.

Jackson, A.S., and M.L. Pollock. 1985. Practical assessment of body composition. *Physician and Sportsmedicine* 13:76-90.

Jacobs, I. 1980. The effects of thermal dehydration on performance on the Wingate anaerobic test. *International Journal of Sports Medicine* 1:21-4.

Jacobs, I., M. Esbjornsson, C. Sylven, I. Holm, E. Jansson. 1987. Sprint training effects on muscle myoglobin, enzymes, fiber types, and blood lactate. *Medicine and Science in Sports and Exercise* 19:368-74.

Jakeman, P., and S. Maxwell. 1993. Effect of antioxidant vitamin supplementation on muscle function after eccentric exercise. *European Journal of Applied Physiology* 67:426-30.

Janeway, C.A., and P. Travers. 1996. *Immunobiology: The immune system in health and disease.* 2nd ed. London: Current Biology Ltd.

Jansson, E., M. Esbjornsson, I. Holm, and I. Jacobs. 1990. Increases in the proportion of fast-twitch muscle fibres in sprint training in males. *Acta Physiologica Scandinavica* 140:359-63.

Jansson, E., B. Sjodin, and P. Tesch. 1978. Changes in muscle fibre type distribution in man after physical training. *Acta Physiologica Scandinavica* 104:235-7.

Jansson, E., C. Sylven, and B. Sjodin. 1983. Myoglobin content and training in humans. In *Biochemistry of exercise.* Vol. 13, ed. H.G. Knuttgen, J.A. Vogel, and J. Poortmans, 821-5. Champaign, IL: Human Kinetics.

Jenkins, D.J.A., T.M.S. Wolever, R.H. Taylor, H. Baker, H. Fielden, J.M. Baldwin, A.C. Bowling, H.C. Newman, A.L. Jenkins, and D.V. Goff. 1981. Glycemic index of foods: A physiological basis for carbohydrate exchange. *American Journal of Clinical Nutrition* 34:362-6.

Jeukendrup, A.E., F. Brouns, A.J.M. Wagenmakers, and W.H.M. Saris. 1997. Carbohydrate-electrolyte feedings improve 1 h time trial cycling performance. *International Journal of Sports Medicine* 18:125-9.

Jezova, D., M. Vigas, P. Tatar, R. Kevtnansky, K. Nazar, H. Kaciuba-Uscilko, and S. Kozlowski. 1985. Plasma testosterone and catecholamine responses to physical exercise of different intensities in men. *European Journal of Applied Physiology* 54:62-6.

Johns, R.J., and V. Wright. 1962. Relative importance of various tissues in joint stiffness. *Journal of Applied Physiology* 17:824-8.

Johnson, B.L., K.J.W. Adamczy, K.O. Tennoe, and S.B. Stromme. 1976. A comparison of concentric and eccentric muscle training. *Medicine and Science in Sports and Exercise* 8:35-8.

Jost, J., M. Weiss, and H. Weicker. 1989. Comparison of sympatho-adrenergic regulation at rest and of the adrenoreceptor system in swimmers, long-distance runners, weight lifter, wrestlers, and untrained men. *European Journal of Applied Physiology* 58:596-604.

Juhn, M.S., and M. Tarnopolsky. 1998. Oral creatine supplementation and athletic performance: A critical review. *Clinical Journal of Sports Medicine* 8:286-97.

Kaminski, M., and R. Boal. 1992. An effect of ascorbic acid on delayed-onset muscle soreness. *Pain* 50:317-21.

Kanakis, C., and R.C. Hickson. 1980. Left ventricular responses to a program of lower-limb strength training. *Chest* 78:618-21.

Kanehisa, H. and M. Miyashita. 1983. Specificity of velocity in strength training. *European Journal of Applied Physiology* 52:104-6.

Kaneko, M., T. Fuchimoto, H. Toji, and K. Suei. 1983. Training effect of different loads on the force-velocity relationship and mechanical power output in human muscle. *Scandinavian Journal of Sports Sciences* 5:50-5.

Kang, J., J.R. Hoffman, H. Walker, and E.C. Chaloupka. 2001. Regulating intensity of exercise using ratings of perceived exertion during treadmill exercise. *Medicine and Science in Sports and Exercise* 33:S84 (abstract).

Kanj, H., S.H. Schneider, and N.B. Ruderman. 1988. Exercise and diabetes mellitus. In *Exercise, nutrition and energy metabolism,* ed. E.S. Horton and R.L. Terjung, 228-41. New York: Macmillan.

Kanter, M. 1995. Free radicals and exercise: Effects of nutritional antioxidant supplementation. *Exercise and Sport Sciences Reviews* 23:375-98.

Kanter, M.M., L.A. Nolte, and J.O. Holloszy. 1993. Effects of an antioxidant vitamin mixture on lipid peroxidation at rest and postexercise. *Journal of Applied Physiology* 74:965-9.

Karagiorgos, A., J.F. Garcia, and G.A. Brooks. 1979. Growth hormone response to continuous and intermittent exercise. *Medicine and Science in Sports and Exercise* 11:302-7.

Karslon, J., B. Dumont, and B. Saltin. 1971. Muscle metabolites during submaximal and maximal exercise in man. *Scandinavian Journal of Clinical Laboratory Investigations* 26:385-94.

Katch, V., A. Weltman, R. Martin, and L. Gray. 1977. Optimal test characteristics for maximal anaerobic work on the bicycle ergometer. *Research Quarterly for Exercise and Sport* 48:319-27.

Katz, A., D.L. Costill, D.S. King, M. Hargreaves, and W.J. Fink. 1984. Maximal exercise tolerance after induced alkalosis. *International Journal of Sports Medicine* 5:107-10.

Kaufman, W.C., and D.J. Bothe. 1986. Wind chill reconsidered, Siple revisited. *Aviation, Space and Environmental Medicine* 57:23-6.

Keen, P., D.A. McCarthy, L. Passfield, H.A.A. Shaker, and A.J. Wade. 1995. Leucocyte and erythrocyte counts during a multi-stage cycling race ("The Milk Race"). *British Journal of Sports Medicine* 29:61-5.

Keizer, H.A. 1998. Neuroendocrine aspects of overtraining. In *Overtraining in sport,* ed. R.B. Kreider, A.C. Fry, and M.L. O'Toole, 145-68. Champaign, IL: Human Kinetics.

Kerr, R. 1982. *The practical use of anabolic steroids with athletes*. San Gabriel, CA: Kerr.

Kindermann, S., A. Schnabel, W.M. Schmitt, G. Biro, J. Cassens, and F. Weber. 1982. Catecholamines, growth hormone, cortisol, insulin and sex hormones in anaerobic and aerobic exercise. *European Journal of Applied Physiology* 49:389-99.

King, D.S., R.L. Sharp, M.D. Vukovich, G.A. Brown, T.A. Reifenrath, N.L. Uhi, and K.A. Parsons. 1999. Effect of oral androstenedione on serum testosterone and adaptations to resistance training in young men. *Journal of the American Medical Association* 281:2020-8.

Kirby, R.L., F.C. Simms, V.J. Symington, and J.B. Garner. 1981. Flexibility and musculoskeletal symptomatology in female gymnasts and age-matched controls. *American Journal of Sports Medicine* 9:160-4.

Kirkendall, D.T. 2000. Physiology of soccer. In *Exercise and sport science,* ed. W.E. Garrett and D.T. Kirkendall, 875-84. Philadelphia: Lippincott, Williams, and Wilkins.

Kirsch, K.A., H. von Ameln, and H.J. Wicke. 1981. Fluid control mechanisms after exercise dehydration. *European Journal of Applied Physiology* 47:191-6.

Kirwin, J.P., D.L. Costill, M.G. Flynn, J.B. Mitchell, W.J. Fink, P.D. Neufer, and J.A. Houmard. 1988. Physiological responses to successive days of intense training in competitive swimmers. *Medicine and Science in Sports and Exercise* 20:255-9.

Kjaer, M. 1989. Epinephrine and some other hormonal responses to exercise in man: With special reference to physical training. *International Journal of Sports Medicine* 10:2-15.

Kjaer, M., J. Bangsbo, G. Lortie, and H. Galbo. 1988. Hormonal response to exercise in humans: Influence of hypoxia and physical training. *American Journal of Physiology* 254:R197-203.

Kjaer, M., and H. Galbo. 1988. Effect of physical training on the capacity to secrete epinephrine. *Journal of Applied Physiology* 64:11-6.

Klinzing, J., and E. Karpowicz. 1986. The effects of rapid weight loss and rehydration on wrestling performance test. *Journal of Sports Medicine and Physical Fitness* 26:149-56.

Knapik, J.J., C.L. Bauman, B.H. Jones, J.M. Harris, and L. Vaughan. 1991. Preseason strength and flexibility imbalances associated with athletic injuries in female collegiate athletes. *American Journal of Sports Medicine* 19:76-81.

Knapik, J.J., R.H. Mawdsley, and M.V. Ramos. 1983. Angular specificity and test mode specificity of isometric and isokinetic strength testing. *Journal of Orthopedic Sports Physical Therapy* 5:58-65.

Knuttgen, H.G., and W.J. Kraemer. 1987. Terminology and measurement in exercise performance. *Journal of Applied Sports Science Research* 1:1-10.

Koivisto, V., R. Hendler, E. Nadel, and P. Felig. 1982. Influence of physical training on the fuel-hormone response to prolonged low intensity exercise. *Metabolism* 31:192-7.

Koivisto, V., V. Soman, E. Nadel, W.V. Tamborlane, and P. Felig. 1980. Exercise and insulin: Insulin binding, insulin mobilization and counterregulatory hormone secretion. *Federation Proceedings* 39:1481-6.

Kokkonen, J., and A.G. Nelson. 1996. Acute stretching exercises inhibit maximal force performance. *Medicine and Science in Sports and Exercise* 28:S1130.

Komi, P.V. 1986. Training of muscle strength and power: Interaction of neuromotoric, hypertrophic and mechanical factors. *International Journal of Sports Medicine* 7:10-5.

Komi, P.V., J. Karlsson, P. Tesch H. Souminen, and E. Hakkinen. 1982. Effects of heavy resistance and explosive-type strength training methods on mechanical, functional and metabolic aspects of performance. In *Exercise and sport biology,* ed. P.V. Komi. 99-102. International Series on Sports Sciences, vol. 12. Champaign, IL: Human Kinetics.

Kozak-Collins, K., E. Burke, and R.B. Schoene. 1994. Sodium bicarbonate ingestion does not improve performance in women cyclists. *Medicine and Science in Sports and Exercise* 26:1510-5.

Koziris, L.P., W.J. Kraemer, J.F. Patton, N.T. Triplett, A.C. Fry, S.E. Gordon, and H.G. Knuttgen. 1996. Relationship of aerobic power to anaerobic performance indices. *Journal of Strength and Conditioning Research* 10:35-9.

Kraemer, W.J., 1988. Endocrine responses to resistance exercise. *Medicine and Science in Sports and Exercise* 20:S152-7.

Kraemer, W.J. 1992a. Endocrine responses and adaptations to strength training. In *Strength and power in sport,* ed. P.V. Komi, 291-304. London: Blackwell Scientific.

———. 1992b. Hormonal mechanisms related to the expression of muscular strength and power. In *Strength and power in sport,* ed. P.V. Komi, 64-76. London: Blackwell Scientific.

———. 1994. Neuroendocrine responses to resistance exercise. In *Essentials of strength training and conditioning,* ed. T. Baechle, 86-107. Champaign, IL: Human Kinetics.

———. 1997. A series of studies: The physiological basis for strength training in American football. *Journal of Strength and Conditioning Research* 11:131-42.

Kraemer, W.J., B.A. Aguilera, M. Terada, R.U. Newton, J.M. Lynch, G. Rosendaal, J.M. McBride, S.E. Gordon, and K. Hakkinen. 1995. Responses of IGF-1 to endogenous increases in growth hormone after heavy-resistance exercise. *Journal of Applied Physiology* 79:1310-5.

Kraemer, W.J., M.R. Deschenes, and S.J. Fleck. 1988. Physiological adaptations to resistance exercise. Implications for athletic conditioning. *Sports Medicine.* 6:246-256.

Kraemer, W.J., J.E. Dziados, S.E. Gordon, L.J. Marchitelli, A.C. Fry, and K.L. Reynolds. 1990. The effects of graded exercise on plasma proenkephalin peptide F and catecholamine responses at sea level. *European Journal of Applied Physiology* 61:214-7.

Kraemer, W.J., J.E. Dziados, L.J. Marchitelli, S.E. Gordon, E. Harman, R. Mello, S.J. Fleck, P. Frykman, and N.T. Triplett. 1993. Effects of different heavy-resistance exercise protocols on plasma β-endorphin concentrations. *Journal of Applied Physiology* 74:450-9.

Kraemer, W.J., S.J. Fleck, J.E. Dziados, E. Harman, L.J. Marchitelli, S.E. Gordon, R. Mello, P. Frykman, L.P Koziris, and N.T. Triplett. 1993. Changes in hormonal concentrations after different heavy-resistance exercise protocols in women. *Journal of Applied Physiology* 75:594-604.

Kraemer, W.J., A.C. Fry, B.J. Warren, M.H. Stone, S.J. Fleck, J.T. Kearney, B.P. Conroy, C.M. Maresh, C.A. Weseman, N.T. Triplett, and S.E. Gordon. 1992. Acute hormonal responses in elite junior weightlifters. *International Journal of Sports Medicine* 13:103-9.

Kraemer, W.J., S.E. Gordon, S.J. Fleck, L.J. Marchitelli, R. Mello, J.E. Dziados, K. Friedl, E. Harman, C. Maresh, and A.C. Fry. 1991. Endogenous anabolic hormonal and growth factor responses to heavy resistance exercise in male and females. *International Journal of Sports Medicine* 12:228-35.

Kraemer, W.J., and L.A. Gotshalk. 2000. Physiology of American football. In *Exercise and sport science,* ed. W.E. Garrett and D.T. Kirkendall, 795-813. Philadelphia: Lippincott, Williams, and Wilkins.

Kraemer, W.J., L. Marchitelli, S. Gordon, E. Harmon, J. Dziados, R. Mello, P. Frykman, D. McCurry, and S. Fleck. 1990. Hormonal and growth factor responses to heavy resistance exercise protocols. *Journal of Applied Physiology* 69:1442-50.

Kraemer, W.J., and R.U. Newton. 2000. Training for muscular power. *Physical Medicine and Rehabilitation Clinics of North America* 11:341-368.

Kraemer, W.J., B.J. Noble, M.J. Clark, and B.W. Culver. 1987. Physiological responses to heavy-resistance exercise with very short rest periods. *International Journal of Sports Medicine* 8:247-52.

Kraemer, W.J., B. Noble, B. Culver, and R.V. Lewis. 1985. Changes in plasma proenkephalin peptide F and catecholamine levels during graded exercise in men. *Proceedings of the National Academy of Science* 82:6349-51.

Kraemer, W.J., J.F. Patton, S.E. Gordon, E. Harman, M.R. Deschenes, K. Reynolds, R.U. Newton, N. Travis-Triplett, and J.E. Dziados. 1995. Compatibility of high-intensity strength and endurance training on hormonal and skeletal muscle adaptations. *Journal of Applied Physiology* 78:976-89.

Kraemer, W.J., J.F. Patton, H.G. Knuttgen, C.J. Hannon, T. Kettler, S.E. Gordon, J.E. Dziados, A.C. Fry, P.N. Frykman, and E.A. Harman. 1991. Effects of high-intensity cycle exercise on sympathoadrenal-medullary response patterns. *Journal of Applied Physiology* 70:8-14.

Kraemer, W.J., J.F. Patton, H.G. Knuttgen, L.J. Marchitelli, C. Cruthirds, A. Damokosh, E. Harman, P. Frykman, and J.E. Dziados. 1989. Hypothalamic-pituitary-adrenal responses to short-term high-intensity cycle exercise. *Journal of Applied Physiology* 66:161-6.

Kraemer, W.J., J.S. Volek, K.L. Clark, S.E. Gordon, S.M. Puhl, L.P. Koziris, J.M. McBride, N.T. Triplett-McBride, M. Putukian, R.U. Newton, K. Hakkinen, J.A. Bush, and W.J. Sebastianelli. 1999. Influence of exercise training on physiological and performance changes with weight loss in men. *Medicine and Science in Sports and Exercise* 31:1320-9.

Kreider, R.B., M. Ferreira, M. Wilson, P. Grindstaff, S. Plisk, J. Reinardy, E. Cantler, and A.L. Almada. 1998. Effects of creatine supplementation on body composition, strength, and sprint performance. *Medicine and Science in Sports and Exercise* 30:73-82.

Kreider, R.B., A.C. Fry, and M.L. O'Toole. 1998. *Overtraining in sport.* Champaign, IL: Human Kinetics.

Kruse, P., J. Ladefoged, U. Nielsen, P. Paulev, and J.P. Sorensen. 1986. β-blockade used in precision sports: Effect on pistol shooting performance. *Journal of Applied Physiology* 61:417-20.

Kuipers, H., and H.A. Keizer. 1988. Overtraining in elite athletes: Review and direction for the future. *Sports Medicine* 6:79-92.

Kuoppasalmi, K., and H. Adlercreutz. 1985. Interaction between catabolic and anabolic steroid hormones in muscular exercise. In *Exercise endocrinology,* ed. K. Fotherby and S.B. Pal, 65-98. Berlin: Walter de Gruyter.

Kuoppasalmi, K.H., Naveri, M. Harkonen, and H. Adlercreutz. 1980. Plasma cortisol, testosterone and luteinizing hormone in running exercise of different intensities. *Scandinavian Journal of Clinical Laboratory Investigations* 40:403-9.

Kurz, M.J., K. Berg, R. Latin, and W. DeGraw. 2000. The relationship of training methods in NCAA division I cross-country runners and 10,000-meter performance. *Journal of Strength and Conditioning Research* 14:196-201.

Laakso, M., S.V. Edelman, G. Brechtel, and A.D. Baron. 1992. Impaired insulin-mediated skeletal muscle blood flow in patients with NIDDM. *Diabetes* 41:1076-83.

LaBotz, M., and B.W. Smith. 1999. Creatine supplement use in an NCAA division I athletic program. *Clinical Journal of Sports Medicine* 9:167-9.

Lachowetz, T., J. Evon, and J. Pastiglione. 1998. The effect of an upper body strength program on intercollegiate baseball throwing velocity. *Journal of Strength and Conditioning Research* 12:116-9.

Landers, J. 1985. Maximum based on reps. *National Strength and Conditioning Association Journal* 6:60-1.

Larsson, L.P., P. Hemmingsson, and G. Boethius. 1994. Self-reported obstructive airway symptoms are common in young cross-country skiers. *Scandinavian Journal of Medicine and Science in Sports* 4:124-7.

Latin, R.W., K. Berg, and T. Baechle. 1994. Physical and performance characteristics of NCAA division I male basketball players. *Journal of Strength and Conditioning Research* 8:214-8.

Laubach, L.L. 1976. Comparative muscular strength of men and women: A review of the literature. *Aviation, Space and Environmental Medicine* 47:534-42.

Laure, P. 1997. Epidemiological approach of doping in sport. *Journal of Sports Medicine and Physical Fitness* 37:218-4.

Lehmann, M., C. Foster, N. Netzer, W. Lormes, J.M. Steinacker, Y. Liu, A. Opitz-Gress, and U. Gastmann. 1998. Physiological responses to short- and long-term overtraining in endurance athletes. In *Overtraining in sport,* ed. R.B. Kreider, A.C. Fry, and M.L. O'Toole, 19-46. Champaign, IL: Human Kinetics.

Lehmann, M., E. Jakob, U. Gastmann, J.M. Steinacker, N. Heinz, and F. Brouns. 1995. Unaccustomed high mileage compared to high intensity-related performance and neuromuscular responses in distance runners. *European Journal of Applied Physiology* 70:457-61.

Lehmann, M., H. Mann, U. Gastmann, J. Keul, D. Vetter, J.M. Steinacker, and D. Haussinger. 1996. Unaccustomed high-mileage vs intensity training-related changes in performance and serum amino acid levels. *International Journal of Sports Medicine* 17:187-92.

Lehmann, M., W. Schnee, R. Scheu, W. Stockhausen, and N. Bachl. 1992. Decreased nocturnal catecholamine excretion: Parameter for an overtraining syndrome in athletes? *International Journal of Sports Medicine* 13:236-42.

Leiper, J.B., and R.J. Maughan. 1988. Experimental models for the investigation of water and solute transport in man: Implications for oral rehydration solutions. *Drugs* 36(supp):65-79.

Leithead, C.S., and E.R. Gunn. 1964. The aetiology of cane cutters cramps in British Guiana. In *Environmental physiology and psychology in arid conditions,* 13-7. Liege, Belgium: UNESCO.

Lemon, P.W.R. 1995. Do athletes need more dietary protein and amino acids? *International Journal of Sports Nutrition* 5:S39-61.

Lemon, P.W.R., M.A. Tarnopolsky, J.D. McDougall, and S.A. Atkinson. 1992. Protein requirements and muscle mass/strength changes during intensive training in novice bodybuilders. *Journal of Applied Physiology* 73:767-75.

Lentini, A.C., R.S. McKelvie, N. McCartney, C.W. Tomlinson, and J.D. MacDougall. 1993. Assessment of left ventricular response of strength trained athletes during weightlifting exercise. *Journal of Applied Physiology* 75:2703-10.

Leveritt, M., and P.J. Abernethy. 1999. Acute effects of high-intensity endurance exercise on subsequent resistance activity. *Journal of Strength and Conditioning Research* 13:47-51.

Levine, B.D., and J. Stray-Gundersen. 1997. "Living high-training low": Effect of moderate-altitude acclimatization with low-altitude training on performance. *Journal of Applied Physiology* 83:102-12.

Lewicki, R., H. Tchorzewski, A. Denys, M. Kowalska, and A. Golinska. 1987. Effect of physical exercise on some parameters of immunity in conditioned sportsmen. *International Journal of Sports Medicine* 8:309-14.

Lewicki, R., H. Tchorzewski, E. Majewska, Z. Nowak, and Z. Baj. 1988. Effect of maximal physical exercise on T-lymphocyte subpopulations and on interleukin 1 IL-1 and interleukin 2 IL-2 production in vitro. *International Journal of Sports Medicine* 9:114-7.

Lewis, S.F., W.F. Taylor, R.M. Graham, W.A. Pettinger, J.E. Schutte, and C.G. Blomqvist. 1983. Cardiovascular responses to exercise as functions of absolute and relative work load. *Journal of Applied Physiology* 54:1314-23.

Liesen, H., B. Dufaux, and W. Hollmann. 1977. Modifications of serum glycoproteins on the days following a prolonged physical exercise and the influence of physical training. *European Journal of Applied Physiology* 37:243-54.

Linde, F. 1987. Running and upper respiratory tract infections. *Scandinavian Journal of Sports Sciences* 9:21-3.

Lohman, T.G. 1981. Skinfolds and body density and their relation to body fatness: A review. *Human Biology* 53:181-225.

Lombardo, J. 1993. The efficacy and mechanisms of action of anabolic steroids. In *Anabolic steroids in sport and exercise,* ed. C.E. Yesalis, 89-106. Champaign, IL: Human Kinetics.

Loughton, S.J., and R.O. Ruhling. 1977. Human strength and endurance responses to anabolic steroids and training. *Journal of Sports Medicine and Physical Fitness* 17:285-96.

Lugar, A., B. Watschinger, P. Duester, T. Svoboda, M. Clodi, and G.P. Chrousos. 1992. Plasma growth hormone and prolactin responses to graded levels of acute exercise and to a lactate infusion. *Neuroendocrinology* 56:112-7.

Luhtanen, P., and P.V. Komi. 1978. Mechanical factors influencing running speed. In *Biomechanics VI-B,* ed. E. Asmussen and K. Jorgensen, 25. Baltimore: University Park Press.

Lund, S., H. Vestergaad, P.H. Anderson, O. Schmitz, L.B.H. Gotzsche, and O. Pedersen. 1993. GLUT-4 content in plasma membrane of muscle from patients with non-insulin-dependent diabetes mellitus. *American Journal of Physiology* 265:E889-97.

Lusiani, L., G. Ronsisvalle, A. Bonanome, A. Visona, V. Castellani, C. Macchia, and A. Pagnan. 1986. Echocardiographic evaluation of the dimensions and systolic properties of the left ventricle in freshman athletes during physical training. *European Heart Journal* 7:196-203.

MacConnie, S.E., A. Barkin, R.M. Lampman, M.A. Schork, and I.Z. Beitins. 1986. Decreased hypothalamic gonadotropin-releasing hormone secretion in male marathon runners. *New England Journal of Medicine* 315:411-7.

MacDougall, J.D. 1992. Hypertrophy or hyperplasia. In *Strength and power in sport,* ed. P.V. Komi, 230-8. London: Blackwell Scientific.

———. 1994. Blood pressure responses to resistive, static, and dynamic exercise. In *Cardiovascular response to exercise,* ed. G.F. Fletcher, 155-174. Mount Kisco, NY: Futura.

MacDougall, J.D., H.J. Green, J.R. Sutton, G. Coates, A. Cymerman, P. Young, and C.S. Houston. 1991. Operation Everest II: Structural adaptations in skeletal muscle in response to extreme simulated altitude. *Acta Physiologica Scandinavica* 142:431-27.

MacDougall, J.D., R.S. McKelvie, D.E. Moroz, D.G. Sale, N. McCartney, and F. Buick. 1992. Factors affecting blood pressure response during heavy weightlifting and static contractions. *Journal of Applied Physiology* 73:1590-7.

MacDougall, J.D., W.G. Reddan, C.R. Layton, and J.A. Dempsey. 1974. Effects of metabolic hyperthermia on performance during heavy prolonged exercise. *Journal of Applied Physiology* 36:538-44.

MacDougall, J.D., D.G. Sale, S.E. Alway, and J.R. Sutton. 1984. Muscle fiber number in biceps brachii in bodybuilders and control subjects *Journal of Applied Physiology: Respiratory, Environmental, Exercise Physiology* 57:1399-1403.

MacDougall, J.D., D.G. Sale, G.C.B. Elder, and J.R. Sutton. 1982. Muscle ultrastructural characteristics of elite powerlifters and bodybuilders. *European Journal of Applied Physiology* 48:117-26.

MacDougall, J.D., D.G. Sale, J.R. Moroz, G.C.B. Elder, J.R. Sutton, and H. Howard. 1979. Mitochondrial volume density in human skeletal muscle following heavy resistance exercise. *Medicine and Science in Sports and Exercise* 11:164-6.

MacDougall, J.D., D. Tuxen, D.G. Sale, J.R. Moroz, and J.R. Sutton. 1985. Arterial blood pressure response to heavy resistance exercise. *Journal of Applied Physiology* 58:785-90.

MacDougall, J.D., G.R. Ward, D.G. Sale, and J.R. Sutton. 1977. Biochemical adaptations of human skeletal muscle to heavy resistance training and immobilization. *Journal of Applied Physiology* 43:700-3.

Mackinnon, L.T. 1994. Current challenges and future expectations in exercise immunology: Back to the future. *Medicine and Science in Sports and Exercise* 26:191-4.

———. 1999. *Advances in exercise immunology.* Champaign, IL: Human Kinetics.

Mackinnon, L.T., T.W. Chick, A. van As, and T.B. Tomasi. 1988. Effects of prolonged intense exercise on natural killer cell number and function. In *Exercise physiology: Current selected research.* Vol. 3, ed. C.O. Dotson and J.H. Humphrey, 77-89. New York: AMS Press.

———. 1989. Decreased secretory immunoglobulins following intense endurance exercise. *Sports Medicine, Training, and Rehabilitation* 1:209-18.

Mackinnon, L.T., E. Ginn, and G. Seymour. 1991. Effects of exercise during sports training and competition on salivary IgA levels. In *Behaviour and Immunity*, ed. A.J. Husband, 169-177. Boca Raton, FL: CRC Press.

———. 1993. Decreased salivary immunoglobulin A secretion rate after intense interval exercise training in elite kayakers. *European Journal of Applied Physiology.* 67:180-4.

Mackinnon, L.T., and S.L. Hooper. 1994. Mucosal (secretory) immune system responses to exercise of varying intensity and during overtraining. *International Journal of Sports Medicine* 15:S179-83.

———. 1996. Plasma glutamine concentration and upper respiratory tract infection during overtraining in elite swimmers. *Medicine and Science in Sports and Exercise* 28:285-90.

Mackinnon, L.T., S.L. Hooper., S. Jones, A.W. Bachmann, and R.D. Gordon. 1997. Hormonal, immunological and hematological responses to intensified training in elite swimmers. *Medicine and Science in Sports and Exercise* 29:1637-45.

MacRae, J.C., P.A. Skene, A. Connell, V. Buchan, and G.E. Lobley. 1988. The action of the β_2-agonist clenbuterol on protein and energy metabolism in fattening wether lambs. *British Journal of Nutrition* 59:457-65.

Magazanik, A., Y. Weinstein, R.A. Dlin, M. Derin, and S. Schwartzman. 1988. Iron deficiency caused by 7 weeks of intensive physical exercise. *European Journal of Applied Physiology* 57:198-202.

Magel, J.R., G.F. Foglia, W.D. McArdle, B. Gutin, and G.S. Pechar. 1975. Specificity of swim training on maximal oxygen uptake. *Journal of Applied Physiology* 38:151-5.

Magnusson, S.P., E.B. Simonsen, P. Aagaard, and M. Kjaer. 1996. Biomechanical responses to repeated stretches in human hamstring muscle in vivo. *American Journal of Sports Medicine* 24:622-8.

Mahesh, V.B., and R.B. Greenblatt. 1962. The in vivo conversion of dehydroepiandrosterone and androstenedione to testosterone in the human. *Acta Endocrinologia* 41:400-6.

Maltin, C.A., M.I. Delday, J.S. Watson, S.D. Heys, I.M. Nevison, I.K. Ritchie, and P.H. Gibson. 1993. Clenbuterol, a beta-adrenoceptor agonist, increases relative muscle strength in orthopaedic patients. *Clinical Science* 84:651-4.

Mangine, R.E., F.R. Noyes, M.P. Mullen, and S.D. Baker. 1990. A physiological profile of the elite soccer athlete. *Journal of Orthopedic Sports Physical Therapy* 12:147-52.

Manning, P.J., R.M. Watson, and P.M. O'Byrne. 1993. Exercise-induced refractoriness in asthmatic subjects involving leukotriene and prostaglandin interdependent mechanisms. *American Review of Respiratory Disease* 148:950-4.

Maresh, C.M., B.C. Wang, and K.L. Goetz. 1985. Plasma vasopressin, renin activity, and aldosterone responses to maximal exercise in active college females. *European Journal of Applied Physiology* 54:398-403.

Margaria, R., P. Aghemo, and E. Rovelli. 1966. Measurement of muscular power (anaerobic) in man. *Journal of Applied Physiology* 21:1662-4.

Maron, B.J. 1986. Structural features of the athletic heart as defined by echocardiography. *Journal of the American College of Cardiology* 7:190-203.

Martin, B., M. Heintzelman, and H.I. Chen. 1982. Exercise performance after ventilatory work. *Journal of Applied Physiology* 52:1581-5.

Martin, D.E., and P.N. Coe. 1997. *Better training for distance runners.* 2nd ed. Champaign, IL: Human Kinetics.

Martineau, L., M.A. Horan, N.J. Rothwell, and R.A. Little. 1992. Salbutamol, a β_2-adrenoreceptor agonist, increases skeletal muscle strength in young men. *Clinical Science* 83:615-21.

Maughan, R.J. 1991. Carbohydrate-electrolyte solutions during prolonged exercise. In *Perspectives in exercise science and sports medicine.* Vol. 4. Ergogenic, ed. D.R. Lamb and M.H. Williams, 35-86. Carmel, IN: Benchmark Press.

Maughan, R.J., C.E. Fenn, and J.B. Leiper. 1989. Effects of fluid, electrolyte and substrate ingestion on endurance capacity. *European Journal of Applied Physiology* 58:481-6.

Maughan, R.J., and T.D. Noakes. 1991. Fluid replacement and exercise stress. *Sports Medicine* 12:16-31.

Mayhew, J.L., T.E. Ball, and J.C. Bowen. 1992. Prediction of bench press lifting ability from submaximal repetitions before and after training. *Sports Medicine, Training and Rehabilitation* 3:195-201.

Mayhew, J.L., B. Levy, T. McCormick, and G. Evans. 1987. Strength norms for NCAA division II college football players. *National Strength and Conditioning Association Journal* 9:67-9.

Mayhew, J.L., J.S. Ware, M.G. Bemben, B. Wilt, T.E. Ward, B. Farris, J. Juraszek, and J.P. Slovak. 1999. The NFL-225 test as a measure of bench press strength in college football players. *Journal of Strength and Conditioning Research* 13:130-4.

McArdle, W.D., R.M. Glaser, and J.R. Magel. 1971. Metabolic and cardiorespiratory response during free swimming and treadmill walking. *Journal of Applied Physiology* 30:733-8.

McArdle, W.D., F.I. Katch, and V.L. Katch. 1996. *Exercise physiology: Energy, nutrition, and human performance.* 4th ed. Baltimore: Williams & Wilkins.

McArdle, W.D., J.R. Magel, D.J. Delio, M. Toner, and J.M. Chase. 1978. Specificity of run training on $\dot{V}O_2$ max and heart rate changes during running and swimming. *Medicine and Science in Sports and Exercise* 10:16-20.

McCarthy, D.A., and M.M. Dale. 1988. The leucocytosis of exercise: A review and model. *Sports Medicine* 6:333-63.

McCarthy, J.P., J.C. Agre, B.K. Graf, M.A. Pozniak, and A.C. Vailas. 1995. Compatibility of adaptive responses with combining strength and endurance training. *Medicine and Science in Sports and Exercise* 27:429-36.

McCartney, N. 1999. Acute responses to resistance training and safety. *Medicine and Science in Sports and Exercise* 31:31-7.

McCartney, N., R.S. McKelvie, J. Martin, D.G. Sale, and J.D. MacDougall. 1993. Weight-training-induced attenuation of the circulatory response of older males to weight lifting. *Journal of Applied Physiology* 74:1056-60.

McCartney, N., L.L. Spriet, G.J.F. Heigenhauser, J.M. Kowalchuk, J.R. Sutton, and N.L. Jones. 1986. Muscle power and metabolism in maximal intermittent exercise. *Journal of Applied Physiology* 60:1164-9.

McCole, S.D., K. Claney, J.C. Conte, R. Anderson, and J.M. Hagberg. 1990. Energy expenditures during bicycling. *Journal of Applied Physiology* 68:748-53.

McConnell, G.K., C.M. Burge, S.L. Skinner, and M. Hargreaves. 1997. Influence of ingested fluid volume on physiological responses during prolonged exercise. *Acta Physiologica Scandinavica* 160:149-56.

McCue, B.F. 1953. Flexibility of college women. *Research Quarterly for Exercise and Sport* 24:316-24.

McDowell, S.L., K. Chalos, T.J. Housh, G.D. Tharp, and G.O. Johnson. 1991. The effect of exercise intensity and duration on salivary immunoglobulin A. *European Journal of Applied Physiology* 63:108-11.

McDowell, S.L., R.A. Hughes, R.J. Hughes, D.J. Housh, T.J. Housh, and G.O. Johnson. 1992. The effect of exhaustive exercise on salivary immunoglobulin A. *Journal of Sports Medicine and Physical Fitness* 32:412-5.

McEvoy, K.P., and R.U. Newton. 1998. Baseball throwing speed and base running speed: The effects of ballistic resistance training.*Journal of Strength and Conditioning Research* 12:216-21.

McFadden Jr., E.R., and I.A. Gilbert. 1994. Exercise-induced asthma. *New England Journal of Medicine* 330:1362-7.

McGee, D., T.C. Jessee, M.H. Stone, and D. Blessing. 1992. Leg and hip endurance adaptations to three weight-training programs. *Journal of Applied Sport Science Research* 6:92-5.

McInnes, S.E., J.S. Carlson, C.J. Jones, and M.J. McKenna. 1995. The physiological load imposed on basketball players during competition. *Journal of Sport Sciences* 13:387-97.

McMaster, W.C., S.C. Long, and V.J. Caiozzo. 1991. Isokinetic torque imbalances in the rotator cuff of the elite water polo player. *American Journal of Sports Medicine* 19:72-5.

Melin, B., J.P. Eclache, G. Geelen, G. Annat, A.M. Allevard, E. Jarsaillon, A. Zebidi, J.J. Legros, and C. Gharib. 1980. Plasma AVP, neurophysin, renin activity, and aldosterone during submaximal exercise performed until exhaustion in trained and untrained men. *European Journal of Applied Physiology and Occupational Physiology* 44:141-51.

Menapace, F.J., W.J. Hammer, T.F. Ritzer, K.M. Kessler, H.F. Warner, J.F. Spann, and A.A. Bove. 1982. Left ventricular size in competitive weight lifters: An echocardiographic study. *Medicine and Science in Sports and Exercise* 14:72-5.

Meredith, C.N., M.J. Zackin, W.R. Frontera, and W.J. Evans. 1989. Dietary protein requirements and protein metabolism in endurance-trained men. *Journal of Applied Physiology* 66:2850-6.

Mero, A., P.V. Komi, and R.J. Gregor. 1992. Biomechanics of sprint running: A review. *Sports Medicine* 13:376-92.

Meydani, M. 1992. Protective role of dietary vitamin E on oxidative stress in aging. *Age* 15:89-93.

Meyer, C.R. 1967. Effect of two isometric routines on strength, size and endurance in exercise and non-exercise arms. *Research Quarterly for Exercise and Sport* 38:430-40.

Milledge, J.S., J.M. Beeley, J. Broome, N. Luff, M. Pelling, and D. Smith. 1991. Acute mountain sickness susceptibility, fitness and hypoxic ventilatory response. *European Respiratory Journal* 4:1000-3.

Milledge, J.S., and P.M. Cotes. 1985. Serum erythropoietin in humans at high altitude and its relation to plasma renin. *Journal of Applied Physiology* 59:360-4.

Miller, W.J., W.M. Sherman, and J.L. Ivy. 1984. Effect of strength training on glucose tolerance and post-glucose insulin response. *Medicine and Science in Sports and Exercise* 16:539-87.

Milner-Brown, H.S., R.B. Stein, and R. Yemm. 1975. Synchronization of human motor units: Possible roles of exercise and supraspinal reflexes. *Electroencephalography and Clinical Neurophysiology* 38:245-54.

Minkler, S., and P. Patterson. 1994. The validity of the modified sit-and-reach test in college-age students. *Research Quarterly for Exercise and Sport* 65:189-92.

Montain, S.J., J.E. Laird, W.A. Latzka, and M.N. Sawka. 1997. Aldosterone and vasopressin responses in the heat: Hydration level and exercise intensity effects. *Medicine and Science in Sports and Exercise* 29:661-8.

Montgomery, J.C., and J.A. MacDonald. 1990. Effects of temperature on nervous system: Implications for behavioral performance. *American Journal of Physiology: Regulatory, Integrative, Comparative Physiology* 259:R191-6.

Moore, M.A., and R.S. Hutton. 1980. Electromyographic investigation of muscle stretching techniques. *Medicine and Science in Sports and Exercise* 12:322-9.

Morgan, W.P. 1985. Affective beneficence of vigorous physical activity. *Medicine and Science in Sports and Exercise* 17:94-100.

Morgan, W.P., D.R. Brown, J.S. Raglin, P.J. O'Connor, and K.A. Ellickson. 1987. Psychological monitoring of overtraining and staleness. *British Journal of Sports Medicine* 21:107-14.

Morganroth, J., B.J. Maron, W.L. Henry, and S.E. Epstein. 1975. Comparative left ventricular dimensions in trained athletes. *Annals of Internal Medicine* 82:521-4.

Moritani, M.T., and H.A. deVries. 1979. Neural factors vs. hypertrophy in time course of muscle strength gain. *American Journal of Physical Medicine and Rehabilitation* 58:115-30.

Morton, A.R. 1995. Asthma. In *Science and medicine in sport,* ed. J. Bloomfield, P. Fricker, and K.D. Fitch, 616-27. Victoria, Australia: Blackwell Science.

Mujika, I., J.C. Chatard, L. Lacoste, F. Barale, and A. Geyssant. 1996. Creatine supplementation does not improve sprint performance in competitive swimmers.

Medicine and Science in Sports and Exercise 28:1435-41.

Mulligan, S.E., S.J. Fleck, S.E. Gordon, L.P. Koziris, N.T. Triplett-McBride, and W.J. Kraemer. 1996. Influence of resistance exercise volume on serum growth hormone and cortisol concentrations in women. *Journal of Strength and Conditioning Research* 10:256-62.

Murray, R. 1987. The effects of consuming carbohydrate-electrolyte beverages on gastric emptying and fluid absorption during and following exercise. *Sports Medicine* 4:322-51.

Nadel, E.R., S.M. Fortney, and C.B. Wenger. 1980. Effect of hydration on circulatory and thermal regulation. *Journal of Applied Physiology* 49:715-21.

Nadel, E.R., I. Holmer, U. Bergh, P.O. Astrand, and J.A.J. Stolwijk. 1974. Energy exchanges of swimming men. *Journal of Applied Physiology* 36:465-71.

National Collegiate Athletic Association (NCAA). 1997. *NCAA study of substance use and abuse habits of college student-athletes.* Indianapolis:National Collegiate Athletic Association.

National Oceanic and Atmospheric Administration. 1976. *US standard atmosphere.* Washington, DC: NOAA.

Ndon, J.A., A.C Snyder, C. Foster, and W.B. Wehrenberg. 1992. Effects of chronic intensive exercise training on the leukocyte response to acute exercise. *International Journal of Sports Medicine* 13:176-82.

Nehlsen-Cannarella, S.L., D.C. Nieman, A.J. Balk-Lamberton, P.A. Markoff, D.B.W. Chritton, G. Gusewitch, and J.W. Lee. 1991. The effect of moderate exercise training on immune response. *Medicine and Science in Sports and Exercise* 23:64-70.

Nelson, A.G., J.D. Allen, A. Cornwell, and J. Kokkonen. 1998. Inhibition of maximal torque production by acute stretching is joint-angle specific. *26th Annual Meeting of the Southeastern Chapter of the ACSM.* Destin, FL. January 1998.

Nelson, A.G., and G.D. Heise. 1996. Acute stretching exercises and vertical jump stored elastic energy. *Medicine and Science in Sports and Exercise* 28:S156.

Neufer, P.D., D.L. Costill, W.J. Fink, J.P. Kirwin, R.A. Fielding, and M.G. Flynn. 1986. Effects of exercise and carbohydrate composition on gastric emptying. *Medicine and Science in Sports and Exercise* 18:658-62.

Neufer, P.D., M.N. Sawka, A.J. Young, M.D. Quigley, W.A. Latzka, and L. Levine. 1991. Hypohydration does not impair skeletal muscle glycogen resynthesis after exercise. *Journal of Applied Physiology* 70:1490-4.

Neufer, P.D., A.J. Young, and M.N. Sawka. 1989. Gastric emptying during exercise: Effects of heat stress and hypohydration. *European Journal of Applied Physiology and Occupational Physiology* 58:557-60.

Newton, R.U., and W.J. Kraemer. 1994. Developing explosive muscular power: Implications for a mixed methods training strategy. *Strength and Conditioning* 16:20-31.

Newton, R.U., W.J. Kraemer, and K. Hakkinen. 1999. Effects of ballistic training on preseason preparation of elite volleyball players. *Medicine and Science in Sports and Exercise* 31:323-30.

Newton, R.U., and K.P. McEvoy. 1994. Baseball throwing velocity: A comparison of medicine ball training and weight training. *Journal of Strength and Conditioning Research* 8:198-203.

Nielsen, H.B., N.H. Secher, N.J. Christensen, and B.K. Pedersen. 1996. Lymphocytes and NK cell activity during repeated bouts of maximal exercise. *American Journal of Physiology: Regulatory, Integrative, Comparative Physiology* 271:R222-7.

Nieman, D.C. 1994. Exercise, infection and immunity. *International Journal of Sports Medicine* 15:S131-41.

———. 2000. Exercise, the immune system, and infectious disease. In *Exercise and sport science,* ed. W.J. Garrett Jr. and D.T. Kirkendall, 177-90. Philadelphia: Lippincott, Williams, and Wilkins.

Nieman, D.C., L.S. Berk, M. Simpson-Westerberg, K. Arabatzis, S. Youngberg, S.A. Tan, J.W. Lee, and W.C. Eby. 1989. Effects of long-endurance running on immune system parameters and lymphocyte function in experienced marathoners. *International Journal of Sports Medicine* 10:317-23.

Nieman, D.C., D.A. Hensen, G. Gusewitch, B.J. Warren, R.C. Dotson, and S.L. Nehlsen-Cannarella. 1993. Physical activity and immune function in elderly women. *Medicine and Science in Sports and Exercise* 25:823-31.

Nieman, D.C., D.A. Hensen, R. Johnson, L. Lebeck, J.M. Davis, and S.L. Nehlsen-Cannarella. 1992. Effects of brief, heavy exertion on circulating lymphocyte subpopulations and proliferative response. *Medicine and Science in Sports and Exercise* 24:1339-45.

Nieman, D.C., D.A. Hensen, C.S. Sampson, J.L. Herring, J. Stulles, M. Conley, M.H. Stone, D.E. Butterworth, and J.M. Davis. 1995. The acute immune response to exhaustive resistance exercise. *International Journal of Sports Medicine* 16:322-8.

Nieman, D.C., L.M. Johanssen, and J.W. Lee. 1989. Infectious episodes in runners before and after a roadrace. *Journal of Sports Medicine and Physical Fitness* 29:289-96.

Nieman, D.C., L.M. Johanssen, J.W. Lee, and K. Arabatzis. 1990. Infectious episodes in runners before and after the Los Angeles Marathon. *Journal of Sports Medicine and Physical Fitness* 30:316-28.

Nieman, D.C., A.R. Miller, D.A. Hensen, B.J. Warren, G. Gusewitch, R.L. Johnson, D.E. Butterworth, J.L. Herring, and S.L. Nehlsen-Cannarella. 1994. Effect of high- versus moderate-intensity exercise on lymphocyte subpopulations and proliferative response. *International Journal of Sports Medicine* 15:199-206.

Nieman, D.C., A.R. Miller, D.A. Hensen, B.J. Warren, G. Gusewitch, R.L. Johnson, J.M. Davis, D.E. Butterworth, and S.L. Nehlsen-Cannarella. 1993. Effect of high- versus moderate intensity exercise on natural killer cell activity. *Medicine and Science in Sports and Exercise* 25:1126-34.

Nieman, D.C., S. Simandle, D.A. Hensen, B.J. Warren, J. Suttles, J.M. Davis, K.S. Buckley, J.C. Ahle, D.E. Butterworth, O.R. Fagoaga, and S.L. Nehlsen-Cannarella. 1995. Lymphocyte proliferative response to 2.5 hours of running. *International Journal of Sports Medicine* 16:404-8.

Nieman, D.C., S.A. Tan, J.W. Lee, and L.S. Berk. 1989. Complement and immunoglobulin levels in athletes and sedentary controls. *International Journal of Sports Medicine*. 10:124-8.

Nissen, S.L., and N.N. Abumrad. 1997. Nutritional role of the leucine metabolite β-hydroxy β-methylbutyrate (HMB). *Nutritional Biochemistry* 8:300-11.

Nissen, S., R. Sharp, M. Ray, J.A. Rathmacher, D. Rice, J.C. Fuller, A.S. Connelly, and N. Abumrad. 1996. Effect of leucine metabolite β-hydroxy-β-methylbutyrate on muscle metabolism during resistance-exercise training. *Journal of Applied Physiology* 81:2095-104.

Noakes, T.D. 1993. Fluid replacement during exercise. *Exercise and Sport Sciences Reviews* 21:297-330.

Noakes, T.D., R.J. Norman, R.H. Buck, J. Godlonton, K. Stevenson, and D. Pittaway. 1990. The incidence of hyponatremia during prolonged ultraendurance exercise. *Medicine and Science in Sports and Exercise* 22:165-70.

Norkin, C.C., and D.J. White. 1995. *Measurement of joint motion: A guide to goniometry*. Philadelphia: Davis.

Nosaka, K., and P.M. Clarkson. 1996. Changes in indicators of inflammation after eccentric exercise of the elbow flexors. *Medicine and Science in Sports and Exercise* 28:953-61.

Nose, H., T. Morimoto, and K. Ogura. 1983. Distribution of water loses among fluid compartments after dehydration in humans. *Japan Journal of Physiology* 33:1019-29.

O'Bryant, H.S., R. Byrd, and M.H. Stone. 1988. Cycle ergometer performance and maximum leg and hip strength adaptations to two different methods of weight-training. *Journal of Applied Sport Science Research* 2:27-30.

O'Connor, P.J., and D.B. Cook. 1999. Exercise and pain: Neurobiology, measurement, and laboratory study of pain in relation to exercise in humans. *Exercise and Sport Sciences Reviews* 27:119-66.

Odland, L.M., J.D. MacDougall, M.A. Tarnopolsky, A. Elorriaga, and A. Borgmann. 1997. Effect of oral creatine supplementation on muscle [PCr] and short-term

maximum power output. *Medicine and Science in Sports and Exercise* 29:216-9.

Olefsky, J.M. 1976. The insulin receptor: Its role in insulin resistance in obesity and diabetes. *Diabetes* 25:1154-64.

Olsson, K.E., and B. Saltin. 1970. Variation in total body water with muscle glycogen changes in man. *Acta Physiologica Scandinavica* 80:11-8.

O'Shea, J.P. 1966. Effects of selected weight training programs on the development of strength and muscle hypertrophy. *Research Quarterly for Exercise and Sport* 37:95-102.

———. 1971. The effects of anabolic steroids on dynamic strength levels of weightlifters. *Nutritional Reports International* 4:363-70.

Osterud, B., J.O. Olsen, and L. Wilsgard. 1989. Effect of strenuous exercise on blood monocytes and their relation to coagulation. *Medicine and Science in Sports and Exercise* 21:374-8.

O'Toole, M.L., P.S. Douglas, and W.D.B. Hiller. 1989. Applied physiology of a triathlon. *Sports Medicine* 8:201-25.

Pakarinen, A., M. Alen, K. Hakkinen, and P. Komi. 1988. Serum thyroid hormones, thyrotropin and thyroxin binding globulin during prolonged strength training. *European Journal of Applied Physiology* 57:394-8.

Pandolf, K.B., R.L. Burse, and R.F. Goldman. 1977. Role of physical fitness in heat acclimatization, decay and reinduction. *Ergonomics* 20:399-408.

Park, S.H., J.N. Roemmick, and C.A. Horswill. 1990. A season of wrestling and weight loss by adolescent wrestlers: Effect on anaerobic arm power. *Journal of Applied Sport Science Research* 4:1-4.

Parra, J., J.A. Cadefau, G. Rodas, N. Amigo, and R. Cusso. 2000. The distribution of rest periods affects performance and adaptations of energy metabolism induced by high-intensity training in human muscle. *Acta Physiologica Scandinavica* 169:157-65.

Pattini, A., F. Schena, and G.C. Guidi. 1990. Serum ferritin and serum iron changes after cross-country and roller ski endurance races. *European Journal of Applied Physiology* 61:55-60.

Pearson, A.C., M. Schiff, D. Mrosek, A.J. Labovitz, and G.A. Williams. 1986. Left ventricular diastolic function in weight lifters. *American Journal of Cardiology* 58:1254-9.

Pearson, D.R., D.G. Hamby, W. Russel, and T. Harris. 1999. Long-term effects of creatine monohydrate on strength and power. *Journal of Strength and Conditioning Research* 13:187-92.

Pedegna, L.R., R.C. Elsner, D. Roberts, J. Lang, and V. Farewell. 1982. The relationship of upper extremity strength to throwing speed. *American Journal of Sports Medicine* 10:352-4.

Pedersen, B.K., and H. Ullam. 1994. NK cell response to physical activity: Possible mechanisms of action. *Medicine and Science in Sports and Exercise* 26:140-6.

Pelliccia, A., B.J. Maron, A. Spataro, M.A. Proschan, and P. Spirito. 1991. The upper limit of physiologic cardiac hypertrophy in highly trained elite athletes. *New England Journal of Medicine* 324:295-301.

Pequignot, J.M., L. Peyrin, M.H. Mayet, and R. Flandrois. 1979. Metabolic adrenergic changes during submaximal exercise and in the recovery period in man. *Journal of Applied Physiology: Respiratory, Environmental, Exercise Physiology* 47:701-5.

Perrine, J.J., and V.R. Edgerton. 1978. Muscle force-velocity and power-velocity relationships under isokinetic loading. *Medicine and Science in Sports and Exercise* 10:159-66.

Persson, C.G.A. 1986. Role of plasma exudation in asthmatic airways. *Lancet* 2:1126-9.

Peters, E.M. 1990. Altitude fails to increase susceptibility of ultramarathon runners to postexercise upper respiratory tract infection. *South African Journal of Sports Medicine* 5:4-8.

Peters, E.M., and E.D. Bateman. 1983. Respiratory tract infections: An epidemiological survey. *South African Medical Journal* 64:582-4.

Peters, E.M., J.M. Goetzsche, B. Grobbelaar, and T.D. Noakes. 1993. Vitamin C supplementation reduces the incidence of postrace symptoms of upper-respiratory-tract infection in ultramarathon runners. *American Journal of Clinical Nutrition* 57:170-4.

Petko, M., and G.R. Hunter. 1997. Four-year changes in strength, power and aerobic fitness in women college basketball players. *Strength and Conditioning* 19:46-9.

Pette, D., and R.S. Staron. 1990. Cellular and molecular diversities of mammalian skeletal muscle fibers. *Review of Physiology, Biochemistry and Pharmacology* 116:2-75.

Phillips, B. 1997. *Sports supplement review.* 3rd Issue. Golden, CO: Mile High.

Phillips, S.M., X.X. Han, H.J. Green, and A. Bonen. 1996. Increments in skeletal muscle GLUT-1 and GLUT-4 after endurance training in humans. *American Journal of Physiology* 270:E456-62.

Pierce, E.F., N.W. Eastman, R.W, McGowan, H. Tripathi, W.L. Dewey, and K.G. Olson. 1994. Resistance exercise decreases β-endorphin immunoreactivity. *British Journal of Sports Medicine* 28:164-6.

Piwonka, R.W., S. Robinson, V.L. Gay, and R.S. Manalis. 1965. Preacclimatization of men to heat by training. *Journal of Applied Physiology* 20:379-84.

Plisk, S. 2000. Speed, agility, and speed-endurance development. In *Essentials of strength training and conditioning.* 2nd ed., ed. T. Baechle and R. Earle, 471-90. Champaign, IL: Human Kinetics.

Plisk, S., and V. Gambetta. 1997. Tactical metabolic training: Part I. *Strength and Conditioning* 19:44-53.

Podolsky, A., K.R. Kaufman, T.D. Calahan, S.Y. Aleksinsky, and E.Y. Chao. 1990. The relationship of strength and jump height in figure skaters. *American Journal of Sports Medicine* 18:400-5.

Poliquin, C. 1988. Five ways to increase the effectiveness of your strength training program. *National Strength and Conditioning Association Journal* 10:34-9.

Poortmans, J.R., H. Auquier, V. Renaut, A. Durussel, M. Saugy, and G.R. Brisson. 1997. Effects of short-term creatine supplementation on renal responses in man. *European Journal of Applied Physiology* 76:566-7.

Poortmans, J.R., and M. Francaux. 1999. Long-term oral creatine supplementation does not impair renal function in healthy athletes. *Medicine and Science in Sports and Exercise* 31:1108-10.

Pope, H.G., and D.L. Katz. 1988. Affective and psychotic symptoms associated with anabolic steroid use. *American Journal of Psychiatry* 145:487-90.

Portington, K.J., D.D. Pascoe, M.J. Webster, L.H. Anderson, R.R. Rutland, and L.B. Gladden. 1998. Effect of induced alkalosis on exhaustive leg press performance. *Medicine and Science in Sports and Exercise* 30:523-8.

Potteiger, J.A. 2000. Aerobic endurance exercise training. In *Essentials of strength training and conditioning.* 2nd ed., ed. T. Baechle and R. Earle, 493-508. Champaign, IL: Human Kinetics.

Potteiger, J., L. Judge, J. Cerny, and V. Potteiger. 1995. Effects of altering training volume and intensity on body mass, performance and hormonal concentrations in weight-event athletes. *Journal of Strength and Conditioning Research* 9:55-8.

Poulmedis, P., G. Rondoyannis, A. Mitsou, and E. Tsarouchas. 1988. The influence of isokinetic muscle torque exerted in various speeds of soccer ball velocity. *Journal of Orthopedic Sports Physical Therapy* 10:93-6.

Prather, I.D., D.E. Brown, P. North, and J.R. Wilson. 1995. Clenbuterol: A substitute for anabolic steroids? *Medicine and Science in Sports and Exercise* 27:1118-21.

Prentice, W.E. 1983. A comparison of static stretching and PNF stretching for improving hip joint flexibility. *Athletic Training* 18:56-9.

Pruden, E.L., O. Siggard-Anderson, and N.W. Tietz. 1987. Blood gases and pH. In *Fundamentals of clinical chemistry,* ed. N.W. Tietz, 624-44. Philadelphia: Saunders.

Pugh, L.G.C.E. 1964. Animals in high altitude: Man above 5000 meters—mountain exploration. In *Handbook of physiology: Adaptation to the environment,* 861-8. Washington, DC: American Physiological Society.

————. 1966. Accidental hypothermia in walkers, climbers, and campers: Report to the Medical Commission on Accident Prevention. *British Medical Journal* 1:123-9.

————. 1970. Oxygen uptake in track and treadmill running with observations on the effect of air resistance. *Journal of Physiology* 207:823-35.

Raglin, J.S., W.P. Morgan, and A.E. Luchsinger. 1990. Mood state and self-motivation in successful and unsuccessful women rowers. *Medicine and Science in Sports and Exercise* 22:849-53.

Raglin, J.S., W.P. Morgan, and P.J. O'Connor. 1991. Changes in mood states during training in female and male college swimmers. *International Journal of Sports Medicine* 12:585-9.

Ram, F.S.F., S.M. Robinson, and P.N. Black. 2000. Effects of physical training in asthma: A systematic review. *British Journal of Sports Medicine* 34:162-7.

Read, M.T.F., and M.J. Bellamy. 1990. Comparison of hamstring/quadricep isokinetic strength ratios and power in tennis, squash, and track athletes. *British Journal of Sports Medicine* 24:178-82.

Reeds, P.J., S.M. Hay, P.M. Dorwood, and R.M. Palmer. 1986. Stimulation of growth by clenbuterol: Lack of effect on muscle protein biosynthesis. *British Journal of Nutrition* 56:249-58.

Reeves, J.T., B.M. Groves, J.R. Sutton, P.D. Wagner, A. Cymerman, M.K. Malconian, P.B. Rock, P.M. Young, and C.S. Houston. 1987. Operation Everest II: Preservation of cardiac function at extreme altitude. *Journal of Applied Physiology* 63:531-9.

Reitman, J.S., B. Vasquez, I. Klimes, and M. Naguelsparan. 1984. Improvement of glucose homeostasis after exercise training in non-insulin-dependent diabetes. *Diabetes Care* 7:431-41.

Rejeski, W.J., P.H. Brubaker, R.A. Herb, J.R. Kaplan, and S.B. Manuck. 1988. The role of anabolic steroids and aggressive behavior in cynomolgus monkeys. *Health Psychology* 7:299-307.

Remes, K., K. Kuoppasalmi, and H. Adlercreutz. 1980. Effect of physical exercise and sleep deprivation on plasma androgen levels. *International Journal of Sports Medicine* 6:131-5.

Ricci, G., D. Lajoie, R. Petitclerc, F. Peronnet, R.J. Ferguson, M. Fournier, and A.W. Taylor. 1982. Left ventricular size following endurance, sprint, and strength training. *Medicine and Science in Sports and Exercise* 14:344-7.

Richter, E.A., K.J. Mikines, H. Galbo, and B. Kiens. 1989. Effect of exercise on insulin action in human skeletal muscle. *Journal of Applied Physiology* 66:876-85.

Riddell, D.I. 1984. Is frostnip important? *Journal of Royal Naval Medical Services* 70:140-2.

Rimmer, E., and G. Sleivert. 2000. Effects of a plyometric intervention program on sprint performance. *Journal of Strength and Conditioning Research* 14:295-301.

Roach, R.C., D. Maes, D. Sandoval, R.A. Robergs, M. Icenogle, H. Hinghofer-Szalkay, D. Lium, and J.A. Loeppky. 2000. Exercise exacerbates acute mountain sickness at simulated high altitude. *Journal of Applied Physiology* 88:581-5.

Robertson, R.J., R. Gilcher, K.F. Metz, C.J. Casperson, T.G. Allison, R.A. Abbott, G.S. Skrinar, J.R. Krause, and P.A. Nixon. 1984. Hemoglobin concentration and aerobic work capacity in women following induced erythrocythemia. *Journal of Applied Physiology* 57:568-75.

Rodnick K.J., W.L. Haskell, A.L. Swislocki, J.E. Foley, and G.M. Reaven. 1987. Improved insulin action in muscle, liver and adipose tissue in physically trained human subjects. *American Journal of Physiology* 253:E489-95.

Rogol, A.D. 1989. Growth hormone: Physiology, therapeutic use and potential for abuse. *Exercise and Sport Sciences Reviews* 17:353-77.

Roitt, I., J. Brostoff, and D. Male. 1993. *Immunology,* 3rd ed. St. Louis: Mosby.

Rome, L.C. 1990. Influence of temperature on muscle recruitment and muscle function in vivo. *American Journal of Physiology:* Regulatory, Integrative, Comparative Physiology 259:R210-22.

Rosenbaum, D., and E.M. Henning. 1997. Reaction time and force development after passive stretching and a 10-minute warm-up run. *Deutsch Zeitschrift Sportmedizin* 48:95-9.

Rothstein, A., E.F. Adolph, and J.H. Wells. 1947. Voluntary dehydration. In *Physiology of man in the desert,* ed. E.F. Adolph, 254-70. New York: Interscience.

Rowbottom, D.G., D. Keast, and A.R. Morton. 1998. Monitoring and preventing of overreaching and overtraining in endurance athletes. In *Overtraining in sport,* ed. R.B. Kreider, A.C. Fry, and M.L. O'Toole, 47-68. Champaign, IL: Human Kinetics.

Rowell, L.B. 1986. *Human circulation regulation during physical stress.* New York: Oxford University Press.

Rowland, T.W., and G.M. Green. 1988. Physiological responses to treadmill exercise in females: Adult-child differences. *Medicine and Science in Sports and Exercise* 20:474-8.

Roy, B.D., J.R. Fowles, R. Hill, and M.A. Tarnopolsky. 2000. Macronutrient intake and whole body protein metabolism following resistance exercise. *Medicine and Science in Sports and Exercise* 32:1412-8.

Rozenek, R., C.H. Rahe, H.H. Kohl, D.N. Marple, G.D. Wilson, and M.H. Stone. 1990. Physiological responses to resistance-exercise in athletes self-administering anabolic steroids. *Journal of Sports Medicine and Physical Fitness* 30:354-60.

Rundell, K.W., R.L. Wilber, L. Szmedra, D.M. Jenkinson, L.B. Mayers, and J. Im. 2000. Exercise-induced asthma screening of elite athletes: Field versus laboratory exercise challenge. *Medicine and Science in Sports and Exercise* 32:309-16.

Rupp, N.T., M.F. Guill, and D.S. Brudno. 1992. Unrecognized exercise-induced bronchospasm in adolescent athletes. *American Journal of Diseases in Children.* 146:941-4.

Ryan, A.J., G.P. Lambert, X. Shi, R.T. Chang, R.W. Summers, and C.V. Gisolfi. 1998. Effect of hypohydration on gastric emptying and intestinal absorption during exercise. *Journal of Applied Physiology* 84:1581-8.

Sale, D.G. 1988. Neural adaptation to resistance training. *Medicine and Science in Sports and Exercise* 20:S1135-45.

Sale, D.G., J.D. MacDougall, and S. Garner. 1990a. Comparison of two regimens of concurrent strength and endurance training. *Medicine and Science in Sports and Exercise* 22:348-56.

Sale, D.G., J.D. MacDougall, I. Jacobs, and S. Garner. 1990b. Interaction between concurrent strength and endurance training. *Journal of Applied Physiology* 68:260-70.

Sale, D.G., D.E. Moroz, R.S. McKelvie, J.D. MacDougall, and N. McCartney. 1993. Comparison of blood pressure response to isokinetic and weight-lifting exercise. *European Journal of Applied Physiology* 67:115-20.

———. 1994. Effect of training on the blood pressure response to weight lifting. *Canadian Journal of Applied Physiology* 19:60-74.

Saltin, B., and P.O. Astrand. 1967. Maximal oxygen uptake in athletes. *Journal of Applied Physiology* 23:353-8.

———. 1993. Free fatty acids and exercise. *American Journal of Clinical Nutrition* 57:S752-7.

Saltin, B., J. Henriksson, E. Nygaard, and P. Anderson. 1977. Fiber types and metabolic potentials of skeletal muscles in sedentary man and endurance runners. *Annals of the New York Academy of Sciences* 301:3-29.

Saltin, B., C.K. Kim, N. Terrados, H. Larsen, J. Svedenhag, and C.J. Rolf. 1995. Morphology, enzyme activities and buffer capacity in leg muscles of Kenyan and Scandinavian runners. *Scandinavian Journal of Medicine and Science in Sports* 5:222-30.

Saltin, B., and L.B. Rowell. 1980. Functional adaptations to physical activity and inactivity. *Federation Proceedings* 39:1506-13.

Sapir, D.G., O.E. Owen, T. Pozefsky, and M. Walser. 1974. Nitrogen sparing induced by a mixture of essential amino acids given chiefly as their keto analogs during prolonged starvation in obese subjects. *Journal of Clinical Investigations* 54:974-80.

Sargeant, A. 1987. Effect of muscle temperature on leg extension force and short-term power output in humans. *European Journal of Applied Physiology* 56:693-8.

Sargeant, A.J., E. Hoinville, and A. Young. 1981. Maximum leg force and power output during short-term dynamic exercise. *Journal of Applied Physiology* 26:188-94.

Sawka, M.N. 1992. Physiological consequences of hypohydration: Exercise performance and thermoregulation. *Medicine and Science in Sports and Exercise* 24:657-70.

Sawka, M.N., R.G. Knowlton, and J.B. Critz. 1979. Thermal and circulatory responses to repeated bouts of prolonged running. *Medicine and Science in Sports and Exercise* 11:177-80.

Sawka, M.N., and K.B. Pandolf. 1990. Effects of body water loss on physiological function and exercise performance. In *Fluid homeostasis during exercise: Perspectives in exercise science and sports medicine.* Vol. 3., ed. C.V. Gisolfi and D.R. Lamb, 1-38. Indianapolis: Benchmark Press.

Sawka, M.N., C.B. Wenger, A.J. Young, and K.B. Pandolf. 1993. Physiological responses to exercise in the heat. In *Nutritional needs in hot environments,* ed. B.M. Marriott, 55-74. Washington, DC: National Academy Press.

Sawka, M.A., and A.J. Young. 2000. Physical exercise in hot and cold climates. In *Exercise and sport science,* ed. W.E. Garrett and D.T. Kirkendall, 385-400. Philadelphia: Lippincott, Williams, and Wilkins.

Sawka, M.N., A.J. Young, B.S. Cadarette, L. Levine, and K.B. Pandolf. 1985. Influence of heat stress and acclimation on maximal power. *European Journal of Applied Physiology* 53:294-8.

Sawka, M.N., A.J. Young, R.P. Francesconi, S.R. Muza, and K.B. Pandolf. 1985. Thermoregulatory and blood responses during exercise at graded hypohydration levels. *Journal of Applied Physiology* 59:1394-401.

Sawka, M.N., A.J. Young, S.R. Muza, R.R. Gonzalez, and K.B. Pandolf. 1987. Erythrocyte reinfusion and maximal aerobic power: An examination of modifying factors. *Journal of the American Medical Association* 257:1496-9.

Schabort, E.J., A.N. Bosch, S.M. Weltan, and T.D. Noakes. 1999. The effect of a preexercise meal on time to fatigue during prolonged cycling exercise. *Medicine and Science in Sports and Exercise* 31:464-71.

Schaefer, M.E., J.A. Allert, H.R. Adams, and M.H. Laughlin. 1992. Adrenergic responsiveness and intrinsic sinoatrial automaticity of exercise-trained rats. *Medicine and Science in Sports and Exercise* 24:887-94.

Schilling, B.K., M.H. Stone, A. Utter, J.T. Kearney, M. Johnson, R. Coglianese, L. Smith, H.S. O'Bryant, A.C. Fry, M. Starks, R. Keith, and M.E. Stone. 2001. Creatine supplementation and health variables: A retrospective study. *Medicine and Science in Sports and Exercise* 33:183-8.

Schmidtbleicher, D., A. Gollhofer, and U. Frick. 1988. Effects of a stretch-shortening typed training on the performance capability and innervation characteristics of leg extensor muscles. In *Biomechanics XI-A*, ed. G. de Groot, A. Hollander, P. Huijing, and G. Van Ingen Schenau 185-9. Amsterdam: Free University Press.

Schneider, S.H., L.F. Amorosa, A.K. Khachadurian, and N.B. Ruderman. 1984. Studies on the mechanism of improved glucose control during regular exercise in type 2 (non-insulin-dependent) diabetes. *Diabetologia* 26:355-60.

Schurmeyer, T., K. Jung, and E. Nieschlag. 1984. The effect of an 1100 km run on testicular, adrenal and thyroid hormones. *International Journal of Andrology* 7:276-82.

Schwarz, L., and W. Kindermann. 1989. β-endorphins, catecholamines and cortisol during exhaustive endurance exercise. *International Journal of Sports Medicine* 10:324-8.

Seals, D.R., and J.M. Hagberg. 1984. The effect of exercise training on human hypertension: A review. *Medicine and Science in Sports and Exercise* 16:207-15.

Seminick, D. 1990. The T-test. *National Strength and Conditioning Association Journal* 12:36-7.

———. 1994. Testing protocols and procedures. In *Essentials of strength training and conditioning,* ed. T. Baechle, 258-73. Champaign, IL: Human Kinetics.

Senay, L. 1979. Temperature regulation and hypohydration: A singular view. *Journal of Applied Physiology* 47:1-7.

Shapiro, L. 1997. The morphological consequences of systemic training. *Cardiology Clinics* 15:373-9.

Sharp, R.L., D.L. Costill, W.J. Fink, and D.S. King. 1986. Effects of eight weeks of bicycle ergometer sprint training on human muscle buffer capacity. *International Journal of Sports Medicine* 7:13-7.

Sharp, R.L., J.P. Troup, and D.L. Costill. 1982. Relationship between power and sprint freestyle swimming. *Medicine and Science in Sports and Exercise* 14:53-6.

Shaver, L.G. 1970. Maximum dynamic strength, relative dynamic endurance, and their relationship. *Research Quarterly for Exercise and Sport* 42:460-5.

Shek, P.N., B.H. Sabiston, A. Buguet, and M.W. Radomski. 1995. Strenuous exercise and immunological changes: A multiple-time-point analysis of leukocyte subsets, CD4/CD8 ratio, immunoglobulin production and NK cell response. *International Journal of Sports Medicine* 16:466-74.

Shephard, R.J., T. Kavanaugh, D.J. Mertens, S. Qureshi, and M. Clark. 1995. Personal health benefits of masters athlete competition. *British Journal of Sports Medicine* 29:35-40.

Sherman, W., D.L. Costill, W.J. Fink, F. Hagerman, L. Armstrong, and T. Murray. 1983. Effect of a 42.2-km foot race and subsequent rest or exercise on muscle glycogen and enzymes. *Journal of Applied Physiology* 55:1219-24.

Sherman, W., D.L. Costill, W.J. Fink, and J.M. Miller. 1981. Effect of exercise-diet manipulation on muscle glycogen and its subsequent utilization during performance. *International Journal of Sports Medicine* 2:114-8.

Sherman, W.M., G. Brodowicz, D.A. Wright, W.K. Allen, J. Simonsen, and A. Dernback. 1989. Effect of 4 h pre-exercise carbohydrate feedings on cycling performance. *Medicine and Science in Sports and Exercise* 21:598-604.

Shvartz, E., Y. Shapiro, A. Magazanik, A. Meroz, H. Birnfeld, A. Mechtinger, and S. Shibolet. 1977. Heat acclimation, physical training, and responses to exercise in temperate and hot environments. *Journal of Applied Physiology: Respiratory, Environmental, Exercise Physiology* 43:678-83.

Sica, D.A., and S. Johns. 1999. Androstenedione: Just another natural substance? *American Journal of the Medical Sciences* March/April:58-63.

Silvester, L.J. 1995. Self-perceptions of the acute and long-range effects of anabolic-androgenic steroids. *Journal of Strength and Conditioning Research* 9:95-8.

Simoneau, J.A., G. Lortie, M.R. Boulay, M. Marcotte, M.C. Thibault, and C. Bouchard. 1985. Human skeletal muscle fiber type alteration with high-intensity intermittent training *European Journal of Applied Physiology* 54:240-53.

———. 1987. Effects of two high-intensity intermittent training programs interspaced by detraining on human skeletal muscle and performance. *European Journal of Applied Physiology* 56:516-21.

Singh, A., E. Moses, and P. Deuster. 1992. Chronic multivitamin-mineral supplementation does not enhance performance. *Medicine and Science in Sports and Exercise* 24:726-32.

Singh, M.V., S.B. Rawal, and A.K. Tyagi. 1990. Body fluid status on induction, reinduction and prolonged stay at high altitude on human volunteers. *International Journal of Biometeorology* 34:93-7.

Sjodin, A.M., A.H. Forslund, K.R. Westerterp, A.B. Andersson, J.M. Forslund, and L.M. Hambraeus. 1996. The influence of physical activity on BMR. *Medicine and Science in Sports and Exercise* 28:85-91.

Sjogaard, G. 1986. Water and electrolyte fluxes during exercise and their relation to muscle fatigue. *Acta Physiologica Scandinavica* 128 (supp):129-36.

Smith, B.W., and M. LaBotz. 1998. Pharmacologic treatment of exercise-induced asthma. *Clinics in Sports Medicine* 17:343-63.

Smith, D.J., and D. Roberts. 1991. Aerobic, anaerobic and isokinetic measures of elite Canadian male and female speed skaters. *Journal of Applied Sport Science Research* 5:110-5.

Smith, J.A., R.D. Telford, M.S. Baker, A.J. Hapel, and M.J. Weidemann. 1992. Cytokine immunoreactivity in plasma does not change after moderate endurance-exercise. *Journal of Applied Physiology* 71:1396-401.

Smith, J.A., R.D. Telford, I.B. Mason, and M.J. Weidemann. 1990. Exercise, training and neutrophil microbicidal activity. *International Journal of Sports Medicine* 11:179-87.

Smolander, J., P. Kolari, O. Korhonen, and R. Ilmarinen. 1986. Aerobic and anaerobic responses to incremental exercise in a thermoneutral and a hot dry environment. *Acta Physiologica Scandinavica* 128:15-21.

Snow, R.J., M.J. McKenna, S.E. Selig, J. Kemp, C.G. Stathis, and S. Zhao. 1998. Effect of creatine supplementation on sprint exercise performance and muscle metabolism. *Journal of Applied Physiology* 84:1667-73.

Snow, T.K., M. Millard-Stafford, and L.B. Rosskopf. 1998. Body composition profile of NFL football players. *Journal of Strength and Conditioning Research* 12:146-9.

Sole, C.C., and T.D. Noakes. 1989. Faster gastric emptying for glucose-polymer and fructose solutions than for glucose in humans. *European Journal of Applied Physiology* 58:605-12.

Spirito, P., A. Pelliccia, M. Proschan, M. Granata, A. Spataro, P. Bellone, G. Caselli, A. Biffi, C. Vecchio, and B.J. Maron. 1994. Morphology of the "athlete's heart" assessed by echocardiography in 947 elite athletes representing 27 sports. *American Journal of Cardiology* 74:802-6.

Sprenger, H., C. Jacobs, M. Nain, A.M. Gressner, H. Prinz, W. Wesemann, and D. Gemsa. 1992. Enhanced release of cytokines, interleukin-2 receptors, and neopterin after long-distance running. *Clinical Immunology and Immunopathology* 53:188-95.

Spriet, L.L. 1991. Blood doping and oxygen transport. In *Perspectives in exercise science and sports medicine.* Vol. 4. Ergogenic, ed. D.R. Lamb and M.H. Williams, 213-48. Carmel, IN: Benchmark Press.

———. 1995a. Anaerobic metabolism during high-intensity exercise. In *Exercise metabolism,* ed. M. Hargreaves, 1-40. Champaign, IL: Human Kinetics.

———. 1995b. Caffeine and performance. *International Journal of Sports Nutrition* 5:S84-99.

Spriet, L.L., D.A. MacLean, D.J. Dyck, E. Hultman, G. Cederblad, and T.E. Graham. 1992. Caffeine ingestion and muscle metabolism during prolonged exercise in humans. *American Journal of Physiology* 262:E891-98.

Sproles, C.B., D.P. Smith, R.J. Byrd, and T.E. Allen. 1976. Circulatory responses to submaximal exercise after dehydration and rehydration. *Journal of Sports Medicine and Physical Fitness* 16: 98-105.

Stamford, B.A., and T. Moffatt. 1974. Anabolic steroid: Effectiveness as an ergogenic aid to experienced weight trainers. *Journal of Sports Medicine and Physical Fitness* 14:191-7.

Stanley, D.C., W.J. Kraemer, R.L. Howard, L.E. Armstrong, and C.M. Maresh. 1994. The effects of hot water immersion on muscle strength. *Journal of Strength and Conditioning Research* 8:134-8.

Staron, R.S., and R.S. Hikida. 1992. Histochemical, biochemical, and ultrastructural analyses of single human muscle fibers with special reference to C fiber protein population. *Journal of Histochemistry and Cytochemistry* 40:563-8.

Staron, R.S., and P. Johnson. 1993. Myosin polymorphism and differential expression in adult human skeletal muscle. *Comparative Biochemical Physiology* 106: B463-75.

Staron, R.S., D.L. Karapondo, W.J. Kraemer, A.C. Fry, S.E. Gordon, J.E. Falkel, F.C. Hagerman, and R.S. Hikida. 1994. Skeletal muscle adaptations during early phase of heavy resistance training in men and women. *Journal of Applied Physiology* 76(3):1247-55.

Staron, R.S., M.J. Leonardi, D.L. Karapondo, E.S. Malicky, J.E. Falkel, F.C. Hagerman, and R.S. Hikida. 1991. Strength and skeletal muscle adaptations in heavy resistance trained women after detraining and retraining. *Journal of Applied Physiology* 70(2):631-40.

Staron, R.S., E.S. Milicky, M.J. Leonardi, J.E. Falkel, F.C. Hagerman, and G.A. Dudley. 1989. Muscle hypertrophy and fast fiber type conversions in heavy resistance trained women. *European Journal of Applied Physiology* 60:71-9.

Stein, R.B., T. Gordon, and J. Shriver. 1982. Temperature dependence of mammalian muscle contractions and ATPase activities. *Journal of Biophysiology* 40:97-107.

Stone, M.H., S.J. Fleck, N.T. Triplett, and W.J. Kraemer. 1991. Health-and performance-related potential of resistance training. *Sports Medicine* 11:210-31.

Stone, M.H., and A.C. Fry. 1998. Increased training volume in strength/power athletes. In *Overtraining in sport,* ed. R.B. Kreider, A.C. Fry, and M.L. O'Toole, 87-130. Champaign, IL: Human Kinetics.

Stone, M.H., J.K. Nelson, S. Nader, and D. Carter. 1983. Short-term weight training effects on resting and recovery heart rates. *Athletic Training* (Spring): 69-71.

Stone, M.H., H. O'Bryant, and J. Garhammer. 1981. A hypothetical model for strength training. *Journal of Sports Medicine* 21:342-51.

Stone, M.H., G.D. Wilson, D. Blessing, and R. Rozenek. 1983. Cardiovascular response to short-term Olympic style weight-training in young men. *Canadian Journal of Applied Sport Science* 8:134-9.

Stone, W.J., and S.P. Coulter. 1994. Strength/endurance effects from three resistance training protocols with women. *Journal of Strength and Conditioning Research* 8:231-4.

Storms, W.W. 1999. Exercise-induced asthma: Diagnosis and treatment for the recreational or elite athlete. *Medicine and Science in Sports and Exercise* 31:S33-8.

Stowers, T., J. McMillan, D. Scala, V. Davis, D. Wilson, and M. Stone. 1983. The short-term effects of three different strength-power training methods. *National Strength and Conditioning Association Journal* 5:24-7.

Strahan, A.R., T.D. Noakes, G. Kotzenberg, A.E. Nel, and F.C. de Beer. 1984. C reactive protein concentrations during long distance running. *British Medical Journal* 289:1249-51.

Strauss, R.H., J.E. Wright, G.A.M. Finerman, and D.H. Catlin. 1983. Side effects of anabolic steroids in weight-trained men. *Physician and Sportsmedicine* 11:87-96.

Stromme, S.B., H.D. Meen, and A. Aakvaag. 1974. Effects of an androgenic-anabolic steroid on strength development and plasma testosterone levels in normal males. *Medicine and Science in Sports and Exercise* 6:203-8.

Strydom, N.B., and C.G. Williams. 1969. Effect of physical conditioning on state of heat acclimatization of Bantu laborers. *Journal of Applied Physiology* 27:262-5.

Sutton, J.R., M.J. Coleman, J. Casey, and L. Lazarus. 1973. Androgen responses during physical exercise. *British Medical Journal* 1:520-2.

Sutton, J., and L. Lazarus. 1976. Growth hormone in exercise: Comparison of physiological and pharmacological stimuli. *Journal of Applied Physiology* 41:523-7.

Sutton, J.R., J.T. Reeves, P.D. Wagner, B.M. Groves, A. Cymerman, M.K. Malconian, P.B. Rock, P.M. Young, S.D. Walter, and C.S. Houston. 1988. Operation Everest II: Oxygen transport during exercise at extreme simulated altitude. *Journal of Applied Physiology* 64:1309-21.

Suzuki, K., S. Naganuma, M. Totsuka, K.J. Suzuki, M. Mochizuki, M. Shiraishi, S. Nakaji, and K. Sugawara. 1996. Effects of exhaustive endurance exercise and its one-week daily repetition on neutrophil count and functional status in untrained men. *International Journal of Sports Medicine* 17:205-12.

Swirzinski, L., R.W. Latin, K. Berg, and A. Grandjean. 2000. A survey of sport nutrition supplements in high school football players. *Journal of Strength and Conditioning Research* 14:464-69.

Syrotuik, D.G., G.J. Bell, R. Burnham, L.L. Sim, R.A. Calvert, and I. Maclean. 2000. Absolute and relative strength performance following creatine monohydrate supplementation combined with periodized resistance training. *Journal of Strength and Conditioning Research* 14:182-90.

Szlyk, P.C., I.V. Sils, R.P. Francesconi, and R.W. Hubbard. 1990. Patterns of human drinking: Effects of exercise, water temperature, and food consumption. *Aviation, Space and Environmental Medicine* 61:43-8.

Szlyk, P.C., I.V. Sils, R.P. Francesconi, R.W. Hubbard, and L.E. Armstrong. 1989. Effects of water temperature and flavoring on voluntary dehydration in men. *Physiology and Behavior* 45:639-47.

Tabata, I., Y. Atomi, Y. Mutoh, and M. Miyahita. 1990. Effect of physical training on the responses of serum adrenocorticotropic hormone during prolonged exhausting exercise. *European Journal of Applied Physiology* 61:188-92.

Takamata, A., G.W. Mack, C.M. Gillen, and E.R. Nadel. 1994. Sodium appetite, thirst, and body fluid regulation in humans during rehydration without sodium replacement. *American Journal of Physiology: Regulatory, Integrative, Comparative Physiology* 266:R1493-502.

Tan, R.A., and S.L. Spector. 1998. Exercise-induced asthma. *Sports Medicine* 25:1-6.

Tarnopolsky, M.A. 1994. Caffeine and endurance performance. *Sports Medicine* 18:109-25.

Tarnopolsky, M.A., S.A. Atkinson, J.D. MacDougall, A. Chesley, S. Phillips, and H.P. Schwarcz. 1992. Evaluation of protein requirements for trained strength athletes. *Journal of Applied Physiology* 73:1986-95.

Tarnopolsky, M.A., S.A. Atkinson, J.D. MacDougall, B.B. Senor, P.W. Lemon, and H. Schwarcz. 1991. Whole body leucine metabolism during and after resistance exercise in fed humans. *Medicine and Science in Sports and Exercise* 23:326-33.

Tarnopolsky, M.A., J.D. MacDougall, and S.A. Atkinson. 1988. Influence of protein intake and training status on nitrogen balance and lean body mass. *Journal of Applied Physiology* 64:187-93.

Telford, R., E. Catchpole, V. Deakin, A. Hahn, and A. Plank. 1992. The effect of 7-8 months of vitamin/mineral supplementation on athletic performance. *International Journal of Sports Nutrition* 2:135-53.

Terrados, N., M. Mizuno, and H. Andersen. 1985. Reduction in maximal oxygen uptake at low altitudes: Role of training status and lung function. *Clinical Physiology* 5:S75-9.

Tesch, P.A. 1985. Exercise performance and β-blockade. *Sports Medicine* 2:389-412.

Tesch, P.A., P.V. Komi, and K. Hakkinen. 1987. Enzymatic adaptations consequent to long-term strength training. *International Journal of Sports Medicine* 8:66-9.

Tesch, P.A., and L. Larson. 1982. Muscle hypertrophy in bodybuilders. *European Journal of Applied Physiology* 49:301-6.

Tesch, P.A., A. Thorsson, and N. Fujitsuka. 1989. Creatine phosphate in fiber types of skeletal muscle before and after exhaustive exercise. *Journal of Applied Physiology* 66:1756-9.

Tharp, G.D., and M.W. Barnes. 1990. Reduction of saliva immunoglobulin levels by swim training. *European Journal of Applied Physiology* 60:61-4.

Thoden, J.S., G.P. Kenny, F. Reardon, M. Jette, and S. Livingstone. 1994. Disturbance of thermal homeostasis during postexercise hyperthermia. *European Journal of Applied Physiology* 68:170-6.

Thompson, J.L., M.M. Manore, and J.R. Thomas. 1996. Effects of diet and diet-plus-exercise programs on resting metabolic rate: A meta-analysis. *International Journal of Sports Nutrition* 6:41-61.

Thompson, R.L., and J.S. Hayward. 1996. Wet-cold exposure and hypothermia: Thermal and metabolic responses to prolonged exercise in rain. *Journal of Applied Physiology* 81:1128-37.

Thomson, J.M., J.A. Stone, A.D. Ginsburg, and P. Hamilton. 1983. The effects of blood reinfusion during prolonged heavy exercise. *Canadian Journal of Applied Sport Science* 8:72-8.

Thorstensson, A. 1975. Enzyme activities and muscle strength after sprint training in man. *Acta Physiologica Scandinavica* 94:313-18.

Thorstensson, A., and J. Karlson. 1976. Fatigability and fiber composition of human skeletal muscle. *Acta Physiologica Scandinavica* 98:318-22.

Tiryaki, G.R., and H.A. Atterbom. 1995. The effects of sodium bicarbonate and sodium citrate on 600 m running time of trained females. *Journal of Sports Medicine and Physical Fitness* 35:194-98.

Tischler, M.E., M. Desautels, and A.L. Goldberg. 1982. Does leucine, leucyl-tRNA, or some metabolite of leucine regulate protein synthesis and degradation in skeletal and cardiac muscle? *Journal of Biology and Chemistry* 257:1613-21.

Tomasi, T.B., F.B. Trudeau, D. Czerwinski, and S. Erredge. 1982. Immune parameters in athletes before and after strenuous exercise. *Journal of Clinical Immunology* 2:173-78.

Toner, M.M., E.L. Glickman, and W.D. McArdle. 1990. Cardiovascular adjustments to exercise distributed between the upper and lower body. *Medicine and Science in Sports and Exercise* 22:773-8.

Toner, M.M., and W.D. McArdle. 1988. Physiological adjustments of man to the cold. In *Human performance physiology and environmental medicine at terrestrial extremes,* ed. K.B. Pandolf, M.N. Sawka, and R.R. Gonzalez, 361-99. Indianapolis: Benchmark Press.

Toner, M.M., M.N. Sawka, M.E. Foley, and K.B. Pandolf. 1986. Effects of body mass and morphology on thermal responses in water. *Journal of Applied Physiology* 60:521-5.

Tricker, R., and D. Connolly. 1997. Drugs and the college athlete: An analysis of the attitudes of student athletes at risk. *Journal of Drug Education* 27:105-19.

Troup, J.P., J.M. Metzger, and R.H. Fitts. 1986. Effect of high-intensity exercise on functional capacity of limb skeletal muscle. *Journal of Applied Physiology* 60:1743-51.

Turcotte, L.P. 2000. Muscle fatty acid uptake during exercise: Possible mechanisms. *Exercise and Sport Sciences Reviews* 28:4-9.

Turcotte, L.P., J.R. Swenberger, M.Z. Tucker, and A.J. Yee. 1999. Training-induced elevations in $FABP_{pm}$ is associated with increased palmitate use in contractile muscle. *Journal of Applied Physiology* 87:285-93.

Tvede N., M. Kappel, J. Halkjaer-Kristensen, H. Galbo, and B.K. Pedersen. 1993. The effect of light, moderate and severe bicycle exercise on lymphocyte subsets, natural and lymphokine activated killer cells, lymphocyte proliferative response and interleukin 2 production. *International Journal of Sports Medicine.* 15:100-104.

Tvede, N., B.K. Pedersen, F.R. Hansen, T. Bendix, L.D. Christensen, H. Galbo, and J. Halkjaer-Kristensen. 1989. Effect of physical exercise on blood mononuclear cell subpopulations and in vitro proliferative responses. *Scandinavian Journal of Immunology* 29:383-9.

Tvede, N., J. Steensberg, J. Bashlund, J. Halkjaer-Kristensen, and B.K. Pedersen. 1991. Cellular immunity in highly trained elite racing cyclists during periods of training with high and low intensity. *Scandinavian Journal of Medicine and Science in Sport* 3:163-6.

Urhausen, A., H. Gabriel, and W. Kindermann. 1995. Blood hormones as markers of training stress and overtraining. *Sports Medicine* 20:251-76.

Urhausen, A., and W. Kindermann. 1987. Behavior of testosterone, sex hormone binding globulin (SHBG), and cortisol before and after a triathlon competition. *International Journal of Sports Medicine* 8:305-8.

USA Track and Field. 2000. *USA Track & Field coaching manual.* Champaign, IL: Human Kinetics.

Van Handel, P.J., A. Katz, J.P. Troup, and P.W. Bradley. 1988. Aerobic economy and competitive swim performance of U.S. elite swimmers. In *Swimming science V,* ed. B.E. Ungerechts, K. Wilke, and K. Reischle, 219-27. Champaign, IL: Human Kinetics.

Van Helder, W., K. Casey, R. Goode, and W. Radomski. 1986. Growth hormone regulation in two types of aerobic exercise of equal oxygen uptake. *European Journal of Applied Physiology* 55:236-9.

Van Helder, W.P., M.W. Radomski, and R.C. Goode. 1984. Growth hormone responses during intermittent weight lifting exercise in men. *European Journal of Applied Physiology* 53:31-4.

Veicteinas, A., G. Ferretti, and D.W. Rennie. 1982. Superficial shell insulation in resting and exercising men in cold water. *Journal of Applied Physiology* 52:1557-64.

Verma, S.K., S.R. Mahindroo, and D.K. Kansal. 1978. Effect of four weeks of hard physical training on certain physiological and morphological parameters of basketball players. *Journal of Sports Medicine* 18:379-84.

Viitasalo, J.T., H. Kyrolainen, C. Bosco, and M. Allen. 1987. Effects of rapid weight reduction on vertical jumping height. *International Journal of Sports Medicine* 8:281-5.

Vogel, J.A., and C.W. Harris. 1967. Cardiopulmonary responses of resting man during early exposure to high altitude. *Journal of Applied Physiology* 22:1124-8.

Volek, J.S., N.D. Duncan, S.A. Mazzetti, R.S. Staron, M. Putukian, A.L. Gomez, D.R. Pearson, W.J. Fink, and W.J. Kraemer. 1999. Performance and muscle fiber adaptations to creatine supplementation and heavy resistance training. *Medicine and Science in Sports and Exercise* 31:1147-56.

Volek, J.S., K. Houseknecht, and W.J. Kraemer. 1997. Nutritional strategies to enhance performance of high-intensity exercise. *Strength and Conditioning* 19:11-7.

Volek, J.S., and W.J. Kraemer. 1996. Creatine supplementation: Its effect on human muscular performance and body composition. *Journal of Strength and Conditioning Research* 10:200-10.

Voy, R.O. 1986. The U.S. Olympic committee experience with exercise-induced bronchospasm, 1984. *Medicine and Science in Sports and Exercise* 18:328-30.

Wade, C.H., and J.R. Claybaugh. 1980. Plasma renin activity, vasopressin concentration and urinary excretory responses to exercise in men. *Journal of Applied Physiology: Respiratory, Environmental, Exercise Physiology* 49:930-6.

Wagner, P.D., H.A. Saltzman, and J.B. West. 1974. Measurement of continuous distributions of ventilation-perfusion ratios: Theory. *Journal of Applied Physiology* 36:588-99.

Wagner, P.D., J.R. Sutton, J.T. Reeves, A. Cymerman, B.M. Groves, and M.K. Malconian. 1987. Operation Everest II: Pulmonary gas exchange during a simulated ascent of Mt. Everest. *Journal of Applied Physiology* 63:2348-59.

Wahren, J., L. Hagenfeldt, and P. Felig. 1975. Splanchnic and leg exchange of glucose, amino acids, and free fatty acids during exercise in diabetes mellitus. *Journal of Clinical Investigations* 55:1303-14.

Walberg, J.L., M.K. Leidy, D.J. Sturgill, D.E. Hinkle, S.J. Ritchey, and D.R. Sebolt. 1988. Macronutrient content of a hypoenergy diet affects nitrogen retention and muscle function in weight lifters. *International Journal of Sports Medicine* 9:261-6.

Walker, J.B. 1979. Creatine: Biosynthesis, regulation, and function. *Advances Enzymology Related Areas Molecular Biology* 50:177-242.

Wallace, M.B., J. Lim, A. Cutler, and L. Bucci. 1999. Effects of dehydroepiandrosterone vs androstenedione supplementation in men. *Medicine and Science in Sports and Exercise* 31:1788-92.

Wallberg-Henriksson, H. 1992. Exercise and diabetes mellitus. *Exercise and Sport Sciences Reviews* 20:339-68.

Wallberg-Henriksson, H., R. Gunnarsson, J. Henriksson, R. Defronzo, P. Felig, J. Ostman, and J. Wahren. 1982. Increased peripheral insulin sensitivity and muscle mitochondrial enzymes but unchanged blood glucose control in type I diabetics after physical training. *Diabetes* 31:1044-50.

Wallberg-Henriksson, H., R. Gunnarsson, J. Henriksson, J. Ostman, and J. Wahren. 1984. Influence of physical training on formation of muscle capillaries in type I diabetes. *Diabetes* 33:851-7.

Wallberg-Henriksson, H., R. Gunnarsson, R. Rossner, and J. Wahren. 1986. Long-term physical training in female type I (insulin-dependent) diabetic patients: Absence of significant effect on glycaemic control and lipoprotein levels. *Diabetologia* 29:53-7.

Wallin, D., B. Ekblom, R. Grahn, and T. Nordenborg. 1985. Improvement of muscle flexibility: A comparison between two techniques. *American Journal of Sports Medicine* 13:263-8.

Walsh, R.M., T.D. Noakes, J.A. Hawley, and S.C. Dennis. 1994. Impaired high-intensity cycling performance time at low levels of dehydration. *International Journal of Sports Medicine* 15:392-8.

Ward, M.P., J.S. Milledge, and J.B. West. 1995. *High altitude medicine and physiology.* London: Chapman and Hall Medical.

Ward, P. 1973. The effect of an anabolic steroid on strength and lean body mass. *Medicine and Science in Sports and Exercise* 5:277-82.

Ware, J.S., C.T. Clemens, J.L. Mayhew, and T.J. Johnston. 1995. Muscular endurance repetitions to predict bench press and squat strength in college football players. *Journal of Strength and Conditioning Research* 9:99-103.

Wasserman, K., B.J. Whipp, and J.A. Davis. 1981. Respiratory physiology of exercise: Metabolism, gas exchange, and ventilatory control. *International Review of Physiology* 23:149-211.

Wathen, D. 1993. NSCA position stand: Explosive/plyometric exercises. *National Strength and Conditioning Association Journal* 15:16.

———. 1994. Load assignment. In *Essentials of strength training and conditioning,* ed. T. Baechle, 435-46. Champaign, IL: Human Kinetics.

Webster, M.J., M.N. Webster, R.E. Crawford, and L.B. Gladden. 1993. Effect of sodium bicarbonate ingestion on exhaustive resistance exercise performance. *Medicine and Science in Sports and Exercise* 25:960-5.

Webster, S., R. Rutt, and A. Weltman. 1990. Physiological effects of a weight loss regimen practiced by college wrestlers. *Medicine and Science in Sports and Exercise* 22:229-34.

Weight, L.M., D. Alexander, and P. Jacobs. 1991. Strenuous exercise: Analogous to the acute-phase response? *Clinical Science.* 81:677-83.

Weiler, J., T. Layton, and M. Hunt. 1998. Asthma in United States Olympic athletes who participated in the 1996 Summer Games. *Journal of Allergy Clinical Immunology* 102:722-6.

Weltman, A., J.Y. Weltman, R. Schurrer, W.S. Evans, J.D. Veldhuis, and A.D. Rogol. 1992. Endurance training amplifies the pulsatile release of growth hormone: Effects of training intensity. *Journal of Applied Physiology* 72:2188-96.

Weltman, A., J.Y. Weltman, C.J. Womack, S.E. Davis, J.L. Blumer, G.A. Gaesser, and M.L. Hartman. 1997. Exercise training decreases the growth hormone (GH) response to acute constant-load exercise. *Medicine and Science in Sports and Exercise* 29:669-76.

Wenger, C.B. 1988. Human heat acclimatization. In *Human performance physiology and environmental medicine at terrestrial extremes,* ed. K.B. Pandolf, M.N. Sawka, and R.R. Gonzalez, 153-98. Indianapolis: Benchmark Press.

West, J.B. 1962. Diffusing capacity of the lung for carbon monoxide at high altitude. *Journal of Applied Physiology* 17:421-6.

Wheeler, G.D., S.R. Wall, A.N. Belcastro, and D.C. Cumming. 1984. Reduced serum testosterone and prolactin levels in male distance runners. *Journal of the American Medical Association* 252:514-6.

Whipp, B.J. 1994. Peripheral chemoreceptor control of exercise hyperpnea in humans. *Medicine and Science in Sports and Exercise* 26:337-47.

Wickiewicz, T.L., R.R. Roy, P.L. Powell, J.J. Perrine, and B.R. Edgerton. 1984. Muscle architecture and force-velocity relationships in humans. *Journal of Applied Physiology: Respiratory, Environmental, Exercise Physiology* 57:435-43.

Wilber, R.L., K.W. Rundell, L. Szmedra, D.M. Jenkinson, J. Im, and S. Drake. 2000. Incidence of exercise-induced bronchospasm in Olympic winter sport athletes. *Medicine and Science in Sports and Exercise* 32:732-7.

Wilkerson, J.E., S.M. Horvath, and B. Gutin. 1980. Plasma testosterone during treadmill exercise. *Journal of Applied Physiology: Respiratory, Environmental, Exercise Physiology* 49:249-53.

Wilkes, D., N. Gledhill, and R. Smyth. 1983. Effect of acute induced metabolic acidosis on 800-m racing time. *Medicine and Science in Sports and Exercise* 15:277-80.

Williams, M.H. 1991. Alcohol, marijuana, and beta blockers. In *Perspectives in exercise science and sports medicine.* Vol. 4. Ergogenic, ed. D.R. Lamb and M.H. Williams, 331-72. Carmel, IN: Benchmark Press.

———. 1992. Ergogenic aids and ergolytic substances. *Medicine and Science in Sports and Exercise* 22(supp):344-8.

Williams, M.H., S. Wesseldine, T. Somma, and R. Schuster. 1981. The effect of induced erythrocythemia upon 5-mile treadmill run time. *Medicine and Science in Sports and Exercise* 13:169-75.

Willoughby, D.S. 1992. A comparison of three selected weight training programs on the upper and lower body strength of trained males. *Journal of Applied Research in Coaching Athletics* March:124-46.

———. 1993. The effects of meso-cycle-length weight training programs involving periodization and partially equated volumes on upper and lower body strength. *Journal of Strength and Conditioning Research* 7:2-8.

Wilmore, J.H. 1974. Alterations in strength, body composition and anthropometric measurements consequent to a 10-week weight training program. *Medicine and Science in Sports and Exercise* 6:133-8.

Wilmore, J.H., and D.L. Costill. 1999. *Physiology of sport and exercise.* Champaign, IL: Human Kinetics.

Wilmore, J.H., P.R. Stanforth, J. Gagnon, A.S. Leon, D.C. Rao, J.S. Skinner, and C. Bouchard. 1996. Endurance exercise training has a minimal effect on resting heart rate: The HERITAGE study. *Medicine and Science in Sports and Exercise* 28:829-35.

Wilson, G.J., R.U. Newton, A.J. Murphy, and B.J. Humphries. 1993. The optimal training load for the development of dynamic athletic performance. *Medicine and Science in Sports and Exercise* 25:1279-86.

Wilson, J.D. 1988. Androgen abuse by athletes. *Endocrine Reviews* 9:181-99.

Wolfel, E.E., B.M. Groves, G.A. Brooks, G.E. Butterfield, R.S. Mazzeo, L.G. Moore, J.R. Sutton, P.R. Bender, T.E. Dahms, R.E. McCullough, R.G. McCullough, S.-Y. Huang, S.-F. Sun, R.F. Glover, H.N. Hultgren, and J.T. Reeves. 1991. Oxygen transport during steady-state submaximal exercise in chronic hypoxia. *Journal of Applied Physiology* 70:1129-36.

Worrel, T.W., D.H. Perrin, B.M. Gansneder, and J. Gieck. 1991. Comparison of isokinetic strength and flexibility measures between hamstring injured and noninjured athletes. *Journal of Orthopedic Sports Physical Therapy* 13:118-25.

Worrel, T.W., T.L. Smith, and J. Winegardner. 1994. Effect of hamstring stretching on hamstring muscle performance. *Journal of Orthopaedic and Sports Physical Therapy* 20:154-9.

Wright, J.E., and M.H. Stone. 1993. Literature review: Anabolic-androgenic steroid use by athletes. *National Strength and Conditioning Association Journal* 15:10-28.

Wright, V., and R.J. Johns. 1960. Physical factors concerned with the stiffness of normal and diseased joints. *Bulletin of Johns Hopkins Hospital* 106:215-31.

Yaglou, C.P., and D. Minard. 1957. Control of heat casualties at military training centers. *Archives Industrial Health* 16:302-5.

Yang, R.C., G.W. Mack, R.R. Wolfe, and E.R. Nadel. 1998. Albumin synthesis after intense intermittent exercise in human subjects. *Journal of Applied Physiology* 84:584-92.

Yarbrough, B.E., and R.W. Hubbard. 1989. Heat related illnesses. In *Management of wilderness and environmental emergencies,* ed. P.S. Auerbach and E.C. Geehr, 119-43. St. Louis: Mosby.

Yerg II, J.E., D.R. Seals, J.M. Hagberg, and J.O. Holloszy. 1985. Effect of endurance exercise training on ventilatory function in older individuals. *Journal of Applied Physiology* 58:791-4.

Yesalis, C.E. 1993. Incidence of anabolic steroid use: A discussion of methodological issues. In *Anabolic steroids in sport and exercise,* ed. C.E. Yesalis, 49-70. Champaign, IL: Human Kinetics.

Yesalis, C.E., S.P. Courson, and J. Wright. 1993. History of anabolic steroid use in sport and exercise. In *Anabolic steroids in sport and exercise,* ed. C.E. Yesalis, 35-48. Champaign, IL: Human Kinetics.

Yesalis, C.E., J. Vicary, W. Buckley, A. Streit, D. Katz, and J. Wright. 1990. Indications of psychological dependence among anabolic-androgenic steroid abusers. In *Anabolic steroid abuse: NIDA Research Monograph,* ed. G.C. Lin and L. Erinoff, 196-214. Rockville, MD: National Institute on Drug Abuse.

Young, A.J. 1990. Energy substrate utilization during exercise in extreme environments. *Exercise and Sport Sciences Reviews* 18:65-117.

Young, A.J., J.W. Castellani, C. O'Brien, R.L. Shippee, P. Tikuisis, L.G. Meyer, L.A. Blanchard, J.E. Kain, B.S. Cadarette, and M.N. Sawka. 1998. Exertional fatigue, sleep loss, and negative energy balance increase susceptibility to hypothermia. *Journal of Applied Physiology* 85:1210-7.

Young, A.J., S.R. Muza, M.N. Sawka, R.R. Gonzalez, and K.B. Pandolf. 1986. Human thermoregulatory responses to cold air are altered by repeated cold water immersion. *Journal of Applied Physiology* 60:1542-8.

Young, A.J., M.N. Sawka, L. Levine, B.S. Cadarette, and K.B. Pandolf. 1985. Skeletal muscle metabolism during exercise is influenced by heat acclimation. *Journal of Applied Physiology* 59:1929-35.

Young, W.B. 1993. Training for speed/strength: Heavy versus light loads. *National Strength and Conditioning Journal* 15:34-42.

Young, W.B., and G.E. Bilby. 1993. The effect of voluntary effort to influence speed of contraction on strength, muscular power and hypertrophy development. *Journal of Strength and Conditioning Research* 7:172-8.

Zinman, B., S. Zuniga-Guajardo, and D. Kelly. 1984. Comparison of the acute and long-term effects of exercise on glucose control in type I diabetics. *Diabetes Care* 7:515-9.

INDEX

Note: Page numbers followed by *t* or *f* refer to the table or figure on that page.

ABOUT THE AUTHOR

Jay R. Hoffman, PhD, is an associate professor in the Department of Health and Exercise Science at the College of New Jersey in Ewing, New Jersey. A former professional athlete, Dr. Hoffman has coached elite athletes and conducted research on them throughout his professional career.

He has published 50 articles in peer-reviewed journals and has reviewed articles for nine journals in the field. He's won numerous awards, including the Editorial Excellence Award –*2001 Journal of Strength and Conditioning Research* and the NSCA Young Investigator of the Year Award in 2000.

He is a member of the American College of Sports Medicine, American Physiological Society, and the National Strength and Conditioning Association.

He earned a bachelor's degree in athletic administration and pre-med from St. John's University at Jamaica, New York, and a master's degree in exercise physiology from Queens College at Flushing, New York. He earned his PhD in exercise science from the University of Connecticut.